Down to the Hour: Short Time in the Ancient Mediterranean and Near East

Time, Astronomy, and Calendars

TEXTS AND STUDIES

VOLUME 8

The titles published in this series are listed at *brill.com/tac*

Down to the Hour: Short Time in the Ancient Mediterranean and Near East

Edited by

Kassandra J. Miller
Sarah L. Symons

BRILL

LEIDEN | BOSTON

Cover illustration: Original artwork by Kenneth Jackson depicting (top to bottom) Babylonian astronomical tablets, Sol Invictus, a water clock, an ancient Egyptian lunar deity, star clocks, and Ahura Mazda, all within a sundial's grid. © Kenneth Jackson.

Library of Congress Cataloging-in-Publication Data

Names: Miller, Kassandra J., editor. | Symons, Sarah L., editor.
Title: Down to the hour : short time in the ancient Mediterranean and Near East / edited by Kassandra J. Miller, Sarah L. Symons.
Description: Leiden ; Boston : Brill, [2020] | Series: Time, astronomy, and calendars 2211-632X ; volume 8 | Short time: measuring time at the sub-day level (e.g., hours, minutes). | Includes bibliographical references.
Identifiers: LCCN 2019039734 (print) | LCCN 2019039735 (ebook) | ISBN 9789004373471 (hardback) | ISBN 9789004416291 (ebook)
Subjects: LCSH: Time measurements—History. | Statistical astronomy. | Clocks and watches--History.
Classification: LCC QB213 .D77 2020 (print) | LCC QB213 (ebook) | DDC 529/.70938—dc23
LC record available at https://lccn.loc.gov/2019039734
LC ebook record available at https://lccn.loc.gov/2019039735

Typeface for the Latin, Greek, and Cyrillic scripts: "Brill". See and download: brill.com/brill-typeface.

ISSN 2211-632X
ISBN 978-90-04-37347-1 (hardback)
ISBN 978-90-04-41629-1 (e-book)

This book is printed on acid-free paper and produced in a sustainable manner.

Contents

Acknowledgments

The idea for this volume first germinated in 2015 when Kassandra Jackson Miller (then at the University of Chicago) and the late Robert Germany (Haverford) decided to organize a panel on "Representations of Time in the Hellenistic and Roman Worlds" at the 2016 annual meeting of the Society for Classical Studies. This panel, which also featured papers by Alexander Jones and Barbara Sattler, was so stimulating that the four speakers felt a follow-up conference was needed, which would allow them to dig more deeply into the material and involve more voices in the conversation. Thus, in February 2017, they reconvened at the University of Chicago for a conference, organized by Miller, called "Down to the Hour: Perspectives on Short Time in the Ancient Mediterranean." The chapters of the present volume represent revised versions of papers delivered at this conference, as well as pieces specially commissioned from other experts.

We are grateful to the many individuals and institutions without whom this project might never have come to fruition. The Society for Classical Studies and the Franke Institute for the Humanities at the University of Chicago provided us with important venues in which our ideas could be developed. New York University's Institute for the Study of the Ancient World, as well as University of Chicago's Oriental Institute, Fishbein Center for the History of Science and Medicine, Center for the Study of Ancient Religion, Stevanovich Institute for the Formation of Knowledge, and the departments of Classics and of Near Eastern Languages & Literature contributed funding to the conference at the University of Chicago and thereby helped us to bring participants to the United States from Canada, Germany, and the United Kingdom. Many thanks are owed to Prof. Christopher Faraone (University of Chicago), who helped to facilitate the conference and advised the editors of this volume at critical junctures along the way, and to Giulia Moriconi, Assistant Editor of Classical Studies for Brill, who has guided us through the publication process. We would also like to thank the Israel Institute for Advanced Study, which hosted several of our contributors as part of its working group on "The Day Unit in Antiquity and the Middle Ages," led by Profs. Sacha Stern and Jonathan Ben-Dov.

The editors wish to express their gratitude to the contributors to this volume, to the other colleagues who attended or participated in the activities out of which this volume grew, and to our families who supported us throughout the editorial process.

Our final thanks must be reserved for Robert Germany, who did not live to see this volume completed. His sudden death in 2017 saddened the authorial team deeply. His future contributions to the field must surely have enriched it, and we dedicate this volume to his memory.

Illustrations

Tables

Figures

Sources for figures are in italics. If no source is given, the figure was constructed by the chapter author.

Notes on Editors and Contributors

Editor and contributing author *Kassandra Jackson Miller* is a Visiting Assistant Professor of Classics at Bard College, New York. Her research is in the areas of ancient religion, magic, and medicine, with a focus on timekeeping.

Editor and contributing author *Sarah L. Symons* is an Associate Professor in the School of Interdisciplinary Science, McMaster University, Canada. Her research in history of astronomy concentrates on interpreting astronomical texts and instruments from ancient Egypt and investigating how the night sky was perceived, explained, and depicted.

Alexandra von Lieven
is an Extracurricular Professor at Westfälische Wilhelms-Universität Münster. Her research interests focus on ancient Egyptian religion, science, and cultural history, particularly on Religious Astronomy.

Anette Schomberg
is part of a research group in the Topoi Excellence Cluster at the Humboldt-Universität, Berlin, investigating the unequal armed balance in antiquity. She is also developing a database of ancient water clocks.

John Steele
is Professor of the History of the Exact Sciences in Antiquity and Chair of the Department of Egyptology and Assyriology at Brown University, Rhode Island. His research focuses on the history of Babylonian astronomy and astrology and its reception in other cultures.

Alexander Jones
is the Director of the Institute for the Study of the Ancient World (New York University). He studies the history and transmission of the mathematical sciences, especially astronomy, in antiquity, with particular focus on the Greco-Roman world.

Barbara M. Sattler
is a Senior Lecturer in Philosophy at the University of St Andrews, UK. The main area of her research is metaphysics and natural philosophy in the ancient Greek world, especially with the Presocratics, Plato, and Aristotle.

James Ker
is Associate Professor of Classical Studies at the University of Pennsylvania. His work has focused on Roman literature, especially Seneca the Younger, conceptions of time in ancient Rome, and receptions of Roman culture.

Anja Wolkenhauer
is chair of Latin philology at Eberhard-Karls-Universität Tübingen, Germany. Her research focuses on scientific issues concerning the history of mentalities, literature, and epistemology, with a focus on time perception in Rome.

Stephan Heilen
is Professor of Classics at the University of Osnabrück, Germany. His main fields of research are the history of astrology in antiquity and in the Renaissance, as well as Neo-Latin poetry.

Introduction

In the 1879 edition of their *Treatise on Natural Philosophy*, William Thomson (better known as Lord Kelvin) and Peter Guthrie Tait discussed a method of measuring an "absolute unit" of time that was based on the behavior of atoms.[1] Atoms' responses to stimulus are so regular and predictable that our modern definition of time is as follows: "The second is the duration of 9 192 631 770 periods of the radiation corresponding to the transition between the two hyperfine levels of the ground state of the cesium 133 atom."[2] All larger units of time (minutes, hours, days, years) are now based on this atomic standard. The instruments that measure Universal Time today are housed in national laboratories around the world. Although the number of people who understand, design, and maintain these clocks is small, the effect of this level of accuracy and standardization in timekeeping is vital to many of us in our own daily lives. Every GPS satellite carries its own atomic clock on board. Every time we use GPS to find a location, we rely on time-accuracy that is corrected for relativistic effects in order to produce location-accuracies of below one meter.[3] The constellation of GPS satellites above us has become our own set of timekeeping stars, driven by scientific knowledge, military requirements, technical needs, and cultural habits.

Our commonplace notion of time today is an abstract concept of a regular stream of "ticks," manifested in ever-climbing numbers delivered remotely to our smart phones or radio-enabled "atomic" wall clocks. We passively consume these numbers, confident that everyone around us is using the same ones. We note that they do not seem to conflict with our own experience of night and day, summer and winter. Occasionally, we are mildly amused to cope with a leap day (or lack of it in 2000) or even a leap second. We may speak of spring being early this year, without any fear that it is our calendar rather than the weather that is at fault. Conversely, we often wonder, "Is that clock telling the right time?"

1 Kelvin and Tait 1879, Vol. 1. Part 1, section 223.
2 BIPM 1969, Resolution 1.
3 Major 2007, chap. 19.

 | DOI:10.1163/9789004416291_002

1 Scope, Aims, and Themes of This Volume

Even with a wide range of timekeeping tools and outrageous levels of precision and accuracy available to us, we don't utilize the same degree of precision or accuracy in every timekeeping context, and sometimes we choose not to consult a device at all (either ignoring the time, or getting a general impression of it based on the amount of light in the sky, how hungry we are, etc.). Although the ancients did not have tools anywhere near as precise as our modern-day watches and smart phones (much less our atomic clocks), they still had a range of different "short" timekeeping strategies at their disposal.

By "short" time measurement, we mean methods for sub-dividing the span of the visible daily circuit of the Sun. In this volume we will consider various divisions of short time, such as watches, *bēru*, and hours, and a variety of tools for measuring them, including sundials, water clocks, star clocks, mathematical tables, and daily human routines. Our contributors comprise ten experts in the literature, history, papyrology, and material culture of ancient Egypt, Mesopotamia, Greece, and Rome, spanning 2000 BCE to the 6th century CE. Our collective aims are

- to assess key short timekeeping methods and, where possible, to describe their technical and conceptual development;
- to explore why certain timekeeping technologies were preferred to others in different contexts across the ancient Mediterranean;
- to investigate how these technologies influenced and were influenced by existing social and cultural structures, intellectual trends, and modes of expression; and
- to detect "imprints" made by time-related activities, which we now see evidenced in a wide variety of textual and material contexts.

Social historians (for example, those working on the Industrial Revolution and the Medieval Period) appreciate that temporal landscapes are as important and fundamental as physical landscapes for understanding how societies functioned and individual members of those societies made decisions and organized their lives.[4] Social historians of antiquity have only more recently begun to explore this area. Initially, though, most publications have focused on how people measured and marked "long" spans of time (years, seasons, months, days).[5] Meanwhile, despite the fact that over 600 sundials and water clocks

4 Mumford 2010; Thompson 1967; Landes 1983; Jenzen and Glasemann 1989; Dohrn-van Rossum 1992; Glennie and Thrift 2009.

5 Meritt 1928; Pritchett and Neugebauer 1947; Samuel 1972; Steele 2007; Feeney 2008; von Lieven 2007; Rüpke 2011; Stern 2012; Steele 2016; Jones 2017; Steele 2017.

from the ancient Mediterranean have been catalogued, these were studied primarily by historians of science from a technical perspective.[6] Their cultural contexts were largely neglected, but since the early 2000s, that has begun to change.[7] So far, contributions have largely focused on individual texts, objects, or corpora; no single volume has put these in conversation specifically on the topic of short time.

In this volume, chapters progress more or less chronologically, from Pharaonic Egypt through the height of the Roman Empire, in order to make certain trends more visible over time. A number of themes emerge which can conveniently be grouped into two broad categories: conceptual models for studying ancient timekeeping and how hourly timekeeping intersects with other intellectual developments.

1.1 *Conceptual Models for Studying Ancient Timekeeping*

Ancient timekeeping devices all draw on natural sources for time: the movement of celestial bodies (Sun, Moon, and stars) or the behavior of natural terrestrial phenomena (the flow of water and rate of burning of substances). These sources are termed "time indicators" by Nilsson.[8] Several chapters touch on the distinction between time-indication (the observation of these phenomena) and time-reckoning, which is the activity of processing the stream of events into a continuous count of units (60-minute hours in our modern case). The ancient world also has many systems for measuring short, non-consecutive time periods for particular functions—what we could term "egg-timer time." Some attested "egg-timer" strategies include burning lamp oil at night or draining the water from a small vessel (*clepsydra*) to time speeches in a Greek or Roman courtroom. We can also imagine other, as yet unattested, possibilities for measuring short intervals, using the human pulse, perhaps, or the rhythms of music. Such systems, however, are outside of the scope of the present volume, which restricts itself to the consideration of strategies for continuous time-reckoning.

There is a tension in our sources between "clock time" (time as a continuous and abstract stream measured by man-made instruments) and what we might call "natural time" (time as marked by natural phenomena, which may be either external, such as sunrise or a sequence of star rises, or internal, such as

6 Gibbs 1976; Schaldach 2001; 2006; Lodomez 2007; Graßhoff et al. 2016; Jones 2016. More recent catalogues, however (e.g., Winter 2013; Talbert 2017), incorporate social and cultural perspectives, as well.

7 Allen 1996; Reckford 1997; Remijsen 2007; Hannah 2008; Ker 2009; Riggsby 2009; Wolkenhauer 2011; von Lieven 2016; HUJI 2017; Miller 2018.

8 Nilsson 1920, 9.

one's perception of hunger or tiredness). In ancient civilizations, the reasons for privileging one of these two time streams above the other may not always appear to us as obvious or logical. We also see, in the early chapters of this book, how moments of interaction between time-indication and time-reckoning evidently occurred and resulted in implications for timekeeping (a term we will use for time-indication and time-reckoning activities collectively), but the documentation or instruments we would need in order to understand the interaction is frustratingly absent from the archaeological record.

We see this tension manifest in our struggle to define the concept of "hour" throughout this book. The two types of "hour" that historians of science generally reference are fixed duration hours (our present 60-minute, atomically-regulated hour falls into this category) and seasonal hours (the division of time between sunrise and sunset into 12 equal portions for daylight hours, and likewise between sunset and sunrise into another set of 12 portions for nighttime hours, equal to each other in duration but unequal to the daylight hours). The intersection between these two hour-types occurs at the equinoxes: the two days of the year when the daylight hour-length equals the nighttime hour-length. Fixed hours are therefore sometimes referred to as equinoctial hours. All these definitions seem reasonable, but as the early chapters of the book demonstrate, they hide a multitude of other interpretations. Our fixed hour may be based on the concept of the equinoctial hour but it is no longer defined by the Sun. There is no rule binding civilizations to using the equinoctial hour to define a fixed hour length: any duration can be defined to serve as an hour-type unit (for example, the *bēru*).

Seasonal hours also pose definition problems. The first two chapters touch on the problem of when nighttime and daytime begin. Was a "moment of sunset" closely defined, or did astronomical observations (of stars or of Sun shadows) begin when observing conditions were optimal, leaving liminal time spaces around sunset and sunrise during which no "hours" were indicated? In this second case, the defined "hours" would be seasonal in that their length would vary with the time of year, but not equivalent to the often-assumed situation where seasonal hours were initiated at an instant of sunrise and sunset. This "continuous equal" or standard type of seasonal hour is based on using the moments of sunrise and sunset to bisect the solar day into two unequal parts (except at the equinoxes). Each of the two parts is then further subdivided into twelve equal segments per part. This produces a sequence of 12 daylight seasonal hours of equal length followed *immediately* by 12 nighttime seasonal hours of a different but also consistent length. No moment of the daily solar cycle is thus left without an assigned hour. The other type of hour, which we could call the "non-continuous equal seasonal hour" consists of starting the daylight portion of the day at some moment *after* sunrise[9] and stopping, after twelve equal periods, at some moment *before* sunset, then waiting for the night to begin, and starting another sequence of equal-length night hours. There would be a second gap around sunrise before the daylight hours restarted. These hours are "seasonal" in that they vary in length through the year, and equal in that they are a precise division into twelve for each part of the daily cycle, but are non-continuous over the daily cycle because they leave two gaps during which time there are no assigned hours. It is likely that all star-based timekeeping methods fall into this category of hour, because stars cannot be seen directly at sunrise and sunset. It is also probable that certain time-keeping methods or instruments produced hours whose length not only varied seasonally but were also not consistent in length even within a single daylight (or night) period, a category which could be termed "non-continuous unequal seasonal hours." The ancient Egyptian L-shaped sundial (see Chapter 1.1) is a convenient example of this scenario. Thus, even when it is obvious that equal-length modern hours are inappropriate to assume, simply using the term "seasonal hours" still does not remove all ambiguity concerning the intended time-period to be measured.

Another tension which is noted throughout the book is that of observational time versus calculated time. Once a timekeeping theory was accepted (for example, a model of shadow lengths which could then be adapted by varying parameters such as latitude), did the model then supersede the observed phenomena, as has been the case with the present-day "mean time" concept, which averages the motion of the Sun? Allied to this issue, we will see how some methods were intrinsically capable of great precision, but for a variety of reasons either did not realize that potential (for example, in the case of meridian transits in stellar timekeeping methods) or were put to use where their precision was unnecessary, unappreciated, or accidentally degraded (for example, carefully constructed sundials which were then re-located, as described in Chapter 4).

With regard to social frameworks, the latter half of the book distinguishes between chronotypes and chronotopes. A chronotype (Gr. χρόνος, "time," plus τύπος, "form" or "pattern") refers to an individual's preferred daily regimen—i.e., the times of day at which he or she likes to eat, sleep, wake up, etc. A chronotope, on the other hand, (Gr. χρόνος plus τόπος, "place") refers to the intimate relationship between a particular location (e.g., a law court or marketplace) and the temporal structure that organizes activity within that space.[10] Several

9 Or something similar such as the night starting at sunset but finishing before sunrise, etc.

10 Bachtin 2014.

of our authors (e.g., Ker, Wolkenhauer, and Miller) make use of these concepts in order to investigate how temporal norms changed from one physical location and cultural context to another, and how individual temporal habits interacted with larger, socially-fixed temporal frameworks (in other words, how chronotypes interact with chronotopes).

The existence of chronotopes means that it is possible to behave in ways that either conform to or deviate from the temporal norms within a given location or cultural context (such as a genre of writing). Socially, conformation to these norms can read as "appropriate" or "good," while deviation can read as "inappropriate" or "bad"—even "scary" or "insurgent." This can refer both to how people talk about time (e.g., which temporal units they use and when) and how they occupy that time. Thus, both of these practices take on ethical dimensions, and can be used to define in-groups and out-groups and to create hierarchies within communities. Several of our authors (e.g., Wolkenhauer, Ker, Miller) explore the roles of short time within identity politics, addressing questions like: what does it mean to be a Roman or a Greek? a king or a commoner? a city-dweller or a farmer? an expert or a layman?

The place of timekeeping methods in society is, of course, a central theme. We see that timekeepers occupy spaces that are familiar to us because our present culture continues to use them (sundials in public places as focal, decorative, and prestigious objects) and other places that are much more surprising (stellar timekeeping texts in tombs and coffins). Timekeeping objects or understanding of timekeeping processes will often be situated in the context of the prestige, skill, or function of those who commission, design, record, or have access to them.

1.2 *How Hourly Timekeeping Intersects with Other Intellectual Developments*

In analyzing timekeeping methods, several of our authors discuss the "accuracy" and "precision" of the methods, with a view to understanding the capabilities of ancient timekeeping devices. These two concepts—to which a third, "standardization," can well be added—offer windows into the technical success of the objects or processes. Timekeepers can be very precise (producing the same result consistently; for example, the length of the first hour after noon will always be the same) without telling the time accurately at all (the length of the first hour after noon bears little relationship to the length of the second hour), and without any degree of standardization (another similar clock nearby gives a different set of hour-lengths entirely). Yet assumptions are embedded in these notions: that the intention for equal hour-lengths is important, and that people in Village A require their movements to be synchronized with those in Village B. The points at which precision, accuracy, and standardization appear to have become important (and the reasons why) are discussed throughout the book.

Our volume also offers insight into how questions about the practical and social roles of short timekeeping intersect with broader intellectual and cultural trends. The contributions by Steele and Heilen illustrate how timekeeping technologies arose out of a concern for celestial omen interpretation—i.e., the practice of observing the movements of celestial bodies to determine how and when it is best to act. Over time, this developed, in turn, into a fervor for astrology which drove later timekeeping innovations. Jones and Heilen also discuss the extent to which interest in defining and measuring temporal spans correlates with interest in using numerals and mathematics as a way to express and generate knowledge. This interest was driven not simply by intellectual curiosity, but also by practical and political concerns. For example, as the Roman empire grew larger, there was widespread interest across social strata in the standardization of weights and measures, which gave individual communities opportunities to express their allegiance to the imperial power (and thus make a statement of political identity) and offered useful "arbiter" units (including time units and timekeeping methods) for converting between one local measure and another.[11]

Traces of how individuals, rather than communities, engaged with available timekeeping systems are occasionally preserved in certain types of material, such as letters, personal narratives, and commonplace objects. However, the amount of such material varies dramatically across the times and places dealt with in this volume. For example, it is very difficult to assess how the "average" ancient Egyptian or Babylonian interacted with available timekeeping systems, because we lack relevant textual evidence in the former case and relevant material evidence in the latter case. More evidence of both kinds is available for the Greek and Roman periods, but it is not evenly dispersed across times, places, and social contexts. To add to these challenges, the individuals whose actions contributed to each body of evidence represent many different social strata and communities: from the inventors of timekeeping devices to those who use or maintain them; from the wealthy elite to the agricultural worker; and from profession to profession. Therefore, only when sufficient attestations survive can we address questions concerning how individuals' interactions with timekeeping systems were shaped not only by their access to relevant knowledge and technology, but also by their personal preferences, cultural practices, and social situations. If, however, we were to have perfect knowledge

11 On how these processes relate to adoption of the Julian calendar, see Stern 2017.

of how individuals in antiquity dealt with the concept of "short time," we would doubtless be able to perceive an intricate and complex network of interactions. The later chapters in this volume (e.g., Heilen, Miller) zoom in on a few such networks, and investigate how and why individuals made different choices about how to engage with different timekeeping methods within different contexts.

Fundamental to all of these themes is the idea that a unit of short time has a "value." This need not be an economic value; we know of no ancient parallels for the modern practices of paying workers "by the hour" or establishing an hourly minimum wage. Yet, individuals in antiquity had to make value judgments about short time measures, responding to questions like: is an hour of productive activity *worth* more than an hour of leisure? And when is it *worth it* to use a phrase like "in the third hour of the day" instead of simply "in the morning"?

2 Chapter Summaries

2.1 *Sun and Stars: Astronomical Timekeeping in Ancient Egypt*—Sarah L. Symons

This chapter describes hourly timekeeping methods in ancient Egypt (from around 2000 BCE to the 1st century CE) that used astronomical bodies, the Sun and the stars, to report time. Four different families of sundials are described and compared in order to demonstrate a possible path of development for solar methods. This, in turn, provides some insight into which concepts were important or desirable when measuring daylight hours. The measurement of night hours is a more complicated area with many uncertainties about the intent of the texts and tables we have. Evidence for the earliest concepts of stellar timekeeping is discussed with an emphasis on exploring the relationship between texts, contexts, and descriptions of astronomical phenomena.

2.2 *The Ancient Egyptian Water Clock between Religious Significance and Scientific Functionality*—Alexandra von Lieven and Anette Schomberg

The water clock was one of several possible ways of measuring hours in ancient Egypt. It was specifically designed to be independent from weather conditions limiting the visibility of the Sun or the stars, which was necessary for other methods. Nevertheless, the close link to the sphere of astronomy, particularly the night sky, is clearly reflected in the standard decoration of water clocks. This ties in with the fact that measuring hours was of utmost importance for several ritual contexts. The first part of the paper addresses these religious aspects of the hours in general and the decoration of water clocks in particular. The second part assesses the functionality and precision of Egyptian water clocks, based on an analysis of the shape and hour scales of preserved specimens.

2.3 *Short Time in Mesopotamia*—John Steele

This chapter examines the ways in which time at the sub-day level was envisioned, divided, used, and measured in different contexts in Mesopotamia (the late third to the first millennium BCE). Two basic systems of dividing the day are found in Mesopotamian texts: a division of the day and the night into three parts each, the length of which vary with the seasons, and a division of the whole of the day plus night into twelve equal parts, each of which is then subdivided into thirty smaller parts, which remain constant throughout the year. The rougher, seasonally varying time unit is used primarily in divinatory contexts whereas the more precisely defined, seasonally constant units are used more commonly in other contexts.

2.4 *Greco-Roman Clock Technology and Its Limits*—Alexander Jones

Except in technical astronomical contexts, statements of times in Greco-Roman sources rarely specify them to a precision finer than whole hours, occasionally nuanced qualitatively by expressions such as "completed" or "near the beginning." The present paper investigates whether ancient clock technology—primarily sundials and water clocks from the 3rd century BCE to the 6th century CE—contributed to this limit of precision.

2.5 *Cosmology and Ideal Society: the Division of the Day into Hours in Plato's* Laws—Barbara M. Sattler

Small units of time, like hours and other subdivisions of the day, are originally not of interest for ancient philosophers—for the astronomical questions of the Presocratics only long units of time, like years and months, are important. This chapter demonstrates that it is only once natural philosophy is brought together with concerns about the ideal society—a combination we find in Plato's *Laws*—that regular subdivisions of the day into hours become a focus of discussion.

Before the 4th century the Greek term ὥρα refers to the seasons, to a fitting time, to the springtime of life, or to any period fixed by the natural revolutions of the heavens. Understanding ὥρα as "hour" is usually understood to be absent from philosophical writings until after Plato. By contrast, this chapter argues that the first occurrence of ὥρα in the sense of hour in a philosophical text can be found in a passage in Plato's *Laws*, in his discussion of the length of certain council meetings; and it is part of his idea that the new ideal society has to adhere also to clear and regular temporal structures.

2.6 *Diurnal Selves in Ancient Rome*—James Ker

This chapter examines diurnal time in ancient Roman discourses on cultural identity. In Latin literature of the 1st and 2nd centuries CE we frequently observe persons positively or negatively characterized on how they use their day—their time-orientation or time-sense. Roman culture included a wide array of different paradigms by which *dies* (or *lux*) was conceived of and divided, including legal definitions, timekeeping technologies, spaces and itineraries, institutional rhythms, religious dimensions, social etiquettes, historical associations, etc.

After surveying cases from Roman literature of the imperial period in which one or more such paradigms serve as a basis for how the author represents a given person or group's "diurnal pattern," the chapter turns to its main case-study, Seneca's *Epistulae morales* 122, which satirizes elite Romans who sleep through the Roman *dies publicus* and live by night. The letter offers us an opportunity to survey multiple cultural discourses at once, on everything from nocturnal animals to funeral customs.

2.7 *Time, Punctuality, and Chronotopes: Concepts and Attitudes Concerning Short Time in Ancient Rome*—Anja Wolkenhauer

The term "short time" is the starting point for this contribution. It raises the question of what "short time" means from a Roman viewpoint and how its length is dependent on certain concepts, places, and social contexts.

In order to answer this question, first the history of timekeeping in Rome (100 BCE–100 CE) according to Pliny the Elder is reinterpreted with a special focus on the idea of exactness in measuring time. Then, three different examples of organizing time will be discussed (judicial speech, *urbs*, and *villa*), introducing the concept of chronotopes, i.e., fixed cultural connections of certain places with certain temporal structures in order to develop the social dimension of time organization and punctuality in Rome.

2.8 *Short Time in Greco-Roman Astrology*—Stephan Heilen

This contribution analyses three levels of chronographical precision that occur in Greco-Roman astrological texts from the first century BCE to the late fifth century CE: standard, high, and extreme precision. In each of these categories, both theoretical texts and practical examples will be examined. While chronographical precision to the hour turns out to be standard practice, the rare extant cases of extreme precision deserve attention because their authors pursue different, partly surprising goals.

2.9 *Hourly Timekeeping and the Problem of Irregular Fevers*—Kassandra Jackson Miller

This chapter explores how the 2nd century CE physician Galen of Pergamum relied on hourly timekeeping to "solve" an enduring medical controversy: how to diagnose and predict the outcomes of irregular, periodic fevers, like malaria. Such fevers displayed patterns of onset and abatement that defied most ancient predictive models, including the "critical day" theories proposed by the Hippocratics. For ancient physicians, this was a substantial concern, since their professional reputations were tied to the accuracy of their prognoses.

The chapter examines Galen's little-studied treatise *Against Those Who Have Written on Types or On Periods*, a treatise which currently lacks a modern translation or commentary. It demonstrates how Galen incorporates hour-units into his fever-classification system and employs them as a rhetorical tool in his polemics against rival physicians. It argues that, for Galen, numbered hours seem to offer the physician a more quantitative—and therefore more broadly comprehensible—lexicon for describing periodic fevers. Furthermore, it argues that Galen's particular interest in hourly timekeeping is closely related to his understanding of scientific method and to his oft-repeated aim of presenting himself as a synthesizer of logic and empiricism.

References

Allen, Danielle. 1996. "A Schedule of Boundaries: An Exploration, Launched from the Water-Clock, of Athenian Time." *Greece & Rome* 43 (2): 157–68.

Bachtin, Michail M. 2014. *Chronotopos. Aus dem Russischen von M. Dewey, mit einem Nachwort v. M.C. Frank und K. Mahlke*. 3rd ed. Frankfurt am Main: Suhrkamp.

BIPM, ed. 1969. *Comptes rendus des séances de la treizième conférence générale des poids et mesures, Paris, 10–16 octobre 1967 ; 15 octobre 1968*. Paris: Bureau international des poids et mesures.

Bonnin, Jérôme. 2015. *La mesure du temps dans l'antiquité*. Paris: Les Belles Lettres.

Dohrn-van Rossum, Gerhard. 1992. *Die Geschichte der Stunde. Uhren und moderne Zeitordnungen*. Munich: C. Hanser.

Feeney, Denis. 2008. *Caesar's Calendar. Ancient Time and the Beginnings of History*. Berkeley: University of California Press.

Gibbs, Sharon L. 1976. *Greek and Roman Sundials*. New Haven, CT: Yale University Press.

Glennie, Paul, and Nigel Thrift. 2009. *Shaping the Day: A History of Timekeeping in England and Wales 1300–1800*. Oxford: Oxford University Press.

Graßhoff, Gerd, Elisabeth Rinner, Karlheinz Schaldach, Bernhard Fritsch, and Liba Taub. 2016. *Ancient Sundials*. [Online database] accessed June 5, 2019, https://doi.org/10.17171/1-1.

Hannah, Robert. 2008. "Timekeeping." In *The Oxford Handbook of Engineering and Technology in the Classical World*, edited by John P. Oleson, 740–58. Oxford: Oxford University Press.

HUJI. 2017. "The Day Unit in Antiquity and the Middle Ages." [Conference website] accessed June 5, 2019, http://ias.huji.ac.il/dayunit.

Jenzen, Igor A., and Reinhard Glasemann. 1989. *Uhrzeiten: Die Geschichte der Uhr und ihres Gebrauchs*. Marburg: Jonas Verlag.

Jones, Alexander, ed. 2016. *Time and Cosmos in Greco-Roman Antiquity*. Princeton: Princeton University Press.

Jones, Alexander. 2017. *A Portable Cosmos: Revealing the Antikythera Mechanism, Scientific Wonder of the Ancient World*. Oxford: Oxford University Press.

Kelvin, William Thomson, Baron, and Peter Guthrie Tait. 1879. *Treatise on Natural Philosophy. New Edition*. Cambridge: Cambridge University Press.

Ker, James. 2009. "Drinking from the Water-Clock: Time and Speech in Imperial Rome." *Arethusa* 42: 279–302.

Landes, David S. 1983. *Revolution in Time: Clocks and the Making of the Modern World*. Cambridge, MA: Belknap Press of Harvard University Press.

Lieven, Alexandra von. 2007. *The Carlsberg Papyri 8: Grundriss des Laufes der Sterne. Das sogenannte Nutbuch*. 2 vols. Carsten Niebuhr Institute Publications 31. Copenhagen: Museum Tusculanum Press.

Lieven, Alexandra von. 2016. "The Movement of Time. News from the 'Clockmaker' Amenemhet." In *Rich and Great: Studies in Honour of Anthony J. Spalinger on the Occasion of His 70th Feast of Thoth*, edited by Renata Landgráfová and Jana Mynářová, 207–31. Prague: Charles University in Prague.

Lodomez, Guy. 2007. "Les fragments de clepsydre de dynastie des Argéades." *Chronique d'Egypte* 82: 57–76.

Major, Fouad G. 2007. *The Quantum Beat: Principles and Applications of Atomic Clocks*. New York: Springer.

Meritt, Benjamin D. 1928. *The Athenian Calendar in the Fifth Century*. Cambridge, MA: Harvard University Press.

Miller, Kassandra J. 2018. "From Critical Days to Critical Hours: Galenic Refinements of Hippocratic Models." *TAPA* 148: 111–38.

Mumford, Lewis. 2010. *Technics and Civilization*. Chicago, IL: University of Chicago Press.

Nilsson, Martin P. 1920. *Primitive Time-Reckoning: A Study in the Origins and First Development of the Art of Counting Time among the Primitive and Early Culture Peoples*. Lund: C.W.K. Gleerup.

Pritchett, William K., and Otto Neugebauer. 1947. *The Calendars of Athens*. Cambridge, MA: Harvard University Press.

Reckford, Kenneth J. 1997. "Horatius: The Man and the Hour." *American Journal of Philology* 118 (4): 583–612.

Remijsen, Sofia. 2007. "The Postal Service and the Hour as a Unit of Time in Antiquity." *Historia* 56 (2): 127–40.

Riggsby, Andrew M. 2009. "For Whom the Clock Drips." *Arethusa* 42 (3): 271–278.

Rüpke, Jörg. 2011. *The Roman Calendar from Numa to Constantine: Time, History, and the Fasti*. Translated by David Richardson. Malden, MA: Wiley-Blackwell.

Samuel, Alan Edouard. 1972. *Greek and Roman Chronology: Calendars and Years in Classical Antiquity*. Munich: C.H. Beck.

Schaldach, Karlheinz. 2001. *Römische Sonnenuhren. Eine Einführung in die antike Gnomonik*. 3rd ed. Frankfurt am Main: Harri Deutsch.

Schaldach, Karlheinz. 2006. *Die antiken Sonnenuhren Griechenlands: Festland und Peloponnes*. Frankfurt am Main: Harri Deutsch.

Steele, John M. 2007. *Calendars and Years: Astronomy and Time in the Ancient World*. Oxford: Oxbow Books.

Steele, John M. 2016. *The Circulation of Astronomical Knowledge in the Ancient World*. Leiden: Brill.

Steele, John M. 2017. *Rising Time Schemes in Babylonian Astronomy*. New York: Springer.

Stern, Sacha. 2012. *Calendars in Antiquity: Empires, States, and Societies*. Oxford: Oxford University Press.

Stern, Sacha. 2017. "Calendars, Politics, and Power Relations in the Roman Empire." In *The Construction of Time in Antiquity: Ritual, Art, and Identity*, edited by Jonathan Ben-Dov and Lutz Doering, 31–49. Cambridge: Cambridge University Press.

Talbert, Richard J.A. 2017. *Roman Portable Sundials: The Empire in Your Hand*. Oxford: Oxford University Press.

Thompson, Edward P. 1967. "Time, Work-Discipline, and Industrial Capitalism." *Past & Present* 38: 56–97.

Winter, Eva. 2013. *Zeitzeichen: zur Entwicklung und Verwendung antiker Zeitmesser*. 2 vols. Berlin: de Gruyter.

Wolkenhauer, Anja. 2011. *Sonne und Mond, Kalender und Uhr: Studien zur Darstellung und poetischen Reflexion der Zeitordnung in der römischen Literatur*. Berlin: de Gruyter.

CHAPTER 1

Sun and Stars: Astronomical Timekeeping in Ancient Egypt

Sarah L. Symons

This chapter will explore evidence in the archaeological record in ancient Egypt that relates to the direct measurement of units of time of a length less than one day, using astronomical phenomena as "time indicators."[1] This evidence includes objects (sundials and, to be discussed in the next chapter, water clocks) but also texts and tables unique to ancient Egypt that conceptualize celestial time and, debatably, could be used as "time reckoners" or clocks.

Analysis of ancient Egyptian "clocks" in their social and cultural setting requires us to reflect on aspects of clocks in the modern day, and therefore to query our assumptions about their form and function. In English, the word "clock" brings to mind a stationary object with some way of easily reading off (or hearing) the passing of hours, minutes, and often seconds. In contrast, the raw output of an atomic clock, as described in this book's Introduction, would be fairly meaningless to an ordinary person, needing conversion into common time units. Atomic clocks are designed for specialist purposes and are extremely expensive and complex, as were most varieties of timekeeping object from ancient Egypt. Modern clocks are usually mass-produced, but handmade and custom clocks are still fairly common: again, we see both of these notions reflected in the sundials described below. A clock today is often placed somewhere that multiple people can refer to it and use it as a shared standard. This type of "public" clock (in size and location) is not well evidenced in ancient Egypt except in the form of water clocks, all of which we know today would have been too heavy to be easily moved, and to which access would (in all likelihood) be very restricted. No monumental sundials dated much before the Greco-Roman era are known. Portable sundials designed for a personal audience have survived from ancient Egypt. How portable timekeeping methods are, especially in terms of the latitude range in which they were developed and used and the length of time (in months, years, or even centuries) during which they would have remained usable, is an interesting question that will be addressed as each type is described.

1 Nilsson 1920, 9.

 | DOI:10.1163/9789004416291_003

Table 1.1 summarizes the characteristics of timekeeping methods that provide a framework for discussion and comparison that we now use to explore astronomical timekeeping in ancient Egypt. Our evidence comes from two main sources: texts and objects. Our methods can be divided into two groups based on their time sources: those which use the Sun and those which use the stars.

1 Sundials

Four major types of sundial are attested from ancient Egypt. The designs are easily distinguished by their overall design or by the nature of their shadow-catching surfaces. Those with a horizontal surface are usually called L-shaped sundials or shadow clocks. Sloping sundials have an inclined plane surface. Vertical planar sundials can be described as semi-circular sundials, as the semi-circular format with roughly equal segments is the only one attested in ancient Egypt. Finally, sundials with conical or spherical shadow-catching surfaces will be grouped together as concave sundials in this chapter.

Of these four types, L-shaped and sloping sundials are unique to Egypt. The semi-circular format is a design that occurs in a variety of times and places, and

TABLE 1.1 Twelve characteristics of timekeeping methods, divided into three groups, which can be used to compare methodologies.

Format	Name	What is the contemporary designation of the object?
	Construction	What is it made out of?
	Portability	Is it designed to be used in a fixed location or moved around?
	Time Source	Where does the time come from?
Output	Purpose	What is the object used for?
	Audience	Who uses the object?
	Definition	What time divisions (and subdivisions) is it intended to measure?
	Display	How is time reported to the user?
Technicalities	Standardization	Is this object's time supposed to match that told by other objects in its class or beyond its class?
	Accuracy	How well does it adhere to the intended time units?
	Localization	Does the object only work in one place, or can it be moved?
	Additional Features	Does this object or class of objects have any additional functionality (e.g. alarms)?

could have been originally or independently developed in Egypt. It is possible that this category conceals two different types of design, discussed below. The fourth type, concave sundials, is a Greco-Roman import. Examples of this type span the geographical area influenced by Hellenized culture. They are by far the most numerous of the four types when counting across the ancient world and have been studied and catalogued somewhat more than the other types.[2]

1.1 *L-Shaped Sundials*

Sundials with a horizontal shadow-catching surface and a vertical gnomon are currently the earliest surviving sundial type from ancient Egypt, and also the rarest. Only two examples are fairly complete: ÄM 14744, from the time of Tuthmosis III (around 1500 BCE), and ÄM 19743 (c. 1000–600 BCE), both now in the Ägyptisches Museum, Berlin, and both made of stone. The larger is just over 30 cm long. Two component parts are often identified as partial L-shaped sundials: wooden gnomons ÄM 14573, also in Berlin, and N 781, in the Musée du Louvre. Both are inscribed to Amenhotep III, c. 1380 BCE. However, it is unclear whether either or both of these objects should be securely identified as sundial parts, as another instrument know as a *merkhet* and usually made of wood is identical in shape to the L-shaped sundials. The difference between a *merkhet* and a sundial is that *merkhets* do not possess hour marks on their top surfaces.

The hour marks on an L-shaped sundial follow a simple pattern. The distance between the gnomon and the closest mark determines the distance between subsequent marks in series of ratios. If we call the distance from the gnomon to mark one "3 units," then the distance from mark one to mark two will be 6 units, from mark two to mark three 9 units, from three to four 12 units, and finally, from four to five 15 units.

Not only do the two surviving examples follow this rule exactly, we also possess a text which explicitly states it—a remarkable example of standardization at such an early point (reign of Seti I, around 1300 BCE). The text is part of the so-called *Book of Nut* (or *Fundamentals of the Course of the Stars*).[3] This text is central to the understanding of the L-shaped sundials and has been the subject of substantial discussion on the way the instrument was intended to be used. In 1910, before the discovery of the *Book of Nut* text,[4] Borchardt had analyzed the two L-shaped sundials in the Berlin collection and had conjectured that

2 Gibbs 1976; Graßhoff et al. 2016.

3 Von Lieven (2007b) restores the name and treats the text extensively.

4 The discovery of the chamber is recorded in Naville (1914). The inscriptions were published in Frankfort (1933).

an L-shaped sundial would have been used by placing its gnomon due east in the morning, then picking it up and turning it 180° at noon.[5] This arrangement requires an entirely new piece, a crossbar, to be added to the instrument. The present author disagrees with this addition, arguing that the small size of the surviving instruments and the provision of a plumb-bob evidenced in each (and in later hieroglyphs depicting the same instrument, which also do not show a crossbar) indicates that the instruments were portable. Setting up an east-west orientation would therefore be impractical, and a Sun-pointing manner of usage is, I argue, simpler and more contextually consistent. However, other researchers still accept the crossbar method.[6] The *Book of Nut* text does not provide evidence of the crossbar, but does mention an act of turning the instrument.

The longevity of the design is attested by a second text, papyrus EA10673 in the British Museum, containing an illustration of an L-shaped sundial.[7] This papyrus dates to the first to third century CE.

Neither the Sun-pointing method nor the fixed method produce equal seasonal hours[8] at all accurately. The fixed version does significantly better, but not well enough to prove that the crossbar is a necessary addition.[9] The hours told by either version are not equal throughout the course of any day of the year, and both methods require adaptations of gnomon height in order to cope with changing day-lengths throughout the seasons. One hieroglyph could be interpreted as depicting a variable-height gnomon.[10] However, if the instruments do not show any evidence of accuracy in the modern sense, the "rule" for the hour marks and the evidence of craftsmen's adherence to it for at least 500 years is a remarkable example of standardization.

1.2 *Sloping Sundials*

The range of sundials depicted in hieroglyphs[11] includes several types which have not survived to the present but which collectively offer a plausible development path between L-shaped sundials (horizontal flat shadow-catching

5 Borchardt 1910, 12–4. The order in which Borchardt became aware of sundial types helps to explain his thinking. His insistence on cardinal alignment beginning with the L-shaped sundials also led him to argue that portable semi-circular sundials ("Reisesonnenuhren") required strict alignment (Symons and Khurana 2016; Krauss 2018).

6 The debate continues, most recently in Hoffmann (2016) and Symons (2016).

7 Griffith and Petrie 1889, Pl. XV; Leitz 2014, 489–491 (Fragmente 59–63) and Plate 116.

8 In either the continuous or non-continuous interpretations outlined in the Introduction.

9 Bruins 1965; Symons 2002a; Tupikova and Soffel 2012.

10 Daumas 1995.

11 Daumas 1995.

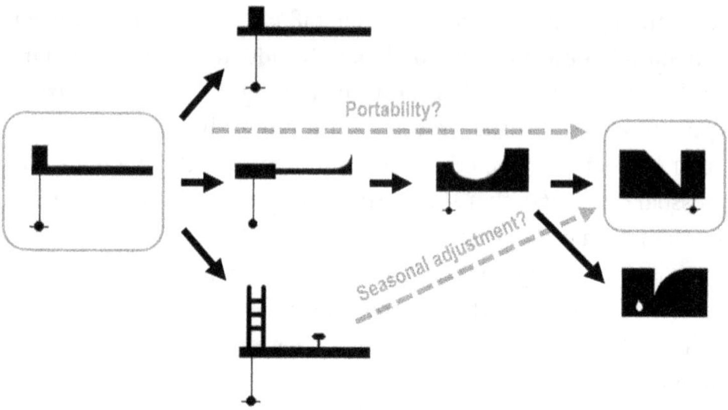

FIGURE 1.1 Arrangement of six attested hieroglyph shapes from Daumas (1995), plus one shape (far left) not attested, but resembling extant L-shaped sundials, showing possible development paths (black arrows) and motives (grey arrows and labels). Only the two shapes in rounded frames are present in the archaeological record. These hieroglyphs suggest that many variants of L-shaped and sloping sundial may have existed.

surface) and sloping sundials (inclined flat shadow-catching surface) (Figure 1.1). More examples of the sloping type have survived, but most are broken and very few have strong indications of date of manufacture.[12] They are made of stone or faience, with wood and bronze versions being rarer. All would have been smaller than 20 cm long when complete. Just like L-shaped sundials, they appear to have originally been fitted with a plumb-bob, and were thus designed to be portable rather than fixed. The present author thinks that it is likely that these sloping sundials were also used by pointing them towards the Sun.

The major innovation, aside from shortening the format by tilting the shadow-catching surface, is that four of these sundials show evidence of seasonally-adjusted timekeeping. Traces of seven scales survive on at least four examples. Each scale is either crossed by slanting lines (Petrie Museum UC16376[13] and Metropolitan Museum of Art 12.181.307[14] are examples) or has

12 Symons and Khurana 2016.
13 Salmas 2014.
14 Scott 1935.

six dots arranged along it (Hoffmann Collection, Paris,[15] and Qantara[16]), which can themselves be joined up by similar slanting lines. In two examples (Hoffmann and Qantara), each scale is accompanied by the name (in Coptic) of one month (those containing one of the two solstices) or two (all the other months in symmetrical pairs, including a central scale marked with the two months of the equinoxes). The slanting lines or dots indicate hours. The month Choiak in the Hoffmann sundial indicates that the Sun made the smallest motions, lowest to the horizon, in this month. Choiak therefore contained the winter solstice and Payni contained the summer solstice at the time this dial was made. Interestingly, the Qantara dial displays Phaophi as the month of the winter solstice and Pharmouthi as that of the summer solstice.[17] Thus, the sloping sundial, at least in its later stages, marked hours that varied in length throughout the year, and furthermore the design was modified to incorporate the change of solstitial month as the Egyptian civil calendar slipped against the solar year.

Of the sixteen examples of sloping sundials known to the present author,[18] only the Qantara instrument contains enough information to be analyzed as a timekeeper. It is the only example that has a complete set of hour dots (or lines), month names, and preserved gnomon. However, its present whereabouts are unknown, and the accuracy of Clédat's drawing[19] cannot be checked. Nevertheless, the intention of measuring hours within seasonally-varying day lengths is very clear. With only one complete example, nothing can be said about any standardization between sundials: the four that retain hour marks collectively do not indicate that a template was used.

1.3 *Semi-Circular Sundials*

Semi-circular sundials from ancient Egypt comprise a group of objects with two shared features: a semi-circle divided into twelve segments and a hole at the intersection of the segment lines. This type of figure has been found in two quarries at Debod and Qertassi,[20] on ostraca from the Valley of the Kings[21] and

15 Legrain 1894.
16 Clédat 1915.
17 It is unknown whether the location of the solstices is an indication of the date of the instrument or the calendrical convention of the time.
18 Symons et al. 2013.
19 Clédat 1915.
20 Roeder 1911.
21 Bickel and Gautschy 2014.

Dendarah,[22] on the back of a concave sundial,[23] and on six hand-sized objects, one of which has not been seen since excavation.[24]

It is not at all clear whether all these objects were themselves ever used as sundials. Several could be construed as representations or models of sundials. It is also not clear whether all of the eleven objects above can be grouped as a single type of sundial. To explain this statement, I shall start with the most secure identification of a semi-circular sundial, SNM 34605, in the Sudan National Museum. This object was found near the Amun Temple in an-Naq'a. It is an otherwise typical Greco-Roman conical sundial, distinguished by being one of the two most southerly examples yet found.[25] Its location puts the sundial south of the Tropic of Cancer, which means that near the summer solstice the Sun passes to the north of the zenith, rendering the shadow-catching bowl and gnomon ineffective. The semi-circular dial occurs on the "back" or north face of the sundial. The face is not vertical but inclined.[26] There is no doubt at all that this is a sundial. There is also no doubt that it was fixed in place.

The two quarry dials are also, by their nature, fixed in place. Roeder's photographs of the Debod (Ptolemaic) example show a gnomon hole very clearly.[27] The hole is not evident in the Qertassi example (perhaps 3rd century CE). The current status of both these quarry dials is unknown, as is their orientation. Roeder identified them as sundials, and this seems a reasonable assumption. Their location at a work site is reminiscent of the context of the most recent addition to the semi-circular sundial corpus in Egypt. The Valley of the Kings sundial's archaeological context is the first to have been thoroughly considered. Bickel and Gautschy[28] conclude that this dial dates from the reign of Seti II (around 1200 BCE) and was used by craftsmen working in the Valley of the Kings. The other example on an ostracon is described by Petrie and placed with objects from the 9th to 12th dynasties from Dendarah.[29] Unfortunately, it is not at all clear whether this was its true archaeological context. Enquiries at the Petrie Museum in 2016 found no trace of the object in the collection there. Both the Dendarah and the Valley of the Kings objects are slightly larger than

22 Petrie 1900, Pl. XIX.18. Petrie did not discuss this object in his text. Borchardt (1920, 49) mentions it in passing.

23 Kroeper and Krzyżaniak 1998.

24 Macalister 1912.

25 Catamo et al. 2000.

26 Catamo et al. 2000.

27 Roeder 1911.

28 Bickel and Gautschy 2014.

29 Petrie 1900.

one would wish to carry, with diameters (when complete) of about 20 cm for the Valley of the Kings example and perhaps up to 40 cm for the Dendarah one (although it is not clear that Petrie's drawing is to scale). In sum, these four "workmen's dials" are most probably sundials, most probably used for timing shifts and breaks, and most probably set up in a fixed location.

The collection of small semi-circular objects could be approached in two ways. First, if one accepts that semi-circular sundials were in secular use from at least the New Kingdom (and potentially much earlier), the small objects could be models of this type of sundial used as decoration or as offerings. In particular, this seems likely to be the case for the small ivory sundial found in Gezer by Macalister.[30] It is the smallest example (diameter 57 mm), making it fiddly to use. It was inscribed to Merneptah and decorated on the opposite side to the dial. Macalister thought it was a pectoral as it has a transverse hole, suggesting threading.

The alternative is that at least some of the small objects are themselves intended to be used as sundials. Borchardt, examining the example in Berlin, called it a "travel sundial" and reconstructed it to have a plumb-bob and hanging loop.[31] The main problem with this interpretation is that if the dial is to work just like a fixed example, the portable dial must be correctly oriented for each use. Instantaneously finding due south at an unknown time of day is not trivial. Borchardt supplied some ingenious ideas to overcome the problem (like remembering north from having observed the stars at night), but the arguments are not convincing. An alternative idea for how these portable sundials could have been used has recently been proposed, suggesting that the sundial was suspended and then twisted until the gnomon's shadow fell on the outer rim.[32]

The timekeeping abilities of a semi-circular sundial depend on a number of factors. One, the incline of the shadow-catching surface, is only adequately addressed in the Naq'a example. The quarry examples seem vertical, but it is hard to tell from photographs. Equally, their orientation to south is not known. Assuming such a dial was set up facing due south, the Sun would rise "behind" it for half the year, meaning that it would be unable to tell the earliest and latest hours of the day while the Sun was not in the southern half of the sky. If the segments are equal, the hours told will not be equal. If the gnomon length does not change during the year, the hours will not be similarly distributed from

30 Macalister 1912.
31 Borchardt 1911.
32 Symons and Khurana 2016.

month to month. In other words, the vertical semi-circular sundial will mark consecutive time periods during most of the day, but their relative lengths will not remain constant. If the sundial is not oriented due south, the hour lengths will not be symmetrical around noon.

The twisting method solves the problem of the missing summer morning and evening shadows: any shadow from sunrise to sunset can be "caught" by the dial. In fact, the twisting method causes the shadow-catching surface to behave like a segment of a fixed cylinder, thus suggesting the portable semi-circular sundial very conjecturally to be a type of pre-cursor to the Greco-Roman concave dials.

1.4 *Concave Sundials*

Whereas L-shaped sundials are unique to Egypt and semi-circular sundials seem to have been developed (perhaps independently of other similar traditions) in Egypt, concave sundials are a reasonably common feature of the entire Greco-Roman world.

Concave sundials are designed to produce seasonal (equal and continuous) hours at any date through the solar year and usually have curves for the solstices and equinoxes as well as hour lines. They are usually monumental, although small versions do occur,[33] but all require their design to be based on latitude and to be set up in the correct orientation. One interesting example from what is now Sudan (Sudan National Museum 34605) was designed especially for its tropical location (around latitude 16°N) and has a semi-circular sundial on its northern face to be used for the days surrounding the summer solstice.

The concave sundials found in modern Egypt and Sudan number at least 14 (Table 1.2). One is spherical, one is an unpublished peculiar conical variant, and the others are conical or (in two cases) of indeterminate type.

All of these concave sundials are made of stone except SC BM EA68475 (a tiny ivory example) and SC GRM 15735 (described as "plaster," which probably indicates this is a modern cast of another object). Their size and weight (again, excepting the ivory one) indicates that this type of sundial is designed to be fixed in place; their use would mainly have been in public gathering places.[34]

This type of sundial uses a combination of shadow length and direction, as cast in the concavity, to allow an observer to tell the time and, to a certain extent, the date. Lines for each of 12 hours are inscribed, as well as three curves indicating shadow lengths throughout the day on each solstice and the

33 Evans and Marée 2008.

34 Schaldach 2006b.

TABLE 1.2 Concave sundials from ancient Egyptian cultural areas, including their identifiers in two databases[a]

Description	AEA Database ID	Dialface (BSDP) ID
Conical Sundial from Basa, possibly in Khartoum	SC Basa	190
Conical Sundial 1936,0309.1 in London	SC BM 1936,0309.1	187
Conical Sundial EA68475 in London	SC BM EA68475	317
Conical Sundial 17483 in Alexandria	SC GRM 17483	525 & 637
Conical Sundial 0235 in Alexandria	SC BAAM 0235	538
Conical Sundial 19810 in Alexandria	SC GRM 19810	429
Conical Sundial 198850 in Warsaw	SC NMWarsaw 198850	660
Conical Sundial 16500 in London	SC Petrie UC16500	188
Conical Sundial 36147 in London	SC Petrie UC36147	N/A
Conical Part of Double Sundial 34605 in Khartoum from an-Naq'a	SC SNM 34605	192 (& 329)
Conical Variant sundial GR.100.1906 in Cambridge	SC Fitz GR.100.1906	N/A
Spherical Sundial 1909,0216.10 in London	SC BM 1909,0216.10	41
Indeterminate Type of Concave Sundial EA68817 in London	SC BM EA68817	N/A
Indeterminate Type of Concave Sundial 15735 in Alexandria	SC GRM 15735	448

a (Symons et al. 2013; Graßhoff 2015).

two equinoxes. All designs required that the sundial be made for a specific latitude,[35] and so display a uniformity of intent, even to the point of displaced sundials being noted to "tell the wrong time."[36]

Concerning the other technical considerations, the quality of manufacture of many of the examples found in Egypt appears to vary, presumably leading to a commensurate variation in accuracy. However, the spread of this type of sundial across the Greco-Roman world can, perhaps, be taken to be a very early example of standardized design, whereby travelers could be assured of understanding local time in any major conurbation.

35 Gibbs 1976.

36 See the passage from Pliny concerning a misplaced sundial in the Roman Forum, addressed later by Jones and Wolkenhauer in their chapters.

2 Star Clocks

Three types of text from ancient Egypt have been called "star clocks." All three types are conveniently gathered together in two volumes of the major work *Egyptian Astronomical Texts*, which will be referred to as *EAT1* and *EAT2*.[37] The earliest attested type is the "diagonal star clock" (*EAT1*), which occurs on at least 27 Middle Kingdom coffin lids, mostly from the Asyut necropolis. A single New Kingdom example is found in a completely different context in the Osireion at Abydos. The next to appear in the archaeological record is the "transit star clock," which is first found in the Osireion at Abydos as well, but survives up to a papyrus copy from the first or second century CE. The final type, the "Ramesside star clock," has a far shorter visible life-span, being confined to just three 20th Dynasty tombs in the Valley of the Kings. These three types will be briefly described and analyzed as timekeeping methods. Finally, the three types will be compared with each other to identify distinguishing features and commonalities.

It is worth noting that the Egyptian word for hour, *wnwt*, is encountered earlier than any of these extant objects, in the *Pyramid Texts*. The word is qualified in its two attestations in the Pyramid of Unas by a determinative consisting of three stars, perhaps indicating that this word was originally associated with periods of the night only.[38]

2.1 *The Diagonal Star Clock*

The "diagonal star clock" consists of a table populated by the names of stars or asterisms we now collectively call "decans," arranged in (usually) 12 rows and up to 40 columns. Each of the first 36 columns represents a 10-day period (decade) of the Egyptian civil year. The final column, if present, is used for the remaining five epagomenal days. There is some variation between the sources, including writing errors, omissions, jumbling of the characteristic diagonal pattern, and curtailment. The tables fall into two main groups based on star names contained within the table (the "decan list")[39] but can be grouped in more fine-grained ways.[40] A detailed explanation of the anatomy of the table can be found in *EAT1* or Symons 2007. *EAT1* calls these tables "clocks" throughout and equates the twelve rows with the twelve hours of the night. This is supported by the single New Kingdom example in the Osireion at Abydos which,

37 Neugebauer and Parker 1960; 1966.

38 Utterances 251 and 320. See Allen (2005, 42 and 61).

39 Symons 2007.

40 E.g. *EAT1*; Locher 1992; Kahl 1993.

although only preserving a very small part of a table, clearly labels the rows with hour names.[41]

The identification of the 12 rows with hours occurs as early as Daressy's publication of the most complete example, that on the coffin of *Msḥti* (CG 28118 in the Egyptian Museum, Cairo),[42] yet in the early literature, this form of coffin decoration was called "calendar,"[43] "table,"[44] and "astrological almanac."[45]

Prior to *EAT1*, Neugebauer and Parker coined the term "diagonal star clock" in the name of the first chapter of the first volume of *EAT*. They consolidate this in the first sentence, "The Egyptian calendar-year on which the diagonal star clocks (hitherto called 'diagonal calendars') were constructed is the well-known civil or 'wandering' year …"[46] and never introduce any doubt into that terminology. This is still a common way to refer to them. The present author prefers the more neutral term "diagonal star table" because Neugebauer and Parker's "clock" terminology supposes a use for these tables which is not fully supported nor fully explored in *EAT1*. The argument for clock rests on two factors which we will examine: that the 12 rows were intended from the first construction of the table to indicate the twelve hours of the night, and that the intention was to use (a copy of) the table in order to determine the hour of night by observing stars at any time of the year.

In order to discuss the first factor, some historiography is required. When Neugebauer and Parker were writing *EAT1* in 1960, they knew of 12 sources on coffins, all but one of which had exactly 12 rows (coffin S9C, their Coffin 10, has 8), plus the Osireion version which does not have a complete column, but the assumption that it would have had twelve rows is reasonable, given that it labels the hours of the night. The conclusion in the 1950s or 1960s, then, would have been very clear: Coffin 10 is an outlier and 12 rows are expected. We now have examples with 6, 8, 9, 10, 12, and even 13 rows. The proportion with 12 rows is now only 57% (again, including the Osireion version in this category).

One of the main reasons for this dramatic swing away from a "magic number" of 12 is that the majority of post-*EAT1* sources fall into the second family of tables (variously called Type 2/II, the late group, or my own designation of type K, referring to the Kenmet constellation which starts their decan list). The K group is far less orderly in general than the other (type 1/I, early group, or type T, which starts with Tjemat decans). Neugebauer and Parker knew of

41 Murray 1904.
42 Daressy 1900.
43 Pogo 1932.
44 Capart 1933.
45 Winlock 1942.
46 *EAT1*, 1.

9 type T coffins, all but one with 12 rows, and the three identified since *EAT1* all have 12 rows. In contrast, they knew of only four type K sources, all of which happen to have (or could be assumed to have, in the Osireion case) 12 rows. However, of the additional eight type K sources now known, only two have 12 rows. Four new sources are of unknown type, with one having six rows, while the other three have an unknown number of rows.

Even accepting that texts copied onto coffins were rarely rendered perfectly, the lack of consistency in number of rows does not suggest that the number of rows was prioritized by the people who painted the tables on the coffins. The tables that contain fewer than 12 rows still start at the top right part of the notional complete table, that is, the 10-row tables drop row 11 and row 12, not row 1 and row 12. This may suggest that no astronomical or chronographical significance of the number of rows was recognized in the painting process. Compare, for example, the fact that the tables are curtailed at a random number of columns dependent on the length of the lid and the choice of the column width. Instead of laying out a whole table by dividing the lid into 40 equal columns, painting started at the right and continued towards the left until the lid was full. It seems that, similarly, the painting of the tables also began from the top row and was simply continued downwards as far as space allowed. Thus, sometimes only the first eight or so rows were included, with no attempt to squash in the remaining ones in order to keep twelve rows, as we would expect if a complete set of 12 was a priority.

Instead, the decorative priority seems to have been the inclusion of the offering text (horizontally, always across the middle of the table regardless of the number of rows) and then the vertical band, which in a full table would occur after the 18th column, but is often brought far earlier (for example, to after the eighth column in K5 X2Y[47]). The surviving coffins demonstrate that the positioning of the vertical band depends strongly on the disposition of the battens (wooden cross-pieces) that hold the planks of the coffin lid together. There are two or, more usually, three battens attached to each lid, and on occasion these are the only parts of the lid to have survived. These battens often contain one or two columns of the table painted over them, starting and ending the table (the "big red dot" is therefore usually located on a batten, see K6 and K7[48]) and continuing the table if there is a third, central batten (for example, K1[49]). However, we have no instances of the vertical band itself being painted on a batten, suggesting that it was seen as important and that it

47 Lapp 1985.

48 Symons 2002b.

49 Dieleman 2014.

needed to be placed flat and undistorted on the planks of the lid itself. In the majority of cases where three battens are present, the vertical band is placed immediately to one side of the central batten.[50]

Certainly, then, in the examples we have, the focus is not on the numbers of rows, and the importance of the number 12 has decreased as the corpus has increased, yet it still maintains its position as the number of rows in the majority of tables.

As for the original intention that these rows represented hours, the Osireion version that labels the rows in this way was created around 700 years after the coffin versions were painted and an unknown number of years beyond that to the point where the tables were originally created. That is a sizeable gap in time in which the understanding of the tables may have changed or evolved. On the other hand, the Osireion seems to have been designed and decorated in a deliberately archaizing style,[51] so using an already-ancient star table would fit with that idea. This type of archaization is also relevant when considering the role of the "transit star clock" in the program of decoration in the same building. It is possible that the interpretation of rows being hours can therefore be projected back to the coffins around 2000 BCE, but it is not completely certain.

The second part of the argument for the diagonal star tables being clocks is more complicated: the intention to use a copy of the table in order to determine the hour of night by observing stars at any time of the year. With no external context available, the true intention of the creator(s) of the tables is impossible to ascertain. Neugebauer and Parker focused instead on whether the tables *could* be used as clocks, and the answer to that in one sense is easy. Given a table of stars in this format that had been prepared for the purpose, the answer is yes, they could. Using the table is rather straightforward: knowing the date, one looks towards the eastern horizon and sees which star is rising (a knowledge of 48 star names is presupposed), then finds that star's name in the column for this 10-day week. The position gives the hour of the night (eighth star in the column = eighth hour, for example). It is not even a stretch

---

50 The "empty" table on coffin T3L in storage at the British Museum is worth noting. It is the only example so far of its type. The table is recognizably a representation of a diagonal star table, but has no star names, only repeated identical star symbols. The arrangement is: 6 rows, no date row, horizontal strip with offering formula in the middle; exactly forty identical columns survive. The lid never had a central batten, and the right-hand end batten is missing. The left batten is present but shows no visible traces of decoration. The vertical band is rendered on the extreme right of the table, next to the missing batten and before the first column of stars. The coffin itself is not from Asyut, and the decorative style is very different to the Siutian coffin tables: the background of the table is rendered as a yellow-and-white checkerboard pattern.

51 Murray 1904; Frankfort 1933; and extensively in von Lieven (2007).

to say that, given the stars had to be learned and memorized, a skillful person would not even need to carry the written table around in order to refer to it, because the diagonal pattern is easy to compute directly from the ordered list of stars.

This is the methodology proposed by Neugebauer and Parker, and it is hard to argue against it in the practical sense of being taught the stars and learning to use the table. Approached from the other direction, that of conceptualizing and constructing the table with this timekeeping aim, it becomes much more difficult.

What sort of hour is measured? We have a choice of two popular models: fixed-length and seasonally adjusted hours. Both are constrained by needing darkness to observe stars, so even if the hours were fixed-length, that length cannot be as long as our 60-minute hours. In fact, if fixed-length is selected, then there have to be 12 in even the shortest night of the year, and so they would be equal in size to the seasonal hours at that point. Seasonal hours at the equinoxes would also still be shorter than 60 minutes, because of the need for "sufficient" darkness. If we require stellar visibility, then this type of clock tells neither seasonal hours nor equinoctial (60-minute) fixed hours, but time periods shorter than either.

If, for simplicity, we were to consider stars arranged around the celestial equator at equal intervals, and did not care whether we could actually observe them, fixed 60-minute hours would require only 24 stars at 15° intervals (360° divided by 24). The stars would tick by at equal intervals marking off the 24 hours of the day, regardless of whether we could see them or not. *Msḥti's* coffin specifies in a label that the "total" of these gods of the sky was 36.[52] These 36 stars of the classical ancient Egyptian decan list would be placed at 10° intervals around or below the celestial equator and would mark 40-minute intervals. A sequence of any 12 of them would therefore span seven hours and 20 minutes.

If instead we wanted seasonal hours, our stars would have to be arranged in such a way that their separations varied around the equator in order to produce the lengthening and shortening of the night hours. Note that if the transition is smooth, the stars would produce slightly different hour lengths over the course of each individual night, which is not the traditional definition of a seasonal hour, but there is no other way of achieving a varying night-length while repeatedly using the same consecutive stars.

Choosing a set of stars that fulfil this goal is not impossible, although there are certainly areas of the sky where the density of bright stars is low. There is

52 *EAT1*, 5.

no need for these stars to lie on the celestial equator. In fact, we know that the real decans do not, because Sirius and stars around the modern Orion area were used in these tables.

Neugebauer and Parker introduced a complicating factor in attempting to identify where these stars lie. They followed the New Kingdom text describing the "transit star clock" described below that specifies that decans follow the model of Sirius and are invisible for 70 days of the year. It is impossible to describe how difficult this second constraint (if applied even somewhat rigidly) on the choice of decan stars is, without diving deeply into naked-eye astronomy. In brief, length of invisibility depends not just on position (declination and right ascension, the star's coordinates relative to the entire celestial sphere) but also on atmospheric conditions, location, weather, horizon features, and even eyesight. The reappearance of a star after invisibility, the heliacal rise, is a very difficult event to predict, and the date can vary year-to-year a little. The same problems apply to the moment of disappearance at the beginning of invisibility, too. With all the varying parameters taken into account, choosing stars at fixed intervals that exhibit this ideal of 70-day invisibility is extremely hard to achieve, even using modern software.

Neugebauer and Parker were very clearly in favor of the diagonal star clock having been constructed on these rigorous principles. Their view of decans was that these were real stars, chosen to fulfil a timekeeping function and with a calendrical (70-day invisibility) duty to resemble Sirius as much as possible (*EAT1*, 97).

Neugebauer (Parker probably not being the author of this section) continues by constructing geometrically the location of the "decanal belt" wherein stars which fulfil these criteria may be found. Bright decans would lie in the southern part of this belt, while fainter ones must lie farther north in order to fulfil the invisibility requirement. Neugebauer does not note that the variation in magnitude (brightness) of the decans would mean, in this schema, that the chosen decans would not be arranged evenly or even smoothly in this belt, as a reader may expect from looking at the beautifully neat figures. The reason is not obvious and has not previously been discussed in the literature, to my knowledge.

In Fig. 25 of *EAT1* (adapted for use in Figure 1.2a), the period of visibility, 70 days, is marked by the movement of the Sun along the ecliptic from *A* to *B* (this interval is fixed for all decans). A star rises heliacally and ends its invisibility when it moves far enough away from the Sun that its light can "push through" the brightness of the sky caused by the nearby Sun. A brighter star can do this in closer proximity to the Sun than a fainter one. The position of the horizon against the starry background when bright star *S* rises heliacally is represented by the curve labelled "Eastern Horizon" passing through *S* and through the ecliptic *near B*. The *nearness* of this curve to *B* is dependent on the brightness of *S*: the brighter *S* is, the closer the horizon curve can be drawn, as a brighter star will remain visible when closer to the Sun than a fainter star could be. A similar curve ("Western Horizon") can be drawn for the time 70 days earlier when the Sun is at *A* and the star is first overwhelmed by its light, causing it to enter its annual period of invisibility. Note that the diagram is already simplifying the situation: heliacal rise does not occur when the star is exactly on the Eastern horizon, and there are many other factors that are missing from this drawn example, but the principles are sound. The Figure 1.2b then considers a fainter star σ. The process is the same except that both the horizon lines must be drawn farther away from *A* and *B*, causing them to intersect much closer to the ecliptic. Faint stars need to be farther north (away from the Sun) in order for them to restrict their invisibility to only 70 days. Iterating this process at 10-day intervals produces the locations of stars of equal brightness and identical behavior to *S*, forming a wavy path across the sky that Neugebauer and Parker called the "decanal belt".[53]

Both diagrams in Figure 1.2a and Figure 1.2b present a roughly symmetrical picture because we are considering both rising and setting equally. Figure 1.2c combines iterations of σ intersections with those of *S* intersections to delineate the "decanal belt", the region between the σ and the *S* curves. It represents the findings of Neugebauer in this regard and was considered sufficient definition to show that such stars could be identified.

However, there are conceptual problems which present real difficulties in translating this invisibility-belt into a belt of time-keeping stars. When we choose a series of stars for a diagonal star clock, selecting from those which obey the 70-day rule, we remain interested only in their risings. We choose stars which rise at 10-day intervals, meaning we are constantly looking at the intersection of the "Eastern Horizon" curve and the ecliptic. If the stars are all of equal magnitude, they would be arranged along a single path (like the σ or *S* curves in Figure 1.2c).

The real sky does not contain such a distribution. Sirius is the brightest star and so would be the only one on the *S* curve. Consider the sunrise on the day *S* rises heliacally. The Sun is at position *B* in the sky and the rise of *S* will occur when the Sun is slightly below the horizon (leftmost dashed line in Figure 1.3a), near sunrise. However, other stars with invisibility periods of 70 days will also have risen heliacally this night, but in their case it happened earlier, when the Sun was farther away from the horizon. The first to become visible would

53 *EAT1*, 97–100.

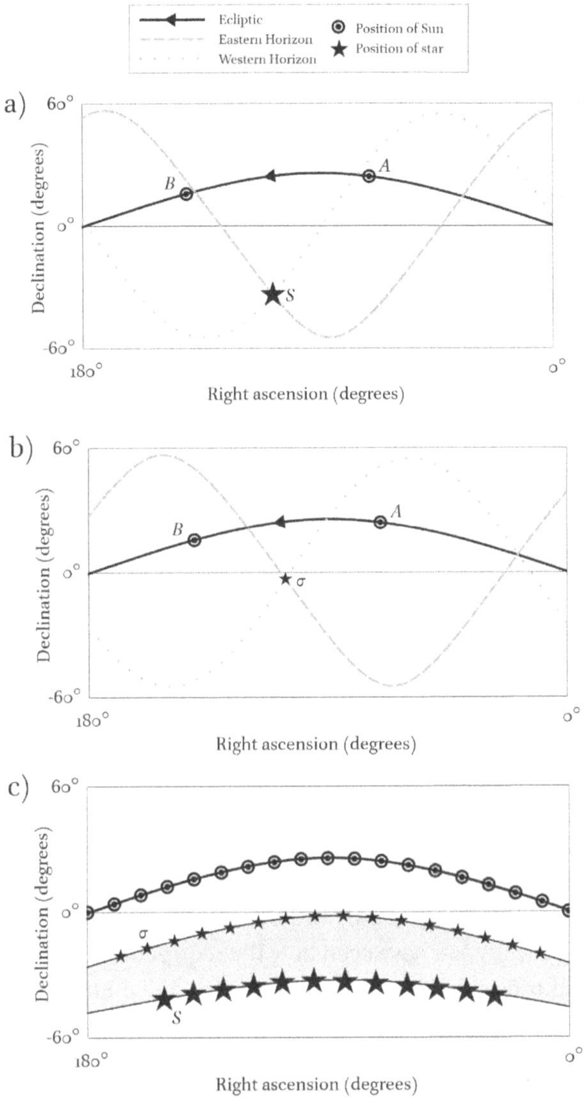

FIGURE 1.2 Neugebauer and Parker's geometric scheme for locating the "decanal belt." a) Locating a single bright star *S* that rises heliacally (Eastern Horizon, gray dashed line) and sets heliacally (Western Horizon, gray dots) 70 days apart as the Sun moves from *A* to *B* along the ecliptic (solid black with arrow) (based on *EAT1* fig. 25). b) Location of a fainter star *σ* which exhibits the same 70-day period of invisibility as *S*. c) The location stream of bright (*S*) and fainter stars (*σ*) which all rise and set heliacally 70 days apart consecutively at 10-day intervals, constructed by repeating the location of *S*- and *σ*-like stars at 10-day intervals. The shaded band defined by the two curves linking the *S* and *σ* stars is one part of Neugebauer and Parker's "decanal belt".

have been the star of magnitude σ, faint, but visible to the naked eye, when the horizon relative to the Sun was in the position of the rightmost dashed line in Figure 1.3a. Other stars of intermediate magnitude may also have reappeared, an example of which is labelled Σ in the Figure. Any of these three stars could be chosen to be a decan, but *S* is the obvious choice. Ten days later, when the Sun has moved to position *C*, a similar situation will occur. *S*, Σ, and σ, and groups of co-heliacally-rising stars like them at 10-day intervals, form sets of decanal candidates arranged along curves crossing the decanal belt.

Let us imagine a complete list of 36 decans where we have one star of *S* magnitude (Sirius), eight stars of intermediate magnitude Σ, and the remaining 27 of our notional 36 stars having the fainter magnitude of σ (Figure 1.3b). Keeping Sirius in its true position and distributing the others in a random sequence of σ and Σ, we can see that variation in magnitude causes the decans to no longer lie on a smooth path. Some will rise farther north than others. There is no smooth line or path of decans. If we removed all the lines from Figure 1.3b, it would be hard to see the order of decans (or, for example, choose a new decan if we had forgotten one) just by observing the night sky, even though we set up this fictional set of decans in a regular pattern.

Neugebauer and Parker treated each decan as a single star in their analysis. Some or even most decans are, instead, likely to have been small groups of stars (asterisms). The stars within each grouping may have behaved slightly differently to one another, further complicating the analysis. However, it seems possible that the visibility of the brightest of each grouping would have counted as heliacal rise, set, or transit for the decan as a whole, allowing us to continue with this model.[54]

The focus on choosing stars has forced us away from the consideration of hours, to which we must now turn. We have seen how the requirements forced Neugebauer and Parker into certain decisions about star choice and these themselves have ramifications for resulting hour length. The stars produced by the method above rise heliacally at 10-day intervals. Their time of rising (by which we mean visibility over the horizon) on days subsequent to their heliacal rise will alter not only by around four minutes earlier per day due to the regular rotation of stars, but also will occur nearer the horizon as the Sun moves away from them. Each night that the Sun moves away, the sky brightness decreases, thus also hastening changing the position (altitude and azimuth) and, critically, the *time* at which they become visible. After some days have passed, their rising occurs in proper darkness and will continue to be observed at the same position and at regular time intervals until the sky brightens again.

In other words, the focus on choosing stars that exhibit certain rising behavior when the Sun is close to them says nothing about how regularly, one after

54 I am grateful to Alexandra von Lieven for her thoughts on this point.

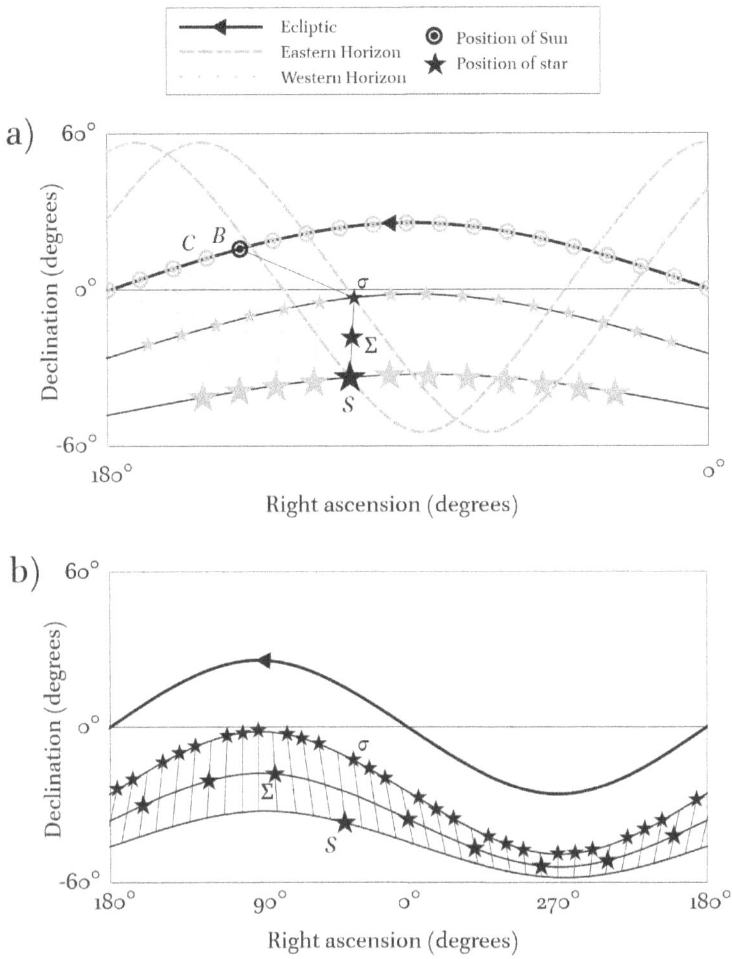

FIGURE 1.3 The problem of mixed-magnitude decans. a) The location of three stars (in black, from Figure 1.2) of different magnitudes *S*, Σ, and *σ* which all rise on the same night, defined by the position of the Sun at *B* (connected by a fine line). At *C*, 10 days later, another set of stars will rise heliacally. Other Sun and star positions at 10-day intervals are shown in gray. b) Extending the schema to a complete set of 36 fictional decans, with one bright decan *S*, eight medium-brightness decans Σ, and 27 faint decans *σ*. The decans no longer appear to form a clear procession across the sky, even though they rise at 10-day intervals and all obey the requirement for 70 days of invisibility.

the other, these same stars will process in sequence in the dark of the night, when they are measuring hours neither just after sunset nor just before dawn.[55] The problem is significant: two stars, *X* and *Y*, of quite different magnitudes can rise heliacally in one order (say *X* rises 10-days before *Y*), but 30 days later,

55 My thanks to Teije de Jong for an interesting conversation on this topic.

when both are rising in a dark sky, *Y* could actually rise before *X*, throwing the order of hours into disarray—the one thing we agree must not happen. The difference in position (and hence time) between heliacal and dark sky rises is more violent for fainter stars than for brighter ones, yet fainter stars are statistically more common and not all decans could be really bright stars.

We do not know how soon after heliacal rise (which happens just before dawn) the last hour of the Egyptian night ended, and therefore when the 12th decan's activity would occur in Neugebauer and Parker's timekeeping model. It is more likely that the hours of the night were counted from the beginning of the night, and Neugebauer and Parker argued that a decan rising marked the *end* of each hour.[56] Leitz later refined this statement to conclude that decans marked the end of the respective night hour at the beginning of a decade.[57] We have no way of knowing, therefore, how closely the end of the 12th hour of the night and the heliacal rise of a star were in time or date, but the farther apart they were, the more devastating would be the disruption to the time separation between normal nightly risings of two stars that rise heliacally 10 days apart.

Neugebauer and Parker themselves hint that it may be impossible to keep all of these requirements in hand in the real sky and what the inevitable conclusion may be when they are just about to start searching for the location of the decanal belt:

> In order to carry out this program we need not aim for high accuracy. Both the diagonal clocks and the texts of (B) [*the Book of Nut*] are obviously schematic simplifications of the appearances. Thus it suffices for our purposes if we can understand qualitatively the contents of these schemes. This has the advantage that we can make use of very simple devices to reach our goal.
>
> *EAT1*, 97

Indeed, the whole notion of a diagonal star clock may be based on these same simplifications. The question of whether one believes the diagonal star table to be a clock or not is a question of where ones draws the line between what is schematically possible and what actually occurred, and also with what intent, four thousand years ago. We can posit several ideas for use intentions, ranked in order of complexity and labelled with letters:

56 *EAT1*, 106.

57 Leitz 1995, 71–73. Also, compare the Ramesside star clock (described in the next section) where 13 stars are used: one for the "beginning of the night" and one for the end of each hour.

A. The table is a representation of the star-sky (a decorative star-pattern).

B. The table is a representation of the **movement** of the star-sky, with dates often included to show that the table covers the whole year.

C. The table is a representation of the movement of **selected stars in 10-day periods.**

D. The table is a representation of the movement of selected stars in 10-day periods and, additionally, shows that they **move sequentially during each night**—i.e., it is a schematic illustration of both the nightly and yearly movement of the stars.

E. The table is designed to **describe** the sequential nightly and yearly movements of certain stars in 10-day periods—i.e., instead of only symbolism, it is an exposition of knowledge.[58]

F. The table is designed to describe the nightly and yearly movements of certain stars, in 10-day periods, which also move sequentially during each night **at roughly equal time intervals.**

G. The table is designed to use the nightly and yearly movements of certain stars, in 10-day periods, which also move sequentially during each night **to define time intervals.**

H. The table is designed, using the nightly and yearly movements of certain stars, in 10-day periods, which also move sequentially during each night, **so that it can be used to determine the passing of defined time intervals.**

The intellectual content of the table as realized can also be categorized, again highlighting the differences at each complexity step in bold:

1. The table collates the movements over the course of one civil year in ten-day units of generalized stars.
2. The table collates the movements over the course of one civil year in ten-day units of **a stream of stars which do something consecutively.**
3. The table collates the movements over the course of one civil year in ten-day units of a stream of stars which **rise consecutively at reasonably even intervals** during the night.
4. The table collates the movements over the course of one civil year in ten-day units of a stream of stars **selected to have 70-day periods of invisibility whose heliacal rise occurs consecutively at ten-day intervals** and whose normal rises occur sequentially.

58 An offshoot of category E (and subsequent categories) would be to **predict**, rather than merely to describe. In this scenario, the table could be used to show where stars will be or to define the future movements of stars. It is possible that the theory behind these tables could have been developed with prediction as an aim, but the more fundamental case is used here as description is a first step towards prediction.

5. The table collates the movements over the course of one civil year in ten-day units of a stream of stars selected to have 70-day periods of invisibility whose heliacal rise occurs consecutively at ten-day intervals and whose normal rises occur sequentially, **marking reasonably even "hours" of the night.**

Combining these categories for intent and execution into a pair of characters, Neugebauer and Parker's methodological analysis would be H5: intent to tell the time, and a reasonably workable realization of that aim. H5 is a useable clock. Perhaps the copies on coffins that have actually survived do not live up to these standards, but the "ideal" source lying behind the decorative implementation would have been a useable clock. Their caveat quoted above suggests that they had misgivings about H5 even while describing it, and perhaps they truly believed the situation was closer to G5 or even G4.[59]

The author's current thoughts fall more towards the E2 locus, with a movement of understanding towards G3 at some point before the Osireion decoration was completed. This scenario indicates that the table is intended as a representation of the moving sky, with actual observation to select stars based on rising behavior. Such observation was attended to well enough to produce the triangle instead of recycling the initial decans. The 70-day invisibility was not rigorously pursued or even used as a criterion at all. The stars are "like Sirius" in a more general sense. There is a notion that the "ideal" behavior is rigid, but the table itself is schematic of this ideal behavior rather than based on recording strict individual observations.

Overall, I expect that A–C are aims too simplistic to have produced the tables as we see them on coffins, although C comes nearer to an explanation for the "transit star clock" below. Execution type 1 is probably too pessimistic: the intriguing "triangle" area of the table would possibly not have been included in a truly abstract schema of the sky. I emphasize that E2 does not mean that the diagonal star table is somehow less important than if it truly is an H5 object. The beauty of 10-day stars falling into place across the year, using the in-between "triangle stars" in the second half of the year, and then resetting automatically to the original decans after the half-decade of epagomenal days is demonstrably clever.

59 It would be interesting to learn where other researchers place their own understanding in this schema. For example, Depuydt (2010) seems to argue for an F3-area stance. The blank table on coffin T3L, in reality probably a "picture" of a table, would be a convenient example of what a B1 level of intent would produce (Symons 2007, 2). This is only slightly more representational than an all-over star pattern, a common decorative theme in temple and tomb ceilings, which could perhaps be denoted as a "A0."

That such a table could then come to be understood as indicating where stars are at consecutive intervals during the night is not a wild leap. The arrival of the hour labels in the Osireion version, as I have argued before,[60] is not sufficient to fix the use of the table as a timekeeping instrument at its inception or even by the New Kingdom. A bus timetable lists hours and minutes, and could (if one's bus company proved particularly reliable) even be used as a clock in a pinch, however its intention is not to mark time but to describe the movements of buses.

2.2 *The Ramesside Star Clock*

The description and explanation of the Ramesside star clock takes up the whole second volume of *EAT*. More recently, Leitz approached the subject with a scheme of identifying the stars used within the Ramesside star clock.[61] The number of sources has not changed to the present day: just four instances of the Ramesside star clock are known, all from 20th Dynasty tombs in the Valley of the Kings. Two occur in the tomb of Ramses V and VI (tomb KV9), and one each occur in the tombs of Ramses VII (KV1) and Ramses IX (KV6). The period of time over which these tombs were decorated spans a maximum of sixty years (1160–1103 BCE, taking an early proposed date for the beginning of the reign of Ramses V and a late estimate for the death of Ramses IX). There are no other references to this type of star clock and no attached instructions.

The limited number of sources and their similarity to one another (all differences can be explained by copying errors and changes in layout between tombs) strongly suggests that the Ramesside star clock derives from a single source. Each of the four instances is located as part of an astronomical ceiling in two parts, either decorating two halves of the same chamber (KV9 pillared hall version and KV1) or decorating two consecutive tomb corridors (KV9 corridor version and KV6). Of these tombs, KV9 is the only one that can be considered to have a complete decorative program. Its ceilings contain a wealth of astronomical texts: the paired astronomical diagrams and Ramesside star clock in the first two corridors; the Ramesside star clock again with the same type of paired diagrams with the same pair of associated decan lists in the pillared hall, flanking content from both the *Book of the Day* (which contains a passage explaining the circumpolar constellations[62]) and the *Book of the Night*; another pair of *Books of the Day and Night* in the burial chamber; and even a

60 Symons 2007, 19.
61 Leitz 1995.
62 Text Pb in Müller-Roth (2008, 282–292).

highly unusual depiction of personified decans on the walls of a niche at the far end of the burial chamber.

The Ramesside star clock consists of 24 tables, each one serving 15 days of the civil year (the 5 epagomenal days are ignored). Each table consists of thirteen rows of text, a grid of seven vertical lines and 13 rows matching the text, with (usually) one star symbol per row placed on one of the seven lines, and a figure of a full-faced, kneeling priest.

The text in each row follows the same pattern, with only the top row of each table differing in that it also contains the date of the table. The text states the time, a star name, and the star's position. The time is "beginning of the night" (in the top row), "end of the 1st hour" (in the next row), "end of the 2nd hour," and so on through to "end of the 12th hour" in the bottom row. The star names are drawn from a set of 47 stars which as a group are otherwise completely unattested. The set, which Neugebauer and Parker called "hour stars," includes stars related to Sopdet/Sirius and Sahu/Orion, like the decans, but it also includes stars from constellations "Mooring Post(s)," "Hippo," and "Lion," which are names strongly associated with the northern circumpolar group in ancient Egyptian astronomical diagrams (*EAT2*, 7).

The possible positions of the stars are limited to seven, relating directly to the seven vertical lines depicted in the grid (ideally, the grid and the text match exactly). The central position is called *r ꜥqꜣ ib*, the nearest positions to that (moving outwards from the center) are *irt wnmy* and *irt iꜣby*, then we have *msḏr wnmy* and *msḏr iꜣby*, and finally the two outermost positions *ḳꜥḥ wnmy* and its mirror twin *ḳꜥḥ iꜣby*. *Irt*, *msḏr*, and *ḳꜥḥ* mean eye, ear, and shoulder respectively, so the positions are related to parts of the body. *Wmny* (or *imny*) means right or west, while *iꜣby* means left or east.

Here, then, we have a direct and explicit relationship between star movements and time. Furthermore, the motions of the "hour stars" do not follow an ordered pattern from table to table like the columns of the diagonal star clock, but instead skip rows or tables. Also, not all of the 47 hour stars appear with the same frequency, and the resulting tables strongly resemble copies of real observational records. The Ramesside star clock, of the three extant types of star clock, lays the strongest claim to being observationally rather than conceptually based, and also the most likely text to describe a process that could be used for telling the time.

As we have seen, diagonal star tables occurred on coffin lids, a funerary context similar to that of the Ramesside star clocks. The development of astronomical themes in royal tombs is worth considering in relation to the sudden appearance (and disappearance) of the Ramesside star clock.

Astronomical material began to be used in the Valley of the Kings in the 19th Dynasty with the famous astronomical ceiling of Seti I in KV17. Two subsequent tombs in the dynasty that extended as far as a burial chamber have astronomical ceilings (Merneptah, KV8, and Tawosret/Setnakht, KV14), although the extensive but badly-damaged tomb of Ramses II has not yet been found to have had one. Tomb corridors' ceilings often bore star patterns, flying vultures, or parts of the *Litany of Re*.[63] The burial chamber ceiling in the 20th Dynasty tomb of Ramses III (KV11) has collapsed, and no astronomical material from the tomb has been published. Ramses IV's tomb (KV2) was not completed at the time of his death, so the room which would have become the pillared hall was used as a burial chamber. Uniquely in the Valley of the Kings, the ceiling of this chamber was decorated with the vignette from the *Book of Nut*, containing the transit star clock (discussed below). The ceilings of corridors leading up to this room bear star patterns, flying vultures, and parts of the *Litany of Re* similar to prior tombs.[64] This ceiling program therefore alters considerably for the next tomb, that of Ramses V and VI, with its use of astronomical diagrams, Ramesside star clock, and the *Books of the Day and the Night*. Incidentally, the *Book of the Night* and the *Book of Nut* both make their first full appearance in the Osireion at Abydos in the reign of Seti I,[65] the apparent initiator of royal astronomical ceilings in the Valley. How this trend may have developed beyond Ramses VI is difficult to say. Ramses VII was buried in haste in what would have been a corridor of a full tomb. The corridor was, however, enlarged for the job, and the pair of astronomical diagrams and Ramesside star clock are of the "pillared hall" format rather than the "corridor" format when compared with those of KV9. The tomb of Ramses VIII is either missing or the unfinished tomb which is now ascribed to Mentuherkhepeshef (KV19). Ramses IX made better progress with KV6, using the astronomical diagrams and Ramesside star clock in two corridors. The pillared hall was excavated but not decorated, so we cannot know the intention for that ceiling, but the burial chamber has a pair of *Books of the Day and Night* (both lacking text[66]) that must suggest that the pattern of KV9 could have been in the process of being followed. Neither KV18 (Ramses X, only partially cut) nor the final royal tomb started in the Valley, KV4 (Ramses XI, cut as far as a burial chamber, but only one corridor decorated) preserve astronomical decoration.

---

63 Weeks 2000.
64 Weeks 2000.
65 Roulin 1996; Hornung 1999.
66 Roberson 2012.

The appearance of the Ramesside star clock in the program of tomb decoration in the 20th Dynasty therefore has to be seen as part of a tradition, dating back to the Old Kingdom, of decorating temple and tomb ceilings with motifs which represent the sky. Quite how the Ramesside star clock is related to the appearance in similar New Kingdom contexts[67] of the Books of the Sky (*Nut*, *Day*, and *Night*) is uncertain, but it is clear that the never-before-seen Ramesside star clock's place among the "new" material in royal tombs during the 19th and 20th Dynasties is not in itself exceptional: the concept of the material fits with the increasing complexity of the sky motifs and explanations. However, the nature of the content—which, as we have seen looks very much like an observational record—is somewhat different compared with the funerary texts introduced around the same time, even that of the *Book of Nut* with its embedded table of astronomical data. Possibly the Ramesside star clock functions as a depiction of the movement of the sky while the astronomical diagrams represent the content of the sky in a more static sense, with the Books of the Sky providing the central motivation of the actions of the gods. The kneeling priests (whether observers or passive participants in the process of observation) are somewhat mundane figures in a royal tomb, but they sit around the edge of the tomb's "sky" resembling a link between the heavens and the human world below, especially in the two "chamber" versions where they form two borders for the central astronomical tableaux and (in the case of KV9) texts.[68]

This is the context in which we see them now, but their journey before arriving on royal tomb ceilings is still unknown. Conjecturally, the original source could have been stored in a temple library alongside papyri containing other astronomical information, from which, eventually, material was drawn. But how, when, and why was this original created?

Neugebauer and Parker placed the Ramesside star clock as the culmination (choosing a word carefully) of astronomical timekeeping efforts. In their model, the central position *r ꜥqꜣ ib* is the north-south meridian and the kneeling priest is positioned due south of an undepicted observer. The priest's body forms a guide to the six outer positions. Stars transit the meridian (culminate) directly above the priest's head, at the highest point in their nightly rotation. Positions each side of the meridian are added to improve timekeeping accuracy. This is all the central thesis of *EAT2*, and undoubtedly works as a

67 The date of origin of many texts, including those with astronomical content, is hard to ascertain. However, textual clues remain. See von Lieven (2007, 223f.) for a discussion of this matter in the textual context of the "transit star clock" (below) and von Lieven (2013) for a methodological argument.

68 Weeks 2000.

timekeeping method. Transits are a much more accurate and reliable way to observe the passing of time. Yet the method has its own set of issues.

North-south directionality is reasonably easy to determine at night, given the visibility of the northern area of the sky, even without Polaris being a clear marker in 1150 BCE (or earlier, when the Ramesside star clock was created). For the kind of accuracy that Neugebauer and Parker suggest, the two participants would have to maintain a precise alignment and distance from one another for each observation. Unlike the diagonal star clock, the Ramesside star clock could not be easily memorized because of the lack of pattern. It becomes clear that this "clock" could only really have been used in a specific location, probably within a temple precinct, as the garb of the kneeling priest indicates. The clock would have remained functional for a maximum of 60 years until the creep of the civil calendar against the solar year would have moved the stars well out of their expected positions for a given date.

Neugebauer and Parker's method is one solution to how the tables were constructed, but it is not the only possibility and it does not provide a full explanation of how the eye, ear, and shoulder positions were ascertained, when stars would have been culminating a long way over the head of the kneeling (not standing!) target figure, unless all the stars used were quite close to the horizon. Perhaps the "shoulder" positions were rather looser: there is no indication that these seven positions have to be equally spaced.[69]

Visualizations of the Ramesside star clock have often included the observer using a *merkhet* instrument (a plumb-bob line to indicate verticals), with which Parker agreed.[70] This would help to extend the meridian, eye, ear, and shoulder lines upwards in the sky a certain distance, but not at altitudes approaching the zenith or further, into the realm of the northern sky where the Hippo and Mooring Posts in the north are located. Neugebauer and Parker deferred the problem of these "northern" names in *EAT2* except to note that the Ramesside star clock Hippo is called Reret and the northern Hippo is only ever called Isis. They could have eased the way for this interpretation by labelling the constellation Reret, not Hippo, but they chose not to.[71] The corollary

69 The vertical lines in the grids are more-or-less evenly spaced, but as we also see in the Osireion sundial text, if distances were specified then there was no reason for drawings to be to scale. Here, the "distances" are specified as center-eye-ear-shoulder, and that may have been enough to suggest their relative positions.

70 Parker 1974.

71 Parker (1974) also points out that in Amenemope's onomasticon (Gardiner 1947), after the first three entries (sky, Sun, Moon, and star), only five constellations are listed: *Sahu*, *Meskhetiu*, *Ianw*, *Nakht*, and *Reret*. *Sopdet* is conspicuously absent, unless we take "star" to be self-evidently the single brightest star, Sirius. The list then moves on immediately with

is that there must also be other Lion and Mooring Post constellations in the south. This argument is convenient for any method of observation that requires a single area of focus, and it is easier to argue for two Hippos than for two Sopdets.

Neugebauer and Parker closed *EAT2* on a pessimistic note. Having described a workable scheme for creating a star clock based on transits, they follow the ramifications of their described method through to a discussion of night length, requiring the sky to be dark enough for thirteen observations to be made each night of the year. But, following stars through the tables, they finally come to the conclusion that the Ramesside star clock as they have described it would not work for seasonal hours: "Thus we see here the results of an ill-conceived scheme which attempted to combine contradictory elements, namely star transits and seasonal hours. Adding to this the effects of a rotating calendar must have given rather disastrous results" (*EAT2*, 74).

The Ramesside star clock may have disappointed Neugebauer and Parker in coming close to, but not achieving, the reproducible accuracy that deserved the word "clock," but they themselves warn against attempting to distill scientific rigor from the tables. What happens if we remove that requirement? We have a priest, sitting somewhere, with a written table of stars to observe. His job is to watch for the passing of hours (the priestly title *ỉmy-wnwt*, "hour watcher,"[72] helps with this visualization). He always sits in the same place, always looks in the same direction where a colleague (or even a statue) sits. The stars move. Somehow he distinguishes between a star in the ideal, straight ahead location, and also either mentally, or physically with his body, or using some apparatus, determines when slightly early stars or slightly late ones are in suitable positions to mark the hour.

As no star appears twice in the same table in different hour rows, it seems likely that we can limit the bounds of the *kꜥḥ* positions such that they do not overlap hour to hour. This constraint is more easily accomplished in the middle of the sky. Stars close to the southern horizon move more slowly, compared with human-centric measurements like finger-widths or hand spans, than stars nearer the celestial equator, and thus the observational error in judging their positions is reduced. Neugebauer and Parker did not discuss this.

storm-cloud, tempest, dawn, and darkness. *Sahu*, the absent *Sopdet*, *Nakht* and *Reret* are all constellations mentioned in the hour stars. The determinative used for *Reret* is a pig, not a hippo, and perhaps can be generalised to a word applicable to female mammals (my thanks to Alexandra von Lieven for this suggestion). *Meskhetiu* is the northern Foreleg (Big Dipper) and *Ianw*, "the baboon," sounds similar to *Anw*, a circumpolar constellation, but that is a god not depicted in baboon form.

72 Literally, "one who is in the hour."

The inclusion of stars in the area of Orion and Sirius narrowed their "belt" of hour stars (cf. the decanal belt discussed above) to the southern direction, well below the celestial equator yet still fairly high in the sky. However, their final comments that transits and seasonal hours did not work optimally did not lead them to question whether the southern orientation was correct. Nor has the obvious question of why the positions are given in a horizontal row rather than with any indication of altitude ever been discussed, when it is true that a star always culminates at the same altitude, and it would really help to distinguish stars to know how "up and down" they are relative to each other.

More recently, Leitz and Depuydt have revisited the tables to discuss what can be gleaned from analyzing the data within.[73] Both treat the tables as observational data. Leitz details the considerable number of assumptions required for any analysis, which remains the case to the present, but demonstrates that modern astronomical methods may be useful for eliminating impossible observational scenarios. Depuydt disagrees with some of Leitz's conclusions, but agrees that the Ramesside tables show progression in the intent of recording star movements that define time periods. He specifically attributes the increased complexity and realism in the Ramesside star clocks (compared to the diagonal star tables) to the need to make records that take into account the changing length of night across the seasons.[74]

There are still clearly many questions to be answered about Ramesside star clocks and, in particular, the interpretation of the seven positions. The anthropocentric vocabulary suggests comparison with body parts that may belong to the observer rather than to a "target figure." For example, using a hand to extend a line from one's eye, ear, or shoulder directly in front of one's body; or extending hands and fingers at arm's length (akin to making a "frame" for pretending to film video); or using a cubit rod held at arm's length horizontally in front of the body, with the middle palm representing the central position, and the three outer palms on each side defining the other six positions.

Observational systems comparable to the Ramesside star clocks are rare, but the Babylonian *ziqpu* stars offer one such system of observing culminating stars. A text describes how the observer should stand:

> If you are to observe the *ziqpu*, on the 20th of Month I, you stand in the morning before sunrise, your right to the west, your left to the east, your

73 Leitz 1995, 117–287; Depuydt 1998.
74 Depuydt 1998, 39.

> face directed towards the south; the Shoulder of the Panther stands in the middle of the sky opposite your chest, and the Crook rises.[75]
>
> MUL.APIN I iv 10–14

No similar contemporary textual reference to the physical act of observing has yet been linked to the Ramesside star clocks. However, a much more modern activity which could usefully be compared to the Ramesside star clock (and potentially diagonal star table) observations is the use of stars for irrigation timing in Oman,[76] where local features such as walls and trees serve as observational reference points.

2.3 *Transit Star Clock*

The name "transit star clock" has been applied[77] to text contained within a much longer composition, the *Book of Nut*, which has more recently been restored to its original name, the *Fundamentals of the Course of the Stars*, by von Lieven.[78] *Fundamentals* is an apposite title for the text, which describes the motion of the stars in comparison to that of the Sun. The work is headed by a mixture of text and graphical elements called the vignette (*EAT1*) or the Nut-picture (von Lieven). The part of this vignette that is known as the transit star clock is a list of dates that runs underneath the body of the sky goddess Nut, along the ground, and up her arms in a cycle. The list is divided into 36 parts, each containing three dates. The three dates are invariably labelled *tpt*, *šn dwꜣt*, and *mst*. A very few star names are included in this list, and they are of decan stars. The intention seems to have been that each set of three dates refers to one decan. The dates themselves are regular except for some writing errors. From set to set, each date progresses in 10-day increments. Within a set, the number of days between *tpt* and *šn dwꜣt* is always 90 days and that between *šn dwꜣt* and *mst* is always 70 days. The entirety of the "clock" is the reconstructed list of 36 stars, each doing three things on these very regular dates.

Neugebauer and Parker knew of and discussed two instances of the vignette: one from the Osireion in Abydos and another from the tomb of Ramses IV (KV2 in the Valley of the Kings). These two New Kingdom versions were treated extensively in the second half of *EAT1*, alongside two 1st or 2nd century CE versions of the *Book of Nut* that did not contain the vignette itself, but which made

75 I am indebted to John Steele for this reference and translation. See his Chapter 3 for more discussion of *ziqpu* stars.

76 Nash and Agius 2011.

77 For example Clagett (1995, 2: 56–59).

78 Von Lieven 2007.

reference to some of the texts. Von Lieven's extensive treatment added several more sources for the *Book*, but only two more partial sources for the vignette: a private Late Period tomb of Mutirdis (TT410 in Thebes) and a few fragments of another papyrus (*pOxford* 79/105) again from the 1st or 2nd century CE.

Neugebauer and Parker used their analysis of the *Book of Nut* to provide a definition and a location for the decans in *EAT1* Chapter 3 which they applied to their treatment of diagonal star clocks, as previously discussed. Their definition of a decan included its function to mark hourly time, but the *Book* taken as a whole focuses far more on *dates* than on *hours*. In itself, this is reminiscent of the Middle Kingdom diagonal star clock sources, which omit hours but often include dates. The vignette describes the annual motion of the stars in a mythical setting,[79] relating to ideal behavior exemplified by Sirius. The text label top center (*EAT1* Text T_2), present only in the Osireion version, states *pꜣ ỉr m 1 ꜣḫt ḫft prt Spdt* "That which is done in *I Akhet* when *Sopdet* rises" which must be taken to be captioning the whole image because it is not placed on the eastern horizon where it "belongs" in the schema of the vignette, but instead above the head of the god Shu in the middle of Nut's body, a locale we would logically expect to represent the zenith.

The beginning of Text V is critical to understanding the annual cycle of the stars. Here, periods of time are expressed in "stars," meaning in 10-day units of time. The event *tpt* is preceded by the word *wnwt* "hour," to make "first hour."[80] The terms referring to the three dates in the transit star clock date sets are rendered in the translation in bold for clarity:

> *ỉr ỉm(y)t(w)ny sbꜣ ỉr wnwt tp(t) <r> sbꜣ šnn(w) dwꜣt pꜣ 9*
> *ỉr grt ỉm(y)t(w)ny sbꜣ n msw r sbꜣ ỉrr tp(t) sbꜣ 20 pꜣ r 29 m ntt ꜥnḫ ḥr bꜣk m pt*[81]

> Between the star making the first hour and the Duat-encircling star, 9.
> Between the star which is **born** and the star making the **first** [**hour**], 20, that makes 29 living and working in the sky.

79 Von Lieven (2007) describes the text as an example of "Religious Astronomy." This separates it from being solely "scientific" in its approach.

80 This is secure because where the Osireion version is damaged, the Ramses IV version is extremely clear, and conversely, where the Ramses IV version stops before the "20," the Osireion version continues. The translation is based on a synthesis of both New Kingdom versions.

81 Transliteration composed from the two New Kingdom versions. For a full discussion with notes and variations in later versions, see von Lieven (2007, 69–70 and 400–1).

The later commentaries go into much more detail, explaining that the period of 200 days between *mst* and *tpt* is divided into 80 days in the east and 120 days of work, then the 90 days in the west (*tpt* to *šn dwꜣt*). The text following the vignette describes the 70 days of death and purification that represents the star's period of invisibility (*šn dwꜣt* to *mst*). The total is 360 days.

The link between the sets of three dates and the idea of a transit star clock is thus a process of deduction. The 120 days of "working" is separated from the horizon events by the periods in the east and west, which seems to place "work" as the part of the stars' yearly journey that the Sun does not interfere with, away from the horizon: culmination. Whether or not the 120 days was a later addition to the understanding of this text, Text V tells us that the event *tpt* is definitely time-related, and this drew Neugebauer and Parker to the conclusion that the vignette describes a system of timekeeping via transits. That system is not made explicit within the text itself. We know that it cannot refer directly to the Ramesside star clock because a) the Ramesside star clock deals in 15-day intervals and b) the Ramesside star clock does not use decans. It also does not, in my opinion, cement the idea of the celestial meridian being a well-defined concept in New Kingdom Egypt.

The *Book of Nut* both hints at the use of the high-points of stars being important and, simultaneously, takes away the ability for decans to perform accurate sequential transits that could count off regular periods of time. The reason for this problem is that we now have a set of criteria for being a decan that the real sky with its random sprinkling of stars cannot fulfil:

1) A decan must rise and set (source: *Book of Nut*, which deals solely with rising and setting stars). This precludes circumpolar stars from being decans.
2) A decan must rise (normally not heliacally) over the eastern horizon in sequence with its neighboring decans (source: diagonal star clocks, in which decans rise sequentially throughout the night). If the tables truly are clocks, this rising must happen with reasonably even separation from those neighboring stars to form hours.
3) A decan must be invisible for 70 days, following the model of Sirius (source: *Book of Nut*). So heliacal rise and heliacal set must be 70 days apart.
4) A decan must culminate near dawn 200 days after heliacal rise (source: *Book of Nut*).
5) A decan must culminate in sequence with, and at reasonably even separation from, its neighboring decans (source: Neugebauer and Parker's deduction from the *Book of Nut*). This is the criterion that turns the *Book of Nut* vignette system into a transit star clock.

Neugebauer and Parker attempted to make criteria 1–3 work to form a diagonal star clock, and there we needed to fall back on the idea that the diagonal star clock was schematic rather than an accurate, observational method, because the diagonal pattern is so regular. I have argued earlier in this chapter that preserving criteria 1 and 2 are sufficient, with 3 being an "ideal" that was not enforced.

Criteria 4 and 5 present much greater difficulties astronomically, because stars which rise together do not transit together, nor do they set together. Stars which rise with a certain time-separation do not transit or set with that same time-separation either. The variance comes from their declination, which maps onto their position on the horizon more to the north or to the south. It also depends a little on their magnitude, as stars which are brighter can be seen slightly lower on the horizon, that is, earlier than their faint brethren. This is only considering normal risings and settings, though. Heliacal rise and set, when the Sun is close by but also moving up and down the horizons over the course of the year, is much more complicated. Whereas culminations are absolutely regular and are independent of declination, magnitude, and latitude of the observer, heliacal rise and set dates are dependent on all those factors and more. Pinning down real stars that behave in even approximately the way that criteria 1–5 require is impossible.

The *Fundamentals of the Course of the Stars*, as we should now call the *Book of Nut*, is a title which may indicate that what is described within is the essence of star movements rather than the observed reality. Criteria 1–5 could indeed represent the fundamental behavior of Sirius, but the 35 other stars will not reach those levels of perfection.

What is certain is that the date texts in the vignette do not, on their own, describe a clock. The absence of most of the star names and the lack of hour-by-hour details (or even a general schema for how such hours would be produced) means that, at best, the text indicates that stars were thought to "work" for 120 days, and that the calendrical aspects of their movements were placed above their potential timekeeping abilities in this particular text.

3 Conclusion

Using "hour" as a translation for the term *wnwt* poses no conceptual difficulties in literary texts where it is used figuratively. However, we have seen that, in the contexts of tables and objects which could have been used to define time, the details and nuances of its meaning are far more difficult to analyze.

Each type of sundial, star clock, or (as we see in the next chapter) water clock maintains its own relationship with time, and the links between these devices, if any, remain obscure. We are equally hampered by the issue of which types of timekeeping traditions have survived and which have been lost. From our surviving artefacts of Egyptian design we find no evidence that a continuous count of 24 hours was measured: day hours and night hours consistently retain their distinct flavors, and the existence of some sort of liminal, no-hour parts of the day around sunrise and sunset seems likely. However, in all cases, our understanding of the nature of marked time periods is colored by our assumptions about methods of construction, use, and, perhaps above all, motivation for timekeeping.

References

Allen, James P. 2005. *The ancient Egyptian pyramid texts*. Atlanta, GA: Society of Biblical Literature.

Bickel, Susanne, and Rita Gautschy. 2014. "Eine ramessidische Sonnenuhr im Tal der Könige." *Zeitschrift für Ägyptische Sprache und Altertumskunde* 141 (1): 3–14.

Borchardt, Ludwig. 1910. "Altägyptische Sonnenuhren." *Zeitschrift für Ägyptische Sprache und Altertumskunde* 48: 9–17.

Borchardt, Ludwig. 1911. "Eine Reisesonnenuhr aus Ägypten." *Zeitschrift für Ägyptische Sprache und Altertumskunde* 49: 66–68.

Borchardt, Ludwig. 1920. *Die Geschichte der Zeitmessung und der Uhren. Band 1. Lief. B. Die altägyptische Zeitmessung*. Berlin: de Gruyter.

Bruins, E.M. 1965. "The Egyptian Shadow Clock." *Janus* 52: 127–37.

Capart, Jean. 1933. "Tables astronomiques d'Assiout." *Chronique d'Égypte* 15: 69–77.

Catamo, Mario, Nicoletta Lanciano, Kurt Locher, Manuel Lombardero, and Manuel Valdés. 2000. "Fifteen Further Greco-Roman Sundials from the Mediterranean Area and Sudan." *Journal for the History of Astronomy* 31: 203–21.

Clagett, Marshall. 1995. *Ancient Egyptian Science: A Source Book. Volume 2: Calendars, Clocks, and Astronomy*. Vol. 2. Philadelphia: American Philosophical Society.

Clédat, Jean. 1915. "Notes sur l'isthme de Suez (monuments divers)." *Recueil de travaux relatifs à la philologie et à l'archéologie égyptiennes et assyriennes* 37: 33–40.

Daressy, Georges. 1900. "Une ancienne liste des décans égyptiens." *Annales du service des antiquités de l'Égypte* 1: 79–90.

Daumas, François. 1995. *Valeurs phonétiques des signes hiéroglyphiques d'époque gréco-romaine*. Vol. 4. 4 vols. Montpellier: Université de Montpellier.

Depuydt, Leo. 1998. "Ancient Egyptian star clocks and their theory." *Bibliotheca Orientalis* 55: 5–44.

Depuydt, Leo. 2010. "Ancient Egyptian star tables: A reinterpretation of their fundamental structure." In *Writings of early scholars in the ancient Near East, Egypt, Rome, and Greece: translating ancient scientific texts*, edited by Annette Imhausen and Tanja Pommerening, 241–276. Beiträge zur Altertumskunde 286. Berlin: de Gruyter.

Dieleman, Jacco. 2014. "A Fragment of a Diagonal Star Table from the Los Angeles County Museum of Art (LACMA M80.202.500)." *Journal of the American Research Center in Egypt* 50: 221–32.

Evans, James, and Marcel Marée. 2008. "A Miniature Ivory Sundial with Equinox Indicator from Ptolemaic Tanis, Egypt." *Journal for the History of Astronomy* 39 (1): 1–17.

Frankfort, Henri. 1933. *The Cenotaph of Seti I at Abydos*. London: Egypt Exploration Society.

Gardiner, Alan Henderson. 1947. *Ancient Egyptian Onomastica*. Oxford: Oxford University Press.

Gibbs, Sharon L. 1976. *Greek and Roman Sundials*. New Haven, CT: Yale University Press.

Graßhoff, Gerd, Elisabeth Rinner, Karlheinz Schaldach, Bernhard Fritsch, and Liba Taub. 2016. *Ancient Sundials*. [Online database] accessed June 5, 2019, https://doi.org/10.17171/1-1.

Griffith, Francis Llewellyn, and W.M. Flinders Petrie. 1889. *Two Hieroglyphic Papyri from Tanis*. London: Egypt Exploration Society.

Hoffmann, Friedhelm. 2016. "Ägyptische astronomische Texte." In *Translating Writings of Early Scholars in the ancient Near East, Egypt, Rome, and Greece—Methodological Aspects with Examples*, edited by Annette Imhausen and Tanja Pommerening, 335–78. Beiträge zur Altertumskunde 344. Berlin: de Gruyter.

Hornung, Erik. 1999. *The Ancient Egyptian Books of the Afterlife*. Translated by David Lorton. Ithaca, NY: Cornell University Press.

Kahl, Jochem. 1993. "Textkritische Bemerkungen zu den Diagonalsternuhren des Mittleren Reiches." *Studien zur Altägyptischen Kultur* 20: 95–107.

Krauss, Rolf. 2018. "Über die L-förmigen Schattenuhren und die Schlacht von Megiddo." *Studien zur Altaegyptischen Kultur* 47: 149–75.

Kroeper, Karla, and Lech Krzyżaniak. 1998. "The Amun Temple Complex. Preliminary Report, Seasons 1 and 2." *Archéologie du Nil moyen* 8: 204–5.

Lapp, Günther. 1985. *Särge des Mittleren Reiches aus der ehemaligen Sammlung Khashaba*. Wiesbaden: Harrassowitz.

Legrain, Georges Albert. 1894. *Collection H. Hoffmann. Catalogue des antiquités égyptiennes*. Paris.

Leitz, Christian. 1995. *Altaegyptische Sternuhren*. Leuven: Peeters.

Leitz, Christian. 2014. *Die Gaumonographien in Edfu und ihre Papyrusvarianten: ein überregionaler Kanon im spätzeitlichen Ägypten: Soubassementstudien III*. 2 vols. Studien zur spätägyptischen Religion 9. Wiesbaden: Harrassowitz.

Lieven, Alexandra von. 2007. *The Carlsberg Papyri 8: Grundriss des Laufes der Sterne. Das sogenannte Nutbuch*. 2 vols. Carsten Niebuhr Institute Publications 31. Copenhagen: Museum Tusculanum Press.

Lieven, Alexandra von. 2013. "Why Should We Date Texts by Historic Linguistic Dating?" In *"Dating Egyptian Literary Texts": Göttingen, 9–12 June 2010*, edited by Andréas Stauder and Gerald Moers, 161–76. Hamburg: Widmaier.

Lieven, Alexandra von. 2016. "The Movement of Time. News from the 'Clockmaker' Amenemhet." In *Rich and Great: Studies in Honour of Anthony J. Spalinger on the Occasion of His 70th Feast of Thoth*, edited by Renata Landgráfová and Jana Mynářová, 207–31. Prague: Charles University in Prague.

Locher, Kurt. 1992. "Two Further Coffin Lids with Diagonal Star Clocks from the Egyptian Middle Kingdom." *Journal for the History of Astronomy* 23 (3): 201–7.

Macalister, Robert Alexander Stewart. 1912. *The Excavation of Gezer 1902–1905 and 1907–1909*. London: Palestine Exploration Fund.

Müller-Roth, Marcus. 2008. *Das Buch vom Tage*. Orbis Biblicus et Orientalis 236. Göttingen: Vandenhoeck & Ruprecht.

Murray, Margaret A. 1904. *The Osireion at Abydos*. London: B. Quaritch.

Nash, Harriet, and Dionisius A. Agius. 2011. "The Use of Stars in Agriculture in Oman." *Journal of Semitic Studies* 56 (1): 167–82.

Naville, Edouard. 1914. "Excavations at Abydos: the great pool and the tomb of Osiris." *Journal of Egyptian Archaeology* 1 (3): 159–67.

Neugebauer, Otto, and Richard A. Parker. 1960. *Egyptian Astronomical Texts Vol. 1*. Providence, RI: Brown University Press.

Neugebauer, Otto, and Richard A. Parker. 1966. *Egyptian Astronomical Texts Vol. 2*. Providence, RI: Brown University Press.

Neugebauer, Otto, and Richard A. Parker. 1969. *Egyptian Astronomical Texts Vol. 3*. Providence, RI: Brown University Press.

Nilsson, Martin P. 1920. *Primitive Time-Reckoning: A Study in the Origins and First Development of the Art of Counting Time among the Primitive and Early Culture Peoples*. Lund: C.W.K. Gleerup.

Parker, Richard A. 1974. *Ancient Egyptian Astronomy*. London: The Royal Academy.

Petrie, W.M. Flinders. 1900. *Dendereh, 1898*. London: Egypt Exploration Fund.

Pogo, Alexander. 1932. "Calendars on Coffin Lids from Asyût (Second Half of the Third Millennium)." *Isis* 17: 6–24.

Roberson, Joshua Aaron. 2012. *The Ancient Egyptian Books of the Earth*. Atlanta, GA: Lockwood Press.

Roeder, Günther. 1911. *Les temples immergés de la Nubie: Debod bis Bab Kalabsche*. Cairo: Institut français d'archéologie orientale.

Roulin, Gilles. 1996. *Le livre de la nuit: une composition égyptienne de l'au-delà*. 2 vols. Orbis Biblicus et Orientalis 147. Göttingen: Vandenhoeck & Ruprecht.

Salmas, Anne-Claire. 2014. "La mesure du temps de la journée (2). Modules et fonctionnement des horloges à ombre récentes et des cadrans solaires." *Bulletin de l'institut français d'archéologie orientale* 114: 419–46.

Schaldach, Karlheinz. 2006. *Die antiken Sonnenuhren Griechenlands: Festland und Peloponnes*. Frankfurt am Main: Harri Deutsch.

Scott, Nora E. 1935. "An Egyptian Sundial." *Bulletin of the Metropolitan Museum of Art* 30: 88–89.

Symons, Sarah L. 2002a. "Egyptian Shadow Clocks." In *Proceedings of the XXth International Congress of History of Science Vol. XVI*, edited by M. Dorikens, 13–20. Turnhout: Brepols.

Symons, Sarah L. 2002b. "Two Fragments of Diagonal Star Clocks in the British Museum." *Journal for the History of Astronomy* 33 (3): 257–60.

Symons, Sarah L. 2007. "A Star's Year: The Annual Cycle in the Ancient Egyptian Sky." In *Calendars and Years: Astronomy and Time in the Ancient World*, edited by John M. Steele, 1–33. Oxford: Oxbow Books.

Symons, Sarah L. 2016. "Challenges of Interpreting Egyptian Astronomical Texts." In *Translating Writings of Early Scholars in the Ancient Near East, Egypt, Rome, and Greece—Methodological Aspects with Examples*, edited by Annette Imhausen and Tanja Pommerening, 379–401. Beiträge zur Altertumskunde 344. Berlin: de Gruyter.

Symons, Sarah L., Robert Cockcroft, Jesse Bettencourt, and Cody Koykka. 2013. *Ancient Egyptian Astronomy*. [Online database], accessed June 5, 2019, http://aea.physics.mcmaster.ca/.

Symons, Sarah L., and Himanshi Khurana. 2016. "A Catalogue of Ancient Egyptian Sundials." *Journal for the History of Astronomy* 47 (4): 375–85.

Tupikova, Irina, and Michael Soffel. 2012. "Modelling Ancient Sundials: Ancient and Modern Errors." In *Productive Errors: Scientific Concepts in Ancient Sciences*, edited by Klaus Geus and Mark J. Geller, 93–114. Berlin: Max-Planck-Institut für Wissenschaftsgeschichte.

Weeks, Kent R., ed. 2000. *Atlas of the Valley of the Kings: Theban Mapping Project Pt. 1*. Cairo: The American University in Cairo Press.

Winlock, Herbert Eustis. 1942. *In Search of the Woman Pharaoh Hatshepsut: Excavations at Deir El-Bahri 1911–1931*. London: Kegan Paul.

CHAPTER 2

The Ancient Egyptian Water Clock between Religious Significance and Scientific Functionality

Alexandra von Lieven and Anette Schomberg

The water clock was one of several possible ways to measure hours in Ancient Egypt. It was specifically designed to be independent from weather conditions that might limit the visibility of the Sun or the stars, a necessary precondition for other methods. Nevertheless, the close link to the sphere of astronomy, particularly the night sky, is clearly reflected in the standard decoration of water clocks. This ties in with the fact that measuring hours was of utmost importance for several ritual contexts. The first part of this chapter will address these religious aspects of the hours in general and the decoration of water-clocks in particular; the second part will deal with the technical aspects of the clepsydrae.

1 Cultural Aspects of the Hours

Of course, a culture could use units of "short time"—i.e., hours—in different contexts. In Ancient Egypt, the "hour" as the smallest measured time unit had an enormous significance within the religion and its rituals. Outside this sphere, it does not seem to have held an important position, although occasionally the duration of a procedure for a certain number of hours is mentioned.[1] Moreover, there are rare instances where particular points in time are designated as "the hour of doing *x*" (meaning, e.g., noon).[2]

1 For example, in the tale known as *The Pleasures of Fishing and Fowling*, B4,9 (Caminos 1956, 14 and pl. 3A–3), and in the *Satirical Letter of Papyrus Anastasi I*=pBM EA 10247 17.1 and 18.1 (Fischer-Elfert 1992, 118 and 122, translation Fischer-Elfert 1986, 143 and 149).

2 Again in the *Satirical Letter of Papyrus Anastasi I* 4.6 and 20.1 (Fischer-Elfert 1992, 52 and 128, translation Fischer-Elfert 1986, 50–51 [with note a] and 160). However, while the former indicates a precise time, namely noon, the latter just mentions "the hour of departure," which could be any time.

 | DOI:10.1163/9789004416291_004

1.1 *The Religious Significance of the Hours*

In religion, however, days and nights were structured by means of hours in many different contexts. These are usually linked either to the cult of the Sun god Re[3] or to the god of the dead, Osiris.[4] Since Osiris had an astral aspect as Orion (who himself consisted of several decans, for which see below) or as the Moon, and could even be understood as a nocturnal form of Re, it is quite fitting that it is just these two gods and their manifold cults that made prominent use of the hours. Apart from that—and conceptually linked to it—the hours served also in protection rituals for human beings, in particular the reigning king.[5]

Most obvious is the use of the hours in the solar cult, where it structured many temple rituals, foremost among them the so-called *Stundenritual* (Ritual of the Hours).[6] This consisted of hymns being presented to the god Re each hour. At the beginning of each hour text, the respective hour and its tutelary deity are identified. Also, during each hour, Re himself takes different forms and aspects, which the hymns extol. It is unclear whether the *Stundenritual* was first composed just for the hours of the day (this version is attested many times) and only subsequently expanded with a similar, more rarely attested ritual for the hours of the night, as well, or whether both parts belong together from the start. The latter seems more logical, but the daily and nightly parts do exhibit some marked differences in their details. In particular, the text for the night is heavily interspersed with excerpts from texts otherwise known as spells from the *Book of the Dead*. However, this could have been done on purpose. Moreover, as these spells themselves sometimes have long and complicated histories, their ultimate origins are perhaps not as straightforward as they would seem. In general, the relationship of certain spells known from the *Book of the Dead* to certain cosmographical compositions needs to be researched in much greater depth.[7] Interestingly for the history of all these compositions, in some cases it is even not a particular *Book of the Dead* spell, but rather its respective *Coffin Text* antecedent that is incorporated into a cosmographic text.[8] Besides, one should not think that the *Coffin Texts* or the *Book of the Dead* was each one fixed book (they are fluid corpora of spells), nor should the

3 Quirke 2001; von Lieven 2010.
4 Assmann 2008; Kucharek 2010; Smith 2017.
5 Pries 2009; 2010.
6 Graefe 1995; 2001 to present; 2014; 2017; forthcoming; Griffin 2017; 2018.
7 Von Lieven 2002; Leroux 2013–15.
8 Müller-Roth 2008, 536–544.

modern names suggest an inherent funerary nature and origin for these texts, beyond the fact that most preserved attestations were found in tombs.

At any rate, the *Stundenritual* is closely linked to the *Book of the Day*[9] and its counterpart the *Book of the Night*.[10] This composition (or these; again, it is a matter of contention whether they are one book, of which the nightly part is attested much more often, or whether they are two originally separate books), as well as the *Amduat*[11] and the *Book of Gates*[12] are cosmographic texts organized into twelve hours each. The reason is that they are focused on the Sun god Re and his travels over the course of a day or night. The night seems to have held the greater fascination, as of course it was necessary to explain what the Sun does while it is invisible. However, this impression is perhaps simply due to the preservation of the sources, among which tombs, with their penchant for the night, dominate in contrast to temples.

All three cosmographies (i.e., *Amduat*, *Book of the Day* and *Book of the Night*, and *Book of Gates*) share the idea that an hour is not just a temporal unit, but also a spatial one. The Sun has to traverse one hour to reach the next one. As the seeming movement of the Sun over the sky can actually be observed, this idea is in line with natural facts, even though, in reality, it is of course the Earth that moves, and not the Sun.

Each hour is personified by a goddess, and is separated from the next by a gate (this is where the *Book of Gates* takes its modern name from), each of which has at least one guardian. As the different cosmographies exhibit variants in this respect, and there were even further traditions, which relate to other, often unknown texts, the Egyptians themselves at some point had to collect all of this information in lists. Thus, the *Onomastica* from Tebtynis (manuscripts preserved from the second century CE)[13] offer several different lists of such entities.

Apart from these ideas, there were also other, at least ultimately related conceptions that posited that the Sun god would change his form over the course of a day.[14] Not only did he himself transform within each hour, but so did his

---

9 Piankoff 1942; Müller-Roth 2008; Hornung 1997.

10 Piankoff 1942; Roulin 1996; Hornung 1997.

11 Hornung 1963; 1967; 1987; 1992a; 1992b; 1994; 1997, 40–55, 117–128; Warburton 2007; Darnell and Manassa 2018, 127–248.

12 Hornung 1979; 1984; 1992a; 1997, 56–66, 130–141; Zeidler 1999; Darnell and Manassa 2018, 249–336.

13 Osing 1998; Quack (forthcoming).

14 Von Lieven 2010. This concept is also behind his twelve different names, for which see Gasse 1984.

bark, its crew, and even his enemy, Apopis. A very nice depiction of this transformation over the twelve hours of the day is to be found on the astronomical ceiling of the hypostyle hall in the temple of Dendara.[15] The ultimate culmination of these concepts can be found within the *Dodekaoros*, a list of twelve different animal shapes the Sun takes over twelve double-hours of day and night. This theoretical construct is found in some Greek-language magical papyri from Egypt and in pictorial form on the *Zodiaque Daressy*, and it ultimately made its way via the Silk Route to Central Asia and China, where it morphed into what is today known as the Chinese or East Asian Zodiac.[16]

Individual hours that were apparently especially in need of ritual support are mentioned in further texts from the solar cult like, e.g., the *Litany of the Sun*.[17] It is explicitly stated that one should do certain practical actions as well as recite certain given texts at this time.[18]

The whole cycle of 12 hours is not only relevant for the solar cult, but also for the cult of Osiris, the god of the dead. The theological rationale for this is implicitly given in some of the cosmographic texts, where it is stated that Re and Osiris join during the night in a mystical union, although this idea has to be seen in a more nuanced way than is usually done.[19] Thus, Osiris himself can be understood as the form of the Sun at night. As its setting equals death in a way, this is only logical. In the morning, then, the Sun is born again as a rejuvenated being. This is also the reason for the funerary adaptation of the cosmographic texts, to which they largely owe their preservation.

Apart from the cosmographies themselves, the hours are most prominent in the ritual of the vigils for Osiris. However, also for these rites one has to reckon with several different variants, apparently used in different times and places

15 Cauville 2012, 10–19, pl. VI; 2013, 506–8. Unfortunately, the pictures are reproduced at too small a scale to actually see anything.

16 Von Lieven 2018. The changed places of certain animals between the *Dodekaoros* and the East Asian Zodiac might seem to be one of the major obstacles to the validity of this line of transmission, which I could not totally rebut in this paper (originally from 2006). However, at a recent (2018) workshop at the Max Planck Institute for the History of Science in Berlin, there were two papers given by specialists in Chinese Studies, who independently from each other also offered evidence for the same phenomenon in relation to the Western zodiac in Chinese sources, but could not explain it. Whatever the reasons might be in each case, this finally proves that such a development is possible, and thereby adds the last grain of plausibility to my earlier arguments in relation to the *Dodekaoros*.

17 Hornung 1975; 1976.

18 Von Lieven 2002, 52–56; 2010, esp. 49.

19 See the critical appraisal by Smith (2017, 299–336), elaborating on a remark by Quack (2012, 180–182).

and with different purposes. While far from being unravelled in every detail, it is certainly clear that matters are not as simple and one-dimensional as is sometimes assumed in scholarly literature, where everything is associated with one specific text known because of its first modern edition as "*Stundenwachen des Osiris.*"[20]

In reality, it is important to differentiate between the Osirian *Stundenwachen*, the solar *Stundenritual*, and further different concepts. The fact that the *Stundenritual* can also enter the funerary sphere, via Osiris as the nightly Sun, is best demonstrated by some Late Period coffins employing texts and pictures for each hour around the dead body.[21] In such contexts, the compositions used are not only differentiated by the texts themselves, but even by the iconography of the hours. While hours of the day always wear a Sun disk on their heads, and hours of the night always a star (alternatively, they hold them in their hands), even a seemingly irrelevant detail, like the question of whether they are depicted standing or kneeling, does make a difference.[22] This clearly proves that the individual textual and pictorial elements present on such coffins ultimately hark back to fixed templates derived from entirely different traditions.

These texts are not to be confused with the Osirian liturgies,[23] which were performed on certain occasions, especially during the festival of Khoiak, by two priestesses impersonating Isis and Nephthys at the bier of Osiris (occasionally also by other priests in the roles of other deities). There were several varieties of such texts, but most of them are also organized into twelve hours.

A different strand of tradition must be behind the so-called *Book of Hours*,[24] a long offering liturgy addressing forms of Osiris and some other deities during different hours. Because of the fragmentary preservation of the manuscript, its further setting must remain unclear, but the general temporal framework is explicitly given within the text itself.

The latter texts are usually preserved on unillustrated manuscripts. The hours are not named—as in the *Stundenritual* and its illustrated offshoots on coffins—but they are simply numbered from 1 to 12. It seems that the naming of individual hours, as opposed to their numbering, is a propensity of the texts

20 Junker 1910; Pries 2011.

21 Graefe 2018; forthcoming.

22 As demonstrated by Graefe 2018, esp. 112.

23 Von Lieven 2006a; 2006b; Kucharek 2010. A new edition of one of the most important such texts, the *Glorification by the Two Sisters*, by Kucharek is almost finished; further volumes from her edition project are in preparation.

24 Faulkner 1958; von Lieven 2002, 51.

that originate from the solar cult, even though they might have been adapted for a funerary context, while the originally Osirian texts prefer mere numbers.

It should be noted that the personification of temporal units of different length was in general widespread and is not restricted to the hours.[25]

1.2 *Measuring the Hours*

To measure the hours, the Ancient Egyptians had different techniques. The most obvious way was, of course, to observe the movements of the heavenly bodies.[26] During the day, the Sun was used. For this purpose, different types of Sun and shadow clocks were developed.[27] As they had to be moved over the day to be of use, their handling was not exactly trivial. There are original exemplars extant, as well as instructions for their use.

The nightly counterpart were the diagonal star clocks,[28] using the risings and settings of particular constellations, known today as decans[29] because each ten days another one of them had its heliacal rising, while a different one completely disappeared from view for a longer period.[30] According to the model of Sirius, this period of invisibility was canonically set to be 70 days, although this is not astronomically exact for the historical period. For timekeeping, their culmination was also important. Yet there were many problems with this system. Certainly the greatest was the lack of an intercalary day every four years, which led to a rather quick desynchronization between the natural year, with its phenomena (including seasonally changing phenomena in the night sky), and the calendar. This, as well as the effects of precession, might have been the reasons why over time different "decan families" were defined.[31] Still, while religiously of great importance, decans were a very impractical way to measure the hours exactly, at least over a longer period of time.

It is therefore no wonder that a better technique was sought—and eventually found. This was the water clock or clepsydra. It is not clear when exactly the different types of timekeeping methods were invented, but the word for "hour"

---

25 Mendel 2005; Quack 2013; von Lieven 2017.

26 See the contribution by Symons to the present volume (chapter 1).

27 Borchardt 1920, 26B–53B = Wuensch and Sommer 2013, 128–167 (with review, von Lieven 2013a); Frankfort 1933, 76–80, pl. LXXXII–LXXXIII; Quack 2006; Leitz 2014, 489–491, pl. 116.

28 Neugebauer and Parker 1960, 1–35, 95–121, pl. 1–29; Locher 1998.

29 Quack 2002.

30 The concept is explained in detail by the Egyptians themselves in the demotic commentary on the *Fundamentals of the Course of the Stars*, §§ 39–51; see von Lieven 2007, 61–71, 142–150.

31 Neugebauer and Parker 1969, 105–174; Quack 2002.

is, according to currently available evidence, attested from at least the time of the *Pyramid Texts* (i.e., attested from the mid-3rd millennium BCE, but likely composed much earlier).[32]

The invention of the water clock was hitherto always set in the early 18th dynasty (1550 BCE) and connected with the autobiography of a certain high-ranking official by the name of Amenemhet, who was thought to have claimed to be its inventor. However, a recent re-examination of the issue proved that Amenemhet did not claim to have invented the water clock.[33] Instead, he claims to have improved upon it by inventing a super clock that links elements of a water clock, a shadow clock, and other elements—most importantly, movable figures, operated by a simple type of mechanic (possibly hydraulic, linked to the outflow of the water from the water clock). Unfortunately, the inscription is too poorly preserved to make its precise contents clear. At any rate, Amenemhet describes moving figures, which are not the standard decoration of water clocks.

1.3 *The Decorative Programme of the Clepsydrae*

While not all water clocks followed it, there was nevertheless a certain standard template for their external decoration, which clearly links them to the world of astronomy (see Figure 2.1). This template is the *Classical Sky Picture*, which otherwise is often found on the ceilings of temples or tombs, as well as on the inside of sarcophagus and coffin lids.[34] As all of these could be understood as symbolic representations of the sky, it is evident that the water clock was also thereby functionally transformed into a miniature sky. That purpose is made even clearer by a line of inscription that typically accompanies the picture and surrounds the rim of the clepsydra. It clearly states the clepsydra's purpose as marking the hours when it is cloudy, and therefore neither the Sun nor the stars are visible. Thus, it is evident that the water clock is a substitute for the sky, as it operates entirely independently from the weather conditions. Implicitly, it also dispenses with all the other problems inherent in the other types of clocks. Instead, it offers a remarkable degree of accuracy, which only recently has been rediscovered (see below).

32 For attestations, see Allen 2005, Index 457. Interestingly, there the hours are linked in particular with the stars, a fact further brought to the fore by the orthography of the word, which is often written with a star determinative in hieroglyphic script. The whole question is treated in detail by Quack (forthcoming).

33 Von Lieven 2016.

34 Most attestations are to be found in Neugebauer and Parker 1969. The handy designation was coined by Quack (2002a).

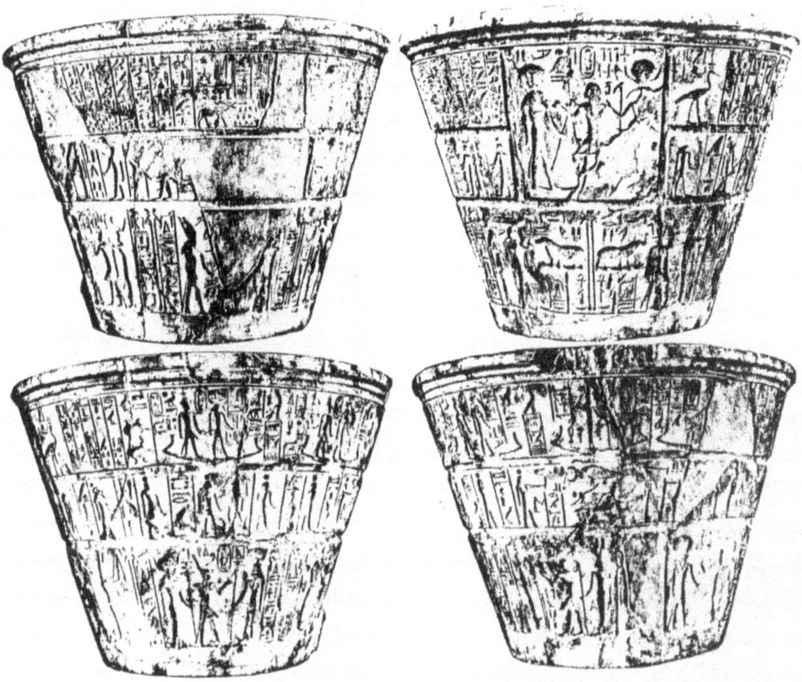

FIGURE 2.1 Different views of the outside decoration of the oldest preserved water clock from Karnak, bearing the names of Amenhotep III. The decoration was originally inlaid with glass paste. It shows the Classical Sky Picture above and the king offering to the deities of the months below.
AFTER NEUGEBAUER AND PARKER 1969, PL. 2

What does the *Classical Sky Picture* depict? It is typically divided into two halves, representing the northern and southern skies, respectively. In this form it would, of course, be problematic to accommodate the spout necessary for letting the water flow out. As the latter is vital for the technical functionality of the clock, the water clocks adapt the decorative scheme of the *Classical Sky Picture* in so far as they add a third register below south and north (compare Figures 2.1 and 2.2). This contains 12 figures of the king, each making an offering to one of 12 deities. The names make it clear that these are deities associated with festivals celebrated in each of the 12 months, so that they can be regarded as early forms of the tutelary deities of the months. Such tutelary deities are well attested later,[35] although their names and iconography then

35 Mendel 2005.

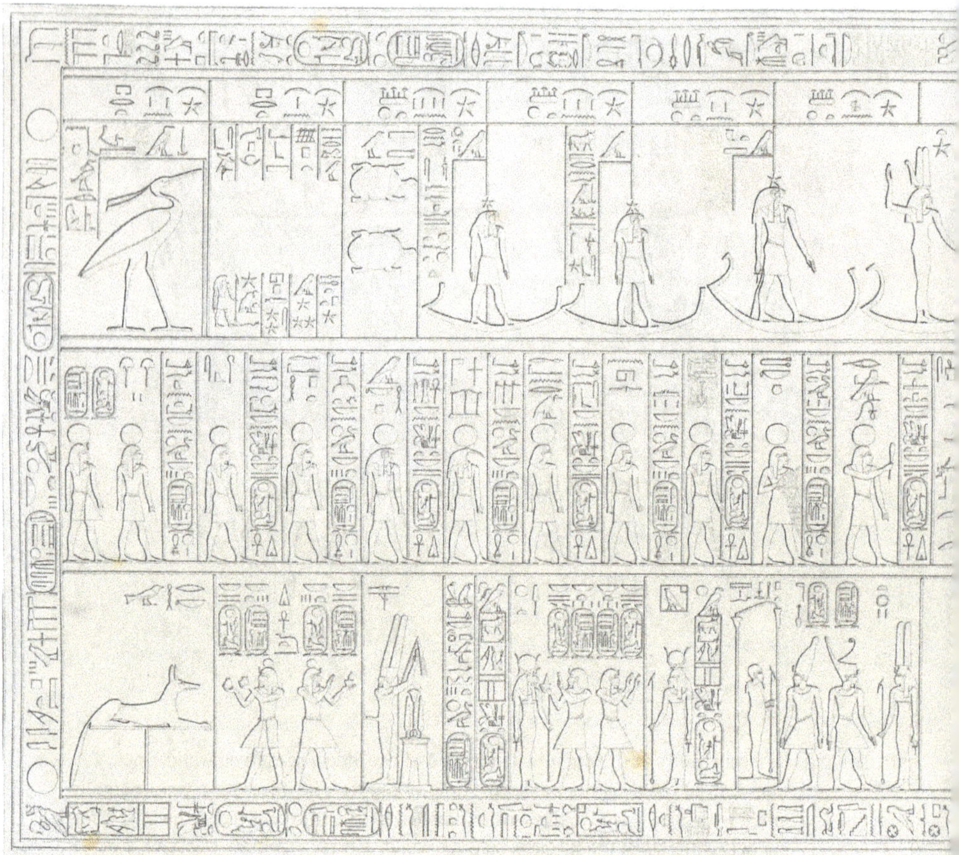

FIGURE 2.2 Ceiling in the Ramesseum, showing the template of a water clock, with the *Classical Sky Picture* above and the king (Ramesses II) offering to the deities of the months below. Apart from the general layout, the presence of a baboon in frontal view in the center of the lowest register indicates the model of a water clock for this depiction.
AFTER NEUGEBAUER AND PARKER 1969, PL. 5

are different—as are, in fact, the names of the months themselves.[36] The water clocks, however, follow the earliest attested list of month names, with six to either side of a central field in which sits a frontally-shown baboon on a *djed* pillar. He guards the water spout. The baboon is, of course, a form of the god Thoth, god of the Moon and thus lord of counting and calculation, as well as of every other science as well. The *djed*-pillar is a hieroglyph with the ideographic and symbolic meaning of "stability" (the word *djed* means "duration,

36 Parker 1950; Depuydt 1997.

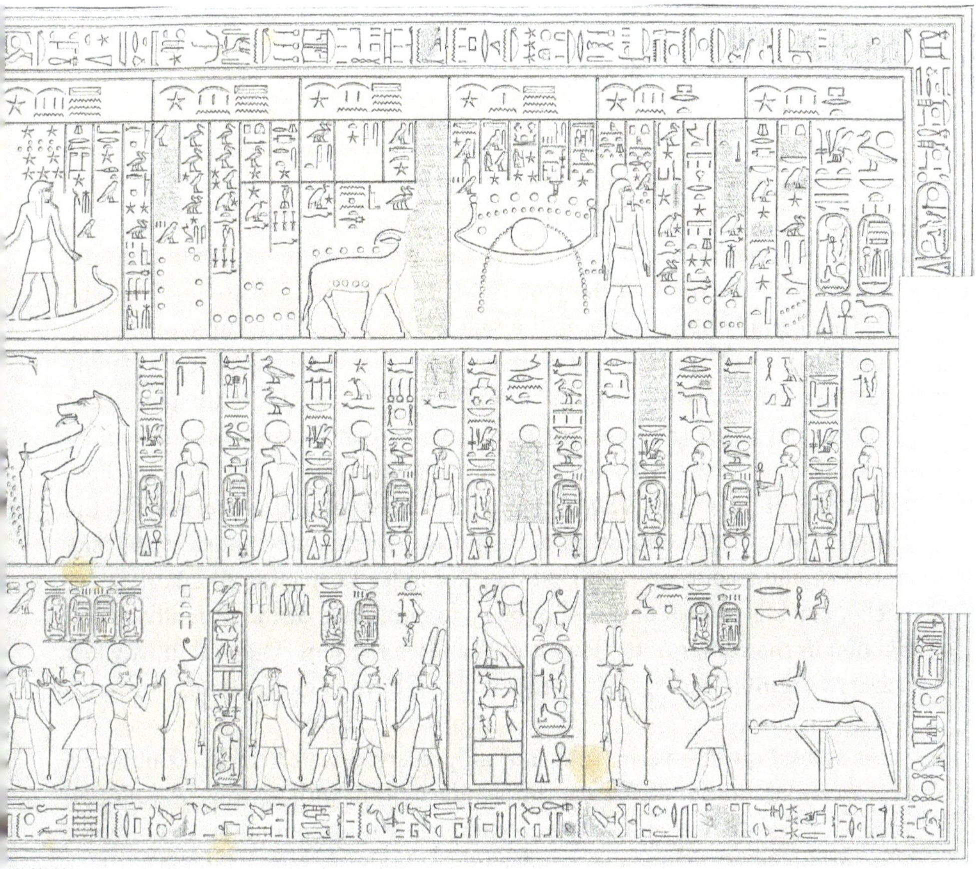

stability"). It is this two- or even three-dimensional depiction of Thoth sitting above the spout of a water clock that is mentioned by Horapollo (1.16),[37] who explicitly connects the baboon with the hours.

The middle register has the constellations of the northern sky in its center, most prominent of which is, of course, the Big Dipper. It is held in place on a chain by a large hippopotamus goddess, who herself must be a circumpolar constellation, although less easily identifiable. To the left and right, there are the gods of the lunar phases. Finally, in the uppermost register, there are the planets and southern constellations (including the decans).

The use of the *Classical Sky Picture* for decorating the outsides of water clocks must be a very ancient practice, although there are few actual clocks with the image preserved. Still, if one looks at all the preserved attestations

37 Thissen 2001, 16–17.

of the *Classical Sky Picture* from (currently) the Middle Kingdom until the Roman Period (early second millennium BCE to first century CE), it is clear that several markedly different strands of tradition can be pinpointed. These differ most significantly in the decan family used, but also in other more or less prominent details.[38] Interestingly, these different strands cluster together in certain contexts. Thus, the preserved water clocks with the *Classical Sky Picture* use the Main Group of the Senmut family of decans. The same decan family, as well as some other elements, are shared by several monuments from the animal cult,[39] the tomb of the famous Senenmut,[40] at least some of the sizable Late Period nobles' tombs in the Asasif,[41] the Meroitic pyramid of a queen called Kheñuwa,[42] and some Ptolemaic coffins.[43] However, the Senmut family of decans also has two subgroups, A and B, and the ceilings of the temples of Ramses II in Luxor (Ramesseum, see Figure 2.2) and Ramses III in Medinet Habu (both in Western Thebes) constitute subgroup A.[44] Remarkably, these two copies of the *Classical Sky Picture* exhibit the typical three registers of the template for a clepsydra instead of the normal version for ceilings, with just a northern and a southern sky. That they are indeed copied from a Vorlage for such a clock is proven beyond doubt by the presence of the frontally-shown baboon in the center of the lowest of the three registers. The reason why just these two temple ceilings follow the model of a water clock remains unfortunately unclear.

As stated, on the other hand, not all preserved water clocks exhibit this type of decoration. Some have a simpler decoration with nothing but offering scenes before deities, and some are completely undecorated on the outside. The insides of all of them are undecorated, but bear marks related to their function (for which see below). The only exception to this pure functionality is the presence of the alternating hieroglyphs for "life" (*ankh*) and, again, "stability" (*djed*) around the bottom in some cases.

---

38 The basic differences in terms of decan families are already listed by Quack (2002). A more thorough investigation of the *Classical Sky Picture* as a whole by von Lieven is in preparation.

39 Up to now, these are Hermopolis B and Abu Yasin, monuments nos. 37 and 41 in Neugebauer and Parker 1969. There may be even more, though; see the list of still unpublished astronomical ceilings in Tuna el-Gebel (Hermopolis) in Schlüter 2017.

40 No. 2 in Neugebauer and Parker 1969.

41 For Karakhamun, see Molinero Polo 2014; Petamenophis and Monthemhet are nos. 32 and 33 in Neugebauer and Parker (1969).

42 Pyramid Begrawiya South 503, Yellin 1984.

43 Namely Ankhhapi and Harendotes (nos. 38 and 46 in Neugebauer and Parker 1969).

44 Neugebauer and Parker 1969, 17–18, 26–27, pls. 5 and 11.

2 Water Clocks as Technical Devices[45]

During the excavation of the Cachette in the temple of Karnak in 1904, a spectacular find was made, although its significance was not immediately clear.[46] Only after the reconstruction of a great number of alabaster fragments did a vessel come to light, open-mouthed but with tapered walls to form a truncated cone. The outside of the vessel was carved with figurative depictions and adorned with inlays of faience and stones (see Figure 2.1). The inside of the vessel displayed 12 columns of 11 drill holes in different intervals, carved into the inner wall of the vessel, each column mounted above an alternating *ankh* or *djed*-pillar; the holes functioned as hour scales (Figure 2.3). The exceptional state of the vessel's preservation enables us not only to understand its function, but also to identify a large number of small and large fragments in different European museums as parts of Egyptian outflow water clocks.[47] These clocks were filled with water at the beginning of each day or night, which then flowed out continuously over a fixed period through a small opening laterally near the bottom, so that the time could be read from the hour scales in the inside of the vessel as the water level dropped.

In the first publication of the Karnak vessel, the instrument was called a "clepsydra," a synonym for a water clock derived from the Greek term κλεψύδρα.[48] In ancient Greek sources, it referred to a device whose external shape, at first glance, resembles the Egyptian water clock. This device was also filled with water, had an opening at the bottom, and served as a time measurer in the law courts of Athens in the 5th and 4th centuries BCE, principally to delimit speaking time within a trial.[49] It would, however, be wrong to classify this device as a "true" water clock comparable to that of the Egyptians, as the Greek devices and the Egyptian water clocks differ in a significant detail: the hour scales. These allowed the Egyptian instruments to serve as literal "clocks," while the Greek devices were only timers.

By dividing day and night into twelve respective hours whose lengths were defined in relation to the calendar, and by visualizing the passing of these hours

---

45 Anette Schomberg would like to thank Hannah Gilb, Ruti Ungar, and Mary Beth Wilson for their invaluable help in terms of translating, reviewing, and proofreading the second half of the chapter.

46 Jambon and Coulon 2017, s.v. CK 937 (clepsydre d'Amenhotep III, Cairo JE 37525); Neugebauer and Parker 1969, 12–14; Desroches-Noblecourt 1976, 139–149; Mengoli 1989; Mills and Symons 2000.

47 Borchardt 1920, 6–22; Lodomez 2007, 57–76; Hölbl 1984, 21–68; Schomberg (forthcoming).

48 Daressy 1915.

49 Young 1939.

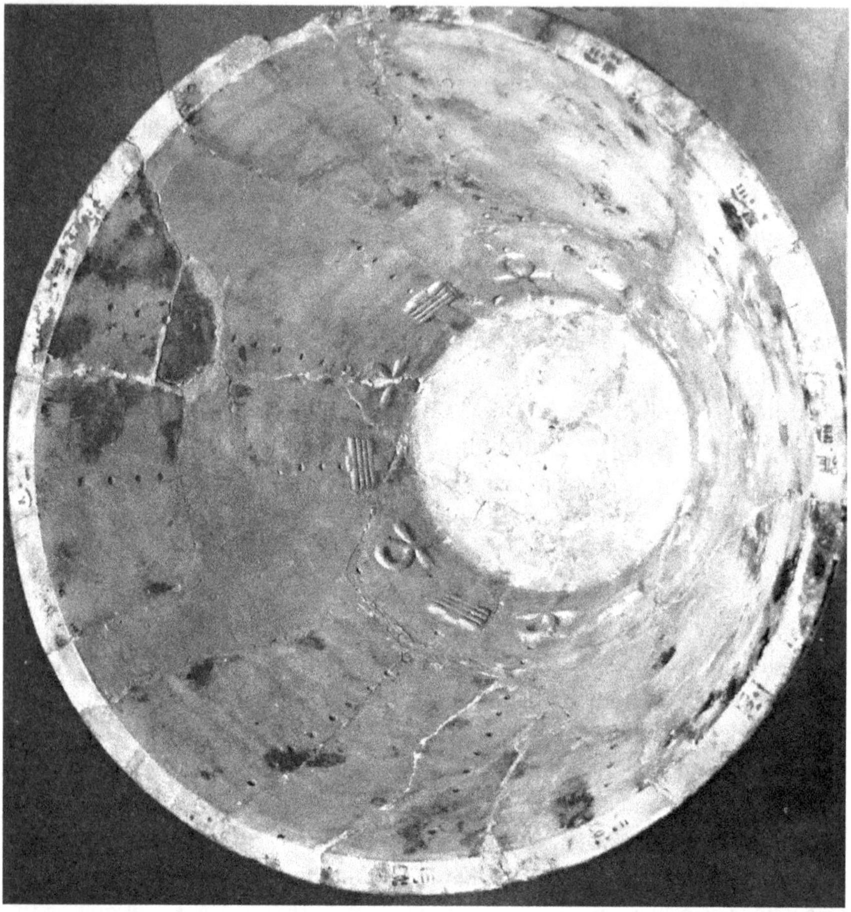

FIGURE 2.3 Interior of the Karnak clock with hour scales.
DESROCHES-NOBLECOURT 1976, 141

via a sinking water level against the hour scales within a vessel, the Egyptians succeeded for the first time in building a device to measure time that was independent from the observation of Sun and stars.

The reason why an hour scale made the difference between a clock and a ceramic vessel/timer was already given by G. Bilfinger in his fundamental study on the ancient hours:

> Das gerichtliche Wassermass [d.h. eine athenische Gerichtsklepsydra] misst einen Zeitraum lediglich nach seiner Länge ohne Rücksicht auf den Zeitpunkt, mit dem derselbe beginnt, also ohne Rücksicht auf den Anfang des (bürgerlichen oder natürlichen) Tages. [...] Man sieht, dass

> beide Messverfahren [gemeint sind Gnomon und Wasseruhr] erst dann zu eigentlichen Zeitmessungen werden, wenn sie im Stande sind, einen Zeitpunkt vermittelst seiner Entfernung vom Tagesanfang zu bestimmen, also wenn sie auf eine Länge des Gesamttages und auf eine bestimmte Einteilung [d.h. Stunden] desselben bezogen werden, nach welcher die verflossenen Teile des Tages ausgedrückt werden können.[50]
>
> The judicial measure (i.e., the Athenian court clepsydra) measures a period only by its length, irrespective of its moment or point in an overarching temporal context, thus, regardless of the beginning of the (civil or natural) day. [...] One sees that both measurement techniques [meaning gnomon and water clock] can only then be considered as a true measurement of time when they are capable of determining a point in time by means of its distance to and from the beginning of the day. In other words, if they refer to the length of the whole day and to a specific division [meaning hours] of this day, in which the parts of the day that have passed can be expressed.

Despite the structural and functional differences between Egyptian water clocks and Greek timers, the term "clepsydra" is still accepted in research. However, in order to avoid any misunderstanding, I will be using the term water clock throughout the following article.

The water clock of Karnak is the first discovered device able to measure the relation of hours to each other as well as to the beginning and end of the day, instead of simply measuring out a specific duration as often as required. Egypt not only played a pioneering role in the development of time measurement, but also in the invention of the water clock for the ancient world. As far as is known today, for more than a thousand years—until the development of the monumental Hellenistic water clocks in Alexandria in the third century BCE[51]—Egypt was the only region in the Mediterranean and the Near East in which time measurement by a water clock is conclusively evidenced.[52]

---

50 Bilfinger 1886, 6.

51 See Schomberg 2017.

52 The analysis of several cuneiform texts has led to the thesis that time measurement with water clocks also existed in the Near East from the middle of the second Millennium onwards. The evidence for this consists of four mathematical calculations of a sinking water level in four different wooden cylinders (in which cylindrical outflow clocks seem discernible) and lists of nightly vigils, with each corresponding to a certain weight. While the calculations, based on the dimensions of the different cylinders and observations of the Egyptian specimens, hardly mean water clocks existed at this time at all, the "weights"

The credit for raising awareness on the subject of Egyptian water clocks in scientific research is most notably due to the Egyptologist Ludwig Borchardt. He became interested in the subject after the finding of the water clock in Karnak, and shortly afterwards he began collecting material on the subject.[53] The first chapter of the groundbreaking book he published in 1920 on time measurement in Egypt is dedicated to the water clock.[54] Hitherto, all research on the subject has still been a dialogue with his approach, theses, and results. However, the conciseness of his work is still an obstacle for any research on these instruments. As he explained in a footnote, the chapter on the water clocks was only a summary of a longer book that he was planning on publishing.[55] His main interest was not the *Classical Sky Picture* on the outside of the Egyptian water clocks (see above in part 1.3), nor was it the context in which these water clocks were developed, used, or even needed (see above in part 1.1). The focus of his research was the technical functionality of the clocks and was based around this accuracy of measurement. Based on his theoretical calculations,[56] and influenced by the later and much more developed Hellenistic water clocks, however, his verdict on their precision was negative.[57] Only newly developed approaches and the use of experimental archeological methods gave us the opportunity to gain new insight into the functionality of the Egyptian water clocks, thus allowing a revision of his view.[58]

of night watches may refer to timers, equivalent to such a device as described by Aeneas Tacticus (22, 24–25). For a discussion of the sources, see Brown, Fermor, and Walker (1999–2000) and Fermor and Steele (2000).

53 The inventory of the Egyptian Museum in Berlin listed for the years 1910–11 a number of plaster casts of fragments of Egyptian water clocks from the Hermitage in St. Petersburg, the Louvre, the Egyptian Museums in Turin and Florence, the Museo Barracco in Rome, the British Museum in London, and the National Museum in Naples, as well as a cast of the clock from Karnak, which Borchardt bought for his research and donated to the museum.

54 Borchardt 1920, 6–26.58–65.

55 Borchardt 1920, 6 n. 1; it is still unknown why this book has not been published. In the estate of Ludwig Borchardt, which is preserved in the *Schweizerisches Institut für Ägyptische Bauforschung und Altertumskunde Kairo*, the handwritten manuscript on water clocks by Borchardt was discovered in 2016, as well as extensive material including photos, calculations, charts, and letters on this topic. A publication of this rich material is in preparation.

56 Nevertheless, his unpublished notes show that Borchardt himself had come to rely not solely on theoretical calculations. He reports in his unpublished book on water clocks about experiments for which he used a plaster copy of the clock of Karnak with a reconstructed outlet.

57 Borchardt 1920, 21.

58 Von Mackensen 1978; Cotterell, Dickson, and Kamminga 1986; Mills and Symons 2000.

2.1 *Shape and Operating Principle of Egyptian Water Clocks*

Since the publication of Borchardt's fundamental study, the research on Egyptian water clocks has been almost limited to the presentation of newly discovered pieces, surveys of objects,[59] or an examination of their functionality and precision.[60] Borchardt was the first to make calculations based on the application of basic laws of fluid mechanics, concerning pressure condition and discharge rates in relation to the special shape of the Egyptian water clocks and their operating principle.[61]

Simply put, ancient water clocks can be divided into two different types: the simpler and older outflow clocks (as represented by the Egyptian water clocks), and the more advanced Hellenistic and Roman inflow clocks. As the name suggests, there is a functional difference between these two types. In the first type, the water flows out of the vessel; in the second, it flows into it. This distinction between outflow and inflow clocks constitutes a key factor regarding ancient water clocks, especially from the perspective of the history of technology. The earliest evidence for the development of inflow clocks dates from the Mouseion in Alexandria. The Roman author Vitruvius (*De Arch.* IX.8) describes monumental water clocks, driven by inflowing water and planned by the engineer Ctesibius of Alexandria under Ptolemy II.[62]

It seems at first glance that the older type, with its inherent weaknesses, was largely replaced by the innovative inflow clocks.[63] Yet, if we consider the outflow clocks, due to their functional principle, from a technological point of view only as a simple predecessor of the elaborate Hellenistic inflow clock, one misses the outstanding accomplishment achieved in their development. As Borchardt was the first to note, the calculation of the behavior of the water level in an outflow clock and the clock's discharge rate is based on the application of "Torricelli's theorem." Applied to the outflow clock, this means that as the water level within a cylindrical vessel decreases (and thereby, the water pressure inside the vessel as well), the speed of the outflow also slows down. The water would not flow out continuously over a period of 12 hours, but would

59 Lembke 1994, 141–142 and 246–248; Beck, Bol, and Bückling 2005, 547f.; Lodomez 2007; Winter 2013; Jones 2016, 51.

60 Von Mackensen 1978; Hölbl 1984; Cotterell, Dickson, and Kamminga 1986; Mengoli 1986; 1989; Mills and Symons 2000.

61 Borchardt 1920, 15 n. 1; this approach was revisited and revised, e.g., in papers by von Mackensen (1978), Mills (1982), Cotterell, Dickson, and Kamminga (1986), and Mills and Symons (2000).

62 Schomberg 2017.

63 Cotterell, Dickson, and Kamminga 1986, 39; Lewis 2000, 361–369; Hannah 2008, 753; Winter 2013, 31.

flow faster at the beginning and then increasingly slower, ending in a trickle. By using the shape of a truncated cone instead of a cylinder, the Egyptians successfully developed an outflow clock even before the 15th century BCE that counteracted this crucial drawback inherent in the outflow clock in comparison to the inflow clock.[64]

As the Italian scientist Evangelista Torricelli stated in his theorem, the velocity, v, of a liquid flowing under the force of gravity out of an opening in a tank is jointly proportional to the square root of the vertical distance, h, between the liquid surface and the center of the opening and to the square root of twice the acceleration caused by gravity, $2g$. In short, $v = \sqrt{2gh}$ where g is the acceleration due to gravity.[65] Simply said, the Egyptians found a solution by using a particular shape for their device: in correspondence with the tapering of the truncated cone, the diameter of the water surface inside consequently becomes smaller. This compensates for the decreasing water level within and stabilizes the water pressure in the vessel. It thus creates both a constant speed of the outflow as well as a constant decrease of the water level inside, which is crucial for the correct display of the hours by an outflow clock.

Considering that this fundamental law in fluid dynamics was discovered by Torricelli in the 17th century CE,[66] the fact that the solution to the problem of the outflow clock was implemented in a clock even before the 15th century BCE is astonishing; even more so because "the angle of the cone must have been determined by empirical experiment, for not until the seventeenth century was fluid mechanics sufficiently advanced to begin to allow mathematical derivation of a theoretical profile for a linear outflow clepsydra."[67]

The reasons for inadequate acknowledgement of the momentousness of this achievement—the development of a water clock with an almost constant outflow—is twofold. Considered against the background of the later development in Greco-Roman times, the Egyptian model created the impression of a technically undemanding forerunner in the mind of modern scholars. Surmounting the basic disadvantage of an outflow clock (adjusting the shape in order to achieve a constant discharge rate) is not seen as anything remarkable, but merely a result of "common observation."[68] Instead, the inflow clock, which simply avoids this problem through a constant water inflow and an overflow valve (this is, however, not attested with certainty at any point in antiquity

64 See von Mackensen 1978; Mills 1982; Cotterell, Dickson, and Kamminga 1986.

65 Borchardt 1920, 145, Anm. 1; von Mackensen 1978, 16f.: Mills 1982; Cotterell, Dickson, and Kamminga 1986, 41–43; Høyrup 1997–98, 195f.

66 Bertoloni Meli 2006, 650–653; Simón Calero 2008, 271–292.

67 Mills 1982, 39.

68 Mills 1982, 39.

and therefore heavily hypothetical) is seen as the advanced and innovative type of water clock.[69]

A further reason for this view is probably the negative results of Ludwig Borchardt's analysis of the functionality and the precision of the water clock from Karnak. Based on his calculations for this clock, its discharge rate was too slow during the first hours. Thus, the clock runs too slowly at the beginning and, after six hours, showed the middle of the day or night almost 45 minutes too late. In the second half of the day or night, the discharge rate was too high, so that after twelve hours the clock was ultimately too fast.[70] The reason for this is the shape of the vessel, as a truncated cone is only a close approximation of the required form. Accordingly, a vessel with a truly constant outflow rate has to have a parabola-shape of the 4th order. This means a smaller diameter in the upper part of the vessel and a larger diameter in the lower part, in order to guarantee the required constant outflow of water and consequently a constant sinking water level inside for the steady display of the hours.[71]

Although Borchardt began his study with a very short collocation of all of the Egyptian water clocks known to him,[72] he touched upon their pictorial decoration and the specific application of these instruments in Egyptian rituals only

69 Drachmann 1948, 17–19; Armstrong 1974, 1f.5f.47f.; Landels 1979, 34; Mills 1982, 39–41; Turner 1984, 2–9.

70 Borchardt 1920, 15.

71 Borchardt 1920, 16 Fig. 3; Mills and Symons 2000, 20 Fig. 4.

72 Borchardt (1920, 6–10) listed 13 clocks altogether, starting with the clock from Karnak. He recognized that, in some cases, fragments now in different museums originally belonged to one clock; for example, his outflow clock "2" (Hermitage, Inv.-No. 2507a, and Naples, National Museum Naples, Inv.-No. 2327) and "3" (British Museum, Inv.-No. 933, and Louvre, Inv.-No. E 30890). His clock "13" and his clock "11" (Egyptian Museum Berlin, Inv.-No. 19556) are now lost. Nevertheless, new fragments have come to light since 1920. Some of them belong to one of Borchardt's clocks (a fragment in the Egyptian Museum Berlin, Inv.-No. 30508, for example, belongs also to his clock "3"). Fragments have been found in the Serapeum of Alexandria, at Tanis, and in Ephesus, or appeared in different museum collections (for example, the Brooklyn Museum, New York; the Oriental Institute, Chicago; the Thorvaldsen's Museum, Copenhagen; the Petrie Museum, London; and the Pushkin Museum, Moscow). Nowadays, two Egyptian water clocks from pre-Hellenistic times are known; four belong to the time of Alexander the Great and Philip III of Macedon, and 14 date to the reigns of Ptolemy I and Ptolemy II and in the early Hellenistic time. The provenance and time of two pieces cannot be determined because they lack pictorial depictions or are now lost. A complete vessel in the Oriental Institute of the University of Chicago was made in Roman times, but served solely as a decorative object; it has dysfunctional hour scales and lacks an outflow opening (see Ritner 2016). In addition, a Roman outflow clock, which was based on the same operating principle, has to be mentioned that is in the Archaeological Museum Frankfurt (see Stutzinger 2001). Schomberg (2019) is a database intended to contain all known water clocks.

briefly. Instead, based on this collocation, he focused on detecting the operating principle of these instruments. His examination of the shape and the scales of the Egyptian outflow clocks led him to the conclusion that these instruments would have utilized a constantly sinking water level in order to divide the day or night into 12 hours of equal length, corresponding to the length of night or day over the 12 months of a year. Consequently, he turned his attention to the question of whether the Egyptians "achieved this, whether their outflow clocks tell the time correctly." However, Borchardt was a "child of his time," and so he continued his sentence with the often overlooked remark, "according to our conception."[73] It was the application of modern standards concerning precision, in combination with the incorrect derivation of Torricelli's theorem that finally led him to the conclusion that the Egyptian water clocks "could not perform this task" of dividing the day/night into 12 parts of equal length.[74]

In 1976, researchers at the Astronomisch-Physikalisches Museum Kassel decided to experimentally investigate the Egyptian water clock and to reconsider Borchardt's results by using a plaster cast of the clock of Karnak with a reconstructed outlet. Based on shape, content, and the length of the hour scales, the diameter of the outlet was calculated to be 0.74 mm. These experiments produced very different practical results.[75]

> Man erkennt [...], dass die Wasseruhr in der ersten Hälfte der Laufdauer bis zu 18 Minuten nachgeht, dann aber die zeitmeßlich wichtige Mitte der Nacht oder des Tages um durchschnittlich 10 Minuten—und nicht wie Borchardt errechnete, um eine dreiviertel Stunde falsch—anzeigt. In der zweiten Hälfte der Laufperiode beginnt die Wasseruhr vorzueilen, so daß sie in der Anzeige der 12. Stunde rund 10–20 Minuten vorgeht, was den Ägypter [...] wenig gekümmert haben dürfte, weil das Ende der Nacht nicht am Laufzeitende der Wasseruhr, sondern wieder am Sonnenaufgang abgelesen wurde [...].[76]

73 Borchardt 1920, 14 ("Bisher haben wir nur erörtert, wie die alten Ägypter das Ziel, durch Auslaufuhren die Zeit in gleiche Teile zu zerlegen und zu messen, zu erreichen suchten, jetzt müssen wir aber auch der Frage näher treten, ob sie dieses Ziel auch erreicht haben, ob ihre Auslaufuhren nach unseren Begriffen richtig gingen").

74 Borchardt 1920, 21 ("Sie haben aber diese Aufgabe nicht erfüllt").

75 Von Mackensen 1978, 17; see also similar experiments by Cotterell, Dickson, and Kamminga (1986, 43f.) and Mills and Symons (2000, 19f.); for the outlet, see also Mills 1982, 47–50.

76 Von Mackensen 1978, 17.

> One recognizes [...] that, in the first half of the period of operation, the water clock loses up to 18 minutes, but then shows the middle of the day/night—the important point in time—by an average of 10 minutes—and not, as calculated by Borchardt, three-quarters of an hour late. In the second half of the period of operation, the water clock starts to rush forward, so that the display of the 12th hour is about 10–20 minutes too fast, which the Egyptians [...] did not mind at all, because the end of the night was not read by the running out of the water clock, but by the rising of the sun [...].

By detecting the error in Borchardt's application of Torricelli's theorem and using experimental archaeology, it was possible to reappraise the functionality of these instruments fundamentally. These experiments showed that—in contrast to Borchardt's opinion—Egyptian water clocks were indeed able to measure 12 temporal hours between sunrise and sunset (and vice versa). Despite the fact that there are inaccuracies of 10 to 20 minutes over 12 hours, the lack of a more precise instrument made the Egyptian clock appear to be correct in the eyes of an observer of that time.[77]

2.2 *Hour Scales*

A crucial factor concerning the functionality of an outflow clock is not only the shape, but also the design of their hour scales. Like all other known ancient water clocks, the Egyptian ones took into account the seasonal variation, as well. Accordingly, the time between sunrise and sunset was divided into 12 hours, and thus the lengths of these hours varied depending on the season/month and the geographical latitude of the site where the clock was installed.[78] Only with the emergence of the mechanical clock—i.e., from the 14th century CE onwards—did it become customary to use the fixed-length hours that are common today.[79]

In relation to ancient water clocks, the exclusive use of seasonal hours made the development of a properly functioning timekeeper even more difficult. Not only did the physical requirements have to be taken into account, but this clock also had to be adaptable to the varying hour lengths over the course of the year. The solution chosen for the Egyptian water clocks consisted of 12 columns of hour scales, comprised of drill holes, lead pins, or hemispherical markers on the inside of the vessel. In addition, the rim of the Karnak vessel bears the

77 Von Mackensen 1978, 18.

78 On the question of the beginnings of night/day, see Luft 1987.

79 Landes 1983, 76–78.

names of the 12 months of the Egyptian civil calendar. These inscriptions are located above the 12 scales, whose sequence corresponds to that of the month gods depicted on the bottom register on the outside. Together, these allow for a reading of the correct temporal hour using month scales of varying length. Here, the longest and shortest scales represent the solstitial months, whereas both of the middle-length scales represent equinoctial months. Together, these four scales compose the fundamental hourly layout of the water clock.

Initially, this may seem like a comparatively straightforward solution for displaying the varying hourly lengths, but a closer look reveals the hidden complexity of the Egyptian water clock. After noticing that day length changes, one has to have a concept of an independent time period (i.e., a calendar with months not determined by the irregular Moon phases). In order to fix each day/night length in a month, and each month over the year (only unnecessary when one uses equinoctial hours; it is an indispensable presupposition for the display of temporal hours), the position of each month in relation to a season or year must not change. While in combination with a lunar calendar, the temporal hour display of a true clock would not work, the use of timers does not represent a problem. Simply said, a clock display based on temporal/seasonal units, as all known water clocks from antiquity up to the Middle Ages were, operates only within the frame of a regular calendar based on the solar year. Consequently, this excludes all societies that used seasons/temporal units and lived according to a lunar calendar.

With regard to the calendar reform of Julius Caesar, only the Egyptian civil calendar of 12 months with 30 days apiece, plus five epagomenal days, provided a closer approximation to the solar year before the first century BCE.[80] Even the 365-day year of this Egyptian civil calendar, however, does not correspond exactly with the astronomical solar year, as the insertion of a leap day every four years was not yet practiced. An attempt to rectify this is for the first time documented for the year 238 BCE, which resulted from the *Decree of Canopus*.[81]

The lack of a leap day had grave consequences for the hour display of the clock from Karnak, which operated on the basis of the 365-day civil year. Because the months of the Egyptian civil year shift by a day every four years compared to the astronomical solar year, these months and their corresponding seasonal variance in hour-lengths gradually "wander" through the solar year. The civil and astronomical New Year fall within a day of each other only after 1,460 years.[82] In the context of a water clock, this means that, after a maxi-

80 Parker 1950, 51–56; von Beckerath 1993, 7–22; Quack 2002, 27–40.

81 Pfeiffer 2004, 249–257.

82 Quack 2002, 50; Gautschy 2011.

mum of 120 years, the hour-length shown on the interior of the clock no longer corresponds to the temporal circumstances because by that time, the slippage between the civil and solar calendars has reached a whole civil month. Each scale would then correspond to the length that was correct for the preceding month. For this reason, the hour scales of an Egyptian water clock are not only determined via latitude (which defines the hour length in relation to sunrise and sunset), but also have a de facto maximum period of 120 years with regard to the timeliness of their hour display. The shifting of the month, and therefore the hour length, within the Egyptian calendar only conditionally opposes the functionality of a water clock if we are to assume that the depictions of the months on the rim of the water clock were either ignored or adjusted as soon as they stopped corresponding to the hour-lengths at the time of their manufacture.

We can therefore anticipate that the length of the hour scales of the Karnak clock correspond to the latitude of Thebes and reflect the calendrical state at the time of Amenhotep III. In this case, both solstices would be recognizable by the longest and shortest hour scales in the 5th and 11th months and both equinoxes by the middle hour scales in the 2nd and 8th months. Since this does not correspond to the hour scales displayed on the Karnak clock, Borchardt doubted that this clock was meant for displaying the time within an Egyptian civil year.[83] An explanation for this seemed to have been provided in the form of a grave inscription, of which Borchardt became aware during the publication process of his book, and which he subsequently treated in an appendix.[84] This inscription from Sheikh Abd el-Qurna gives an account of the grave occupant, Amenemhet, who lived during the time of Pharaohs Ahmose I and Amenhotep I and built a water clock. The inscription concludes thus: "[It was correct] when entering Summer, during Harvest, in the phases of the Moon, who arrives at his hour, each hour according to its day. The water comes out via one pipe."[85] Sethe,[86] using the water clock mentioned by Amenemhet, iden-

83 Borchardt 1920, 20f. ("[…] ganz ausgeschlossen erscheint es aber, dass man zur Zeit Amenophis' III. eine Uhr hergestellt hat, die vor reichlich 100 Jahren einmal richtig gegangen wäre.") ("[…] it seems almost impossible that a clock was produced during the period of Amenhotep III, whose accuracy would have preceded itself by 100 years"); Mengoli 1989, 258f.

84 Borchardt 1920, 60–63; the inscription has been translated, commented on, and extensively discussed lately by von Lieven 2016.

85 Translation by von Lieven (2016, 226).

86 Von Lieven (2016, 212) cites a letter of Sethe to Borchardt: "Ist deine Berechnung der Uhr Amenophis' III. richtig, so kann ich nur auf einen Schluß kommen: die Uhr ist die mechanische Kopie einer älteren Uhr aus der Zeit Amenophis' I., vielleicht der Uhr, […] von der in der Inschrift […] die Rede war." ("If your calculation of the clock from the period of

tified the forerunner of the Karnak clock, which also seemed to serve as an explanation for the "outdated scale-system" of this water clock.[87]

After characterizing the water clock's truncated cone form as inadequate, Borchardt came to a negative conclusion regarding the second factor responsible for the functionality of the clock—namely, the hour scales. One of Borchardt's major points of criticism with regard to the length of the respective hour scales on the Karnak clock concerned their linear increase and decrease over a period of years. In reality, the lengths of a night or a day and, accordingly, the lengths of the temporal hours do not evolve linearly. During the respective solstices, for instance, the lengths of the temporal hours increase or decrease more quickly, while this transition occurs more slowly during the equinoxes. This observation was first documented during the first century BCE.[88] If we follow Borchardt's interpretation, the reception of this knowledge—and, subsequently, an "advance in theory"—can be attested with help from the hour scales of Egyptian outflow clocks from the Hellenistic and Roman periods. While preserving the exterior shape, adjusting the hour scales ensured a more correct display of the time, adapted to the temporal circumstances.[89] This implies, of course, an explanation for reinvestment in an otherwise obviously inadequate and outdated water clock type—especially in light of the advances in water clock construction, which, according to sources like Vitruvius, were developing almost simultaneously in Alexandria.[90]

One look at the 12-hour scales dotting the interior of the Karnak clock makes obvious the difficulty in finding an explanation for their configuration. While 11 scale columns each indicate 11 hour markers, one—the shortest scale of the 10th month (i.e., the scale of the summer solstice, with the shortest hours of the year during the night)—comprises 12 hour markers. Borchardt justifies this by claiming that only the shortest scale of month 10 has enough room at the bottom for 12 hour markers, thus supplementing the remaining 11 scales with an (invisible) 12th hour marker as well as a fill line above the scales, which was not preserved (Figure 2.4).[91] Although no trace of these hypothetical supplements

---

Amenhotep III is correct, I can only conclude that the clock is a mechanical copy of an older clock from the time of Amenophis I, maybe that clock [...] which is described in the inscription.")

87 Parker 1950, 40 § 208, 76 Anm. 73 ("another manifestation of Egyptian conservatism"); Cotterell, Dickson, and Kamminga 1986, 37.

88 Kleomedes, κυκλικὴ θεωρία μετεώρων I, 6; see Bilfinger 1888, 153f.; Borchardt 1920, 13f.; Pogo 1936, 412–412; Hölbl 1984, 33; for Kleomedes see Hübner 1999, 578f.

89 Borchardt 1920, 12–14.

90 Vitruvius, *De Arch.* IX.8.2–15; see Schomberg 2010, 184–197; Schomberg 2017, 300–339.

91 Borchardt 1920, 7 Pl. 3.

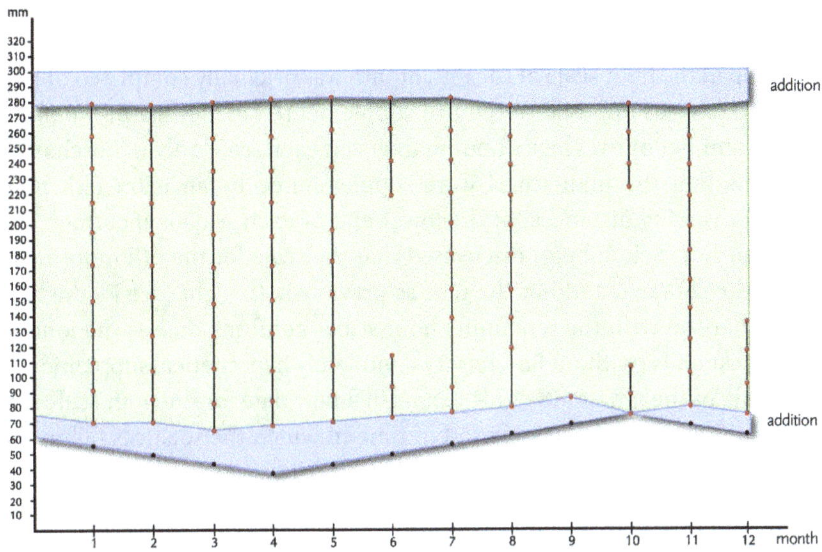

FIGURE 2.4 Hour scales of the Karnak clock

can be found on the water clock, Borchardt's assumptions prove nevertheless to be significant. If we examine the hour scale of the 10th month—which is not only comprised of 12 hour markers, but should also be the shortest scale[92] of the year on account of the temporal hours of the summer solstice—we see that the scale is only fragmentarily preserved and that the hour markers in the middle are missing entirely. Borchardt supplemented the preserved four top and three bottom hour markers with five hypothetical markers in the middle, as he assumed that this water clock measured 12 hours and, consequently, that every monthly scale should have 12 hour markers. The 12th hour marker would have therefore shown the 12th hour and, accordingly, the end of the day or night or sunrise or sunset. The lack thereof was due to an oversight or error in calculating the downward extent of the 11 monthly scales during the manufacture of the water clock, which did not allow sufficient room for the 12th hour marker on 11 of the 12 monthly scales (the shortest month excepted). However, because day and night have the same number of hours, a varied number of hour markers on the clock scales makes no sense at all. As demonstrated in the experimental test of the Karnak clock, a scale with 12 hour markers was unnecessary, as 11 were enough. At the end of the 12th hour, the Sun had risen or set, which ultimately compensated even for the 10–20 minute inaccuracy

92 Borchardt 1920, 20f.

of the clock (caused by the speediness of the water outflow as a result of the insufficient interior diameter discussed above).[93]

The gap in the hour scale of the 10th month was originally comprised of four hour markers—as opposed to the five supplemented by Borchardt—making the total number of the clock's hour markers on each scale only 11. Borchardt's hypothesis that the hour-scales were supplemented by an extra 12th hour marker (at varying distances) at the lower end of each scale is incorrect. The present author (Schomberg) discovered that the scale for the 9th month was actually the shortest, and not the 10th as previously thought (see Figure 2.4). The comparison with the remaining hour scales confirms this, as the longest scale corresponds (without Borchardt's completely hypothetical supplements) to the scale of the 3rd month. This results in a new date for the hour scale system of the Karnak clock. The period of time in which the solstices fell in the 3rd and 9th months and in which the equinoxes fell in the 6th and 12th months coincides neither with the period during which the Karnak clock was created under Amenhotep III, nor with the time of the "inventor," Amenemhet, but rather with the 17th century BCE.[94] The hour scales of this water clock therefore reflect a calendrical situation that precedes the clock itself by 250 years. During the discussion of Amenemhet's grave inscription, von Lieven notes that it follows from the inscription that Amenemhet "did not invent any of the simple clock types, but actually built a sort of super-clock, including elements of different already existing clock types amalgamated with each other" (see above part 1.2).[95] Therefore, the hour scales of the Karnak clock suggest that this water clock emulates an older model of this type.

For the remaining pieces, almost exclusively from the Hellenistic period, Borchardt, on account of their hour scales, concluded that these water clocks present an improved version of the older model and postulated not only a current, calendar-conforming hour system, but also a non-linear increase and decrease in the length of the scales in relation to each other that would have corresponded to the actual increase and decrease in the lengths of day and night.[96] The measurements of the remaining pieces with the help of a 3-D

93 This means that, after one hour, the sinking water level reaches the first hour marker, after two hours the second, and so on, until at the 11th hour, the 11th and last marker was reached. After 12 hours (when the water level has reached the *ankh* or the *djed*-pillar, just below the scale) the Sun will have risen or sunk and indicate unequivocally that 12 hours have passed; see von Mackensen 1978, 14f.

94 For the calculation of the date of the summer solstice, the calculation of the heliacal rising of Sirius was used as a reference; see Gautschy 2012.

95 Von Lieven 2016, 228.

96 Borchardt 1920, 12–14.

scanner in the context of a current research project,[97] however, produced a contradictory result, though Borchardt's claims of the clock's inaccuracy during execution were nevertheless confirmed. Although the absolute length of the scales should increase and decrease in proportion to each other (corresponding to the varying hour-lengths of a given month), and since the internal water level sinks constantly over the course of 12 hours, the intervals of the hour markers on a scale would each have to be the same.[98] This, however, is not the case either with the Karnak clock or with the Hellenistic pieces. If we examine, for example, the completely preserved solstice scale of the Karnak clock (seen in the third month), we see that the intervals between the hour markers differ between 18 and 26 mm.[99] This inaccuracy—especially with regard to the mostly fragmentarily preserved Hellenistic pieces, upon which the "improved scale" theory is based—proves to be significant, as it also exists in these fragments. Individual hour markers/intervals were used here, however, to reconstruct the length of entire hour scales and, on this basis, to draw conclusions concerning the relationship of the length of the scales/the respective lengths of day and night to the different months, as well as to attain the latitude/geographic area in which it was assembled.[100]

Of all the later pieces, the water clock in Museo Barracco in Rome (see Figure 2.5) from the period of Ptolemy II is the most completely preserved, and therefore offers itself as a comparison without needing to refer to the reconstruction of the hour scales. The preserved remains suggest that the hour

97 Schomberg 2019. The aim of this project is the construction of a database that not only contains references and photos of all known water clocks, but also provides 3-D-scans of a number of these instruments.

98 For this reason, it is clear that the hour display of the fully preserved clock at the Oriental Institute in Chicago with a different number of markers on a scale was not made in order to display the time but rather had a decorative purpose; see Ritner 2016.

99 The results of such "scale reconstructions" are shown using a simple calculation. The distance between the two lowest hour markers of scale 3 is 18 mm. This value multiplied by 12, the total length of scale 3 would thus comprise 216 mm. However, the next distance above scale 3 is 23 mm. This value multiplied by 12, the same scale would comprise 276 mm. The true distance between the lowest and the highest hour markers in reality is 215 mm—which, at first glance, would be an argument for the first distance. However, this length is the sum of the distances between the 11 markers—and the factor to multiply with would have been 10. Moreover, only the lowest distance is 18. All the others vary between 21 and 24 mm. If only the lowest distance had been preserved, the total length of scale 3 could not have been reconstructed reliably.

100 See, for example, the fragment of an Egyptian water clock from the period of Ptolemy II, which was found in 1984 in Ephesus. Only the 7 lowest hour markers (from 3 month scales in total) are preserved and used for this type of reconstruction; see Hölbl 1984, 29–39; Dorner 1985.

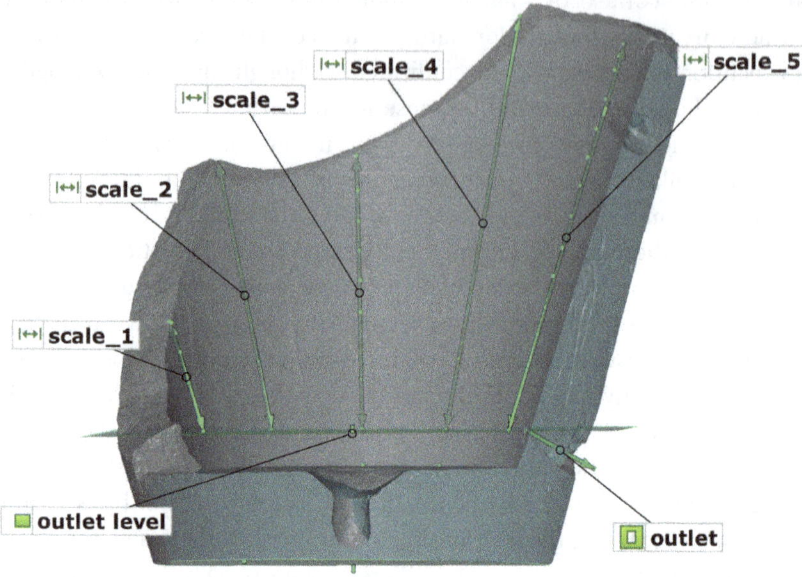

FIGURE 2.5 3-D-model of the clock in the Museo Barracco in Rome

scales of the 3rd and 9th months are the longest and shortest, respectively, and can likely be designated as the solstice scales. The comparison of the remaining scales revealed not only the established inaccuracy of the hour markers/intervals on each scale, but also the linear increase and decrease of each of the scales in relation to one another. However, the similarities with the precursor from Karnak do not end here. Both display the crucial hour scales showing the solstices in the 3rd and 9th months and the equinoxes in the 6th and 12th months. If we assume that the Egyptian civil year, in consequence of the lack of one (leap) day every four years, slips backward continuously from the natural year of 365¼ days, it takes 1,460 years until the solstices and equinoxes fall on the same days. It must follow that a period of time that corresponds to this must have occurred between the manufacture of both hour scale systems: the Karnak clock from the time of Amenhotep III (albeit with an older scale system) and the Barracco clock from the time of Ptolemy II.[101]

Because the hour scales on the Karnak clock reflect the temporal circumstances of the 17th century BCE, these four astronomical occurrences crucial to

101 Gautschy 2011.

time measurement only fall again in the same months from 240 BCE onwards.[102] Nevertheless, Ptolemy II, whose name appears on the clock in Museo Barracco, died in 246 BCE. This means that, either this water clock was equipped with a prospective hour scale system that corresponded to an impending calendrical situation (in marked contrast to the Karnak clock which preserved an utterly outdated scale system), or we are dealing not only with a form and decoration meant to honor a precursor clock type, but also with a display that imitates the hour scales and time measurement of its forerunner. The place where the clock in Museo Barracco was found—in the ancient Iseum Campense in Rome—additionally shows that this clock was exported during the Roman period in order to be used in cultic contexts in Egyptian temples outside of Egypt. Despite the fact that newer outflow clocks of this type could have been produced with hour scales adjusted to the latitude and respective calendar of Rome, the "original" pieces were still used, whose hour scales displayed neither spatial nor temporal context correctly.[103]

At the same time, the water clock in Museo Barracco presents an element that is missing in the Karnak clock. A large, round hole can be found in its base, which is referred to in the literature as the remains of an outlet opening cased in metal.[104] Closer examination of the original shows a second, smaller opening laterally at the bottom end of the wall, in the same place as can be found in the other Egyptian outflow water clocks.[105] It is unlikely that the bigger hole in the base of the clock is a singular phenomenon. Likewise, it is difficult to imagine the purpose of such a large hole in the bottom, which would have to be sealed in order to guarantee a properly working outflow clock. The difference between a vessel in the Oriental Institute in Chicago and the clock in the Museo Barracco in Rome is that the first is a decorative object in the style of a water clock from 3rd century CE Roman Egypt, while the second was a functioning water clock from the time of Ptolemy II, built in Egypt for the

102 Calculated on the basis of the heliacal rising of Sirius at the latitude of Thebes (in relation to the summer solstice) in Gautschy 2011. See also Gautschy 2012.

103 See, for example, the fragment from Ephesus (see Hölbl 1984; Winter 2013, 360f.) and the pieces found in the area of the Iseum Campense in Rome (Museo Egizio, Turin; see Lembke 1994, 246f. No. 52; Winter 2013, 530) (Hermitage, St. Petersburg; see Borchardt 1920, 8 No. 7; Winter 2013, 529f.); a fragment in the British Museum, London, has a secondary, Latin inscription (see Winter 2013, 407f.).

104 Mengoli 1986.

105 Borchardt already recognized the outflow opening at the side, as is clear not only from his drawing (Borchardt 1920, Pl. 5), but also from his unpublished materials.

geographical latitude of Thebes and exported to the Iseum Campense in Rome in order to serve as a true timekeeper in the rituals of an Egyptian sanctuary.[106]

The only preserved Roman outflow clock is stored at the Archaeological Museum in Frankfurt;[107] it is made of bronze and is not in the form of a truncated cone, but the same functional principle underlies the semispherical vessel which tapers towards the bottom. A large, round opening in the center of the base can be seen, as well as a smaller 0.4 mm outlet opening set in gold to the side of the larger opening. An explanation for a second, larger opening in the base is difficult to find. The model of a frustum-shaped outflow clock dating to 1206 CE from the Arabic engineer *al-Jazarī* can be mentioned here,[108] although in this model, the time is no longer read using a scale along the internal walls of the vessel according to a sinking water level. In *al-Jazarī's* clock, the hour display and a figure are attached on top of the clock. This figure sits in the middle on a rotating axis securely anchored to the base and uses a rod to indicate the hour. Amenemhet's grave inscriptions describe moveable additions[109]—traces of which can perhaps be seen on the outflow clocks from Frankfurt and Museo Barracco—which could then be seen as analogous forerunners to *al-Jazarī's* clock.

3 Conclusion

The reciprocal interaction between an Egyptian water clock's physical construction and hour scales determine whether it functions successfully. However, the lack of differentiation between the different types of water clocks and timers, as well as the application of modern standards with regard to the accuracy of the clocks, prevents an unbiased assessment of these instruments.

106 For the Egyptian ritual of the Iseum Campense, see Quack (2003); without question this change of place affected the correct display of the clock negatively. Obviously, with regard to accuracy, the effect of an "original" or a traditional clock imported from Egypt was estimated to be higher than that of a modern and more accurate version adapted to the latitude of Rome. In this case, tradition beats innovation since it would have been easily possible to build a more accurate water clock in the first century CE in Rome. Nevertheless, Pliny's remark (see chapters by Jones and Wolkenhauer) about the Romans using a captured Sicilian sundial for over 100 years without even recognizing that it did not fit the latitude of Rome throws a revealing light on antique standards in this respect.

107 Stutzinger 2001.

108 Hill 1974, 71–74 Pl. 7–8; Sezgin 2002, 194.

109 Von Lieven 2016.

Nevertheless, discoveries from the Late Middle Ages onward[110] of Egyptian outflow water clocks exported into Egyptian temples in the Roman world indicate that these pieces had a special meaning beyond mere time measurement. Even if Borchardt's calculations of the accuracy of the Egyptian water clocks can be corrected through experiments, the inaccuracy of such hour scales seems significant from a modern perspective. However, in the 14th century BCE there was no instrument that could calculate the right time for sacrifices and rituals more accurately.

References

Allen, James P. 2005. *The Ancient Egyptian Pyramid Texts*. Edited by Peter Der Manuelian. Writings from the Ancient World 23. Leiden: Brill.

Armstrong, Joe E. 1974. "The Water Clocks of Ktesibios: Their Origins and Reconstruction." Master's Thesis, San José State University.

Assmann, Jan. 2008. *Altägyptische Totenliturgien. Band 3: Osirisliturgien in Papyri der Spätzeit*. Supplemente zu den Schriften der Heidelberger Akademie der Wissenschaften, Philosophisch-Historische Klasse 20. Heidelberg: Winter.

Beck, Herbert, Peter C. Bol, and Maraike Bückling, eds. 2005. *Ägypten, Griechenland, Rom. Abwehr und Berührung*. Frankfurt am Main: Städelsches Kunstinstitut und Städtische Galerie.

Beckerath, Jürgen von. 1993. "Bemerkungen zum ägyptischen Kalender." *Zeitschrift für Ägyptische Sprache und Altertumskunde* 120: 7–22.

Bertoloni Meli, Domenico. 2006. "Mechanics." In *The Cambridge History of Science Volume 3, Early Modern Science*, edited by Katharine Park and Lorraine Daston, 632–72. Cambridge: Cambridge University Press.

Bilfinger, Gustav. 1886. *Die Zeitmesser der antiken Völker*. Stuttgart: W. Kohlhammer.

Bilfinger, Gustav. 1888. *Die antiken Stundenangaben*. Stuttgart: W. Kohlhammer.

Borchardt, Ludwig. 1920. *Die Geschichte der Zeitmessung und der Uhren. Band 1. Lief. B. Die altägyptische Zeitmessung*. Berlin: de Gruyter.

Brown, David, John Fermor, and Christopher Walker. 1999–2000. "The Water Clock in Mesopotamia." *Archiv für Orientforschung* 47: 130–148.

Caminos, Ricardo. 1956. *Literary Fragments in the Hieratic Script*. Oxford: Oxford University Press.

110 See, for example, the depiction of two Egyptian water clock fragments in Kircher 1654 Pl. 8, which nowadays belong to the collection of the Hermitage in St. Petersburg.

Cauville, Sylvie. 2012. *Dendara XV. Traduction. Le pronaos du temple d'Hathor: Plafond et parois extérieures*. Orientalia Lovaniensia Analecta 213. Leuven: Peeters.

Cauville, Sylvie. 2013. *Dendara. Le pronaos d'Hathor: Analyse de la décoration*. Orientalia Lovaniensia Analecta 221. Leuven: Peeters.

Cotterell, Bryan, Frank P. Dickson, and Johan Kamminga. 1986. "Ancient Egyptian Water-Clocks: A Reappraisal." *Journal of Archaeological Science* 13: 31–50.

Daressy, Georges. 1915. "Deux clepsydres antiques." *Bulletin de l'Institut Égyptien* 9: 5–16.

Darnell, John Coleman, and Colleen Manassa Darnell. 2018. *The ancient Egyptian Netherworld Books*. Writings from the Ancient World 39. Atlanta, GA: SBL Press.

Depuydt, Leo. 1997. *Civil Calendar and Lunar Calendar in Ancient Egypt*. Orientalia Lovaniensia Analecta 77. Leuven: Peeters.

Desroches-Noblecourt, Christiane, ed. 1976. *Ramsès le Grand. Catalogue de l'exposition du Grand Palais*. Paris: Galeries Nationales du Grand Palais.

Dorner, Josef. 1985. "Tagesuhr, Nachtuhr. Ein Beitrag zur altägyptischen Zeitmessung." *Jahreshefte des österreichischen archäologischen Institutes* 56: 27–37.

Drachmann, Aage G. 1948. *Ktesibios, Philon and Heron. A Study in Ancient Pneumatics*. Copenhagen: E. Munksgaard.

Faulkner, Raymond O. 1958. *An Ancient Egyptian Book of Hours (Pap. Brit. Mus. 10569)*. Oxford: Oxford University Press.

Fermor, John, and John M. Steele. 2000. "The Design of Babylonian Waterclocks: Astronomical and Experimental Evidence." *Centaurus* 42: 210–222.

Fischer-Elfert, Hans-Werner. 1986. *Die satirische Streitschrift des Papyrus Anastasi I: Übersetzung und Kommentar*. Ägyptologische Abhandlungen 44. Wiesbaden: Harrassowitz.

Fischer-Elfert, Hans-Werner. 1992. *Die satirische Streitschrift des Papyrus Anastasi I*. 2nd revised ed. Kleine ägyptische Texte 7. Wiesbaden: Harrassowitz.

Frankfort, Henri. 1933. *The Cenotaph of Seti I at Abydos*. London: Egypt Exploration Society.

Gasse, Annie. 1984. "La litanie des douze noms de Rê-Horakhty." *Bulletin de l'Institut Français d'Archéologie Orientale* 84: 189–227.

Gautschy, Rita. 2011. "Der Stern Sirius in Ägypten." *Zeitschrift für ägyptische Sprache und Altertumskunde* 138 (2): 116–31.

Gautschy, Rita. 2012. *The star Sirius in ancient Egypt and Babylonia*. [Online article] accessed June 5, 2019, http://www.gautschy.ch/~rita/archast/sirius/siriuseng.htm.

Graefe, Erhart. 1995. "Das Stundenritual in thebanischen Gräbern der Spätzeit (Über den Stand der Arbeit an der Edition." In *Thebanische Beamtennekropolen. Neue Perspektiven archäologischer Forschung. Internationales Symposion Heidelberg 9.–13.6.1993*, edited by Jan Assmann, Eberhard Dziobek, Heike Guksch, and Friederike Kampp, 85–93. Studien zur Archäologie und Geschichte Altägyptens 12. Heidelberg: Heidelberger Orientverlag.

Graefe, Erhart. 2001 to present. *Das Stundenritual.* [Online database] accessed June 5, 2019, https://www.uni-muenster.de/IAEK/forschen/aeg/proj/laufend/stundenritual.html.

Graefe, Erhart. 2014. "Report on the Work on the Fragments of the 'Stundenritual' (Ritual of the Hours of the Day) in TT 223." In *Thebes in the First Millennium BC*, edited by Elena Pischikova, Julia Budka, and Kenneth Griffin, 295–305. Newcastle upon Tyne: Cambridge Scholars Publishing.

Graefe, Erhart. 2017. "Second Report on Work on the Fragments of the Stundenritual (Ritual of the Hours of the Day) in TT 223." In *Tombs of the South Asasif Necropolis. New Discoveries and Research 2012–14*, edited by Elena Pischikova, 91–95. Cairo: The American University in Cairo Press.

Graefe, Erhart. 2018. "Die Stundentexte zu Seiten der Nut auf der Innenseite des Deckels des inneren Sarges Milano E 0.9.40147 eines p3=f-č3w-(m)-ᶜwwỉ-3śt Peftjau(em)-auuiaset." *Zeitschrift für ägyptische Sprache und Altertumskunde* 145 (2): 111–135.

Graefe, Erhart. Forthcoming. "The Ritual of the Hours of the Day on the Inner Vault of the ḳrsw-Coffin of Nes(pa)qashuty from Deir El-Bahari." In *Deir El Bahari Studies II*, edited by Zbigniew E. Szafrański. Polish Archaeology in the Mediterranean Special Studies. Warsaw: Warsaw University.

Griffin, Kenneth. 2017. "Toward a Better Understanding of the Ritual of the Hours of the Night (Stundenritual)." In *Tombs of the South Asasif Necropolis: New Discoveries and Research 2012–2014*, edited by Elena Pischikova, 97–134. Cairo: The American University in Cairo Press.

Griffin, Kenneth. 2018. "A Preliminary Report on the Hours of the Night in the Tomb of Karakhamun (TT 223)." In *Thebes in the First Millennium BC: Art and Archaeology of the Kushite Period and Beyond*, edited by Elena Pischikova, Julia Budka, and Kenneth Griffin, 59–70. London: Golden House Publications.

Hannah, Robert. 2008. "Timekeeping." In *The Oxford Handbook of Engineering and Technology in the Classical World*, edited by John P. Oleson, 740–58. Oxford: Oxford University Press.

Hill, Donald R., trans. 1974. *The Book of Knowledge of Ingenious Mechanical Devices, by Ibn Al-Razzāz Al-Jazarī.* New York: Springer.

Hölbl, Günther. 1984. "Die ägyptische Wasserauslaufuhr aus Ephesos. III. Die ägyptische Wasseruhr Ptolemaios' II." *Jahreshefte des österreichischen archäologischen Institutes* 55: 21–68.

Hornung, Erik. 1963. *Das Amduat: die Schrift des verborgenen Raumes. Teil 1 und 2.* Ägyptologische Abhandlungen 7. Wiesbaden: Harrassowitz.

Hornung, Erik. 1967. *Das Amduat: die Schrift des verborgenen Raumes. Teil 3: Die Kurzfassung. Nachträge.* Ägyptologische Abhandlungen 13. Wiesbaden: Harrassowitz.

Hornung, Erik. 1975. *Das Buch der Anbetung des Re im Westen (Sonnenlitanei): nach den Versionen des Neuen Reiches. Teil 1: Text.* Aegyptiaca Helvetica 2. Geneva: Éditions de Belles Lettres.

Hornung, Erik. 1976. *Das Buch der Anbetung des Re im Westen (Sonnenlitanei): nach den Versionen des Neuen Reiches. Teil 2: Übersetzung und Kommentar.* Aegyptiaca Helvetica 3. Geneva: Éditions de Belles Lettres.

Hornung, Erik. 1979. *Das Buch von den Pforten des Jenseits: nach den Versionen des Neuen Reiches. Teil 1: Text.* Aegyptiaca Helvetica 7. Geneva: Éditions de Belles Lettres.

Hornung, Erik. 1984. *Das Buch von den Pforten des Jenseits: nach den Versionen des Neuen Reiches. Teil 2: Übersetzung und Kommentar.* Aegyptiaca Helvetica 8. Geneva: Éditions de Belles Lettres.

Hornung, Erik. 1987. *Texte zum Amduat. Teil 1: Kurzfassung und Langfassung, 1. bis 3. Stunde.* Aegyptiaca Helvetica 13. Geneva: Éditions de Belles Lettres.

Hornung, Erik. 1992a. *Die Unterweltsbücher der Ägypter.* 3rd ed. Düsseldorf: Artemis.

Hornung, Erik. 1992b. *Texte zum Amduat. Teil 2: Langfassung, 4. bis 8. Stunde.* Aegyptiaca Helvetica 14. Geneva: Éditions de Belles Lettres.

Hornung, Erik. 1994. *Texte zum Amduat. Teil 3: Langfassung, 9. bis 12. Stunde.* Aegyptiaca Helvetica 15. Geneva: Éditions de Belles Lettres.

Hornung, Erik. 1997. *Altägyptische Jenseitsbücher: ein einführender Überblick.* Darmstadt: Wissenschaftliche Buchgesellschaft.

Høyrup, Jens. 1997–98. "A Note on Water-Clocks and on the Authority of Texts." *Archiv für Orientforschung* 44–45: 192–94.

Hübner, Wolfgang. 1999. "Kleomedes." *Der neue Pauly.* Stuttgart: J.B. Metzler.

Jambon, Emmanuel and Laurent Coulon. 2017. *Cachette de Karnak.* [Online database] accessed June 5, 2019. https://www.ifao.egnet.net/bases/cachette/.

Jones, Alexander, ed. 2016. *Time and Cosmos in Greco-Roman Antiquity.* Princeton: Princeton University Press.

Junker, Hermann. 1910. *Die Stundenwachen in den Osirismysterien nach den Inschriften von Dendera, Edfu und Philae.* Vol. 1. Denkschriften der Kaiserlichen Akademie der Wissenschaften in Wien, Philosophisch-Historische Klasse 54. Vienna: A. Hölder.

Kircher, Athanasius. 1654. "Oedipus Aegyptiacus." In *Oedipus Aegyptiacus.* Vol. III. Rome.

Kucharek, Andrea. 2010. *Altägyptische Totenliturgien 4: Die Klagelieder von Isis und Nephthys in Texten der Griechisch-Römischen Zeit.* Supplemente zu den Schriften der Heidelberger Akademie der Wissenschaften, Philosophisch-Historische Klasse 22. Heidelberg: Winter.

Landels, John G. 1979. "Water-Clocks and Time Measurement in Classical Antiquity." *Endeavour* 3 (1): 32–37.

Landes, David S. 1983. *Revolution in Time: Clocks and the Making of the Modern World.* Cambridge, MA: Belknap Press of Harvard University Press.

Leitz, Christian. 2014. *Die Gaumonographien in Edfu und ihre Papyrusvarianten: ein überregionaler Kanon im spätzeitlichen Ägypten: Soubassementstudien III*. 2 vols. Studien zur spätägyptischen Religion 9. Wiesbaden: Harrassowitz.

Lembke, Katja. 1994. *Das Iseum Campense in Rom: Studie über den Isiskult unter Domitian*. Heidelberg: Verlag Archäologie und Geschichte.

Leroux, Nicolas. 2013–15. "Nestanebetisherou et les heures de la nuit: arrêt sur le texte 'mythologique' du papyrus Greenfield—pBM EA 10554–71." *Cahiers de Recherches de l'Institut de Papyrologie et d'Égyptologie de Lille* 30: 201–16.

Lewis, Michael J.T. 2000. "Theoretical Hydraulics, Automata, and Water Clocks." In *Handbook of Ancient Water Technology*, edited by Örjan Wikander, 343–69. Leiden: Brill.

Lieven, Alexandra von. 2002. "Mysterien des Kosmos: Kosmographie und Priesterwissenschaft." In *Ägyptische Mysterien?*, edited by Jan Assmann and Martin Bommas, 47–58. Munich: Fink.

Lieven, Alexandra von. 2006a. "Eine punktierte Osirisliturgie (pCarlsberg 589+PSI inv. I 104+pBerlin 29022)." In *Hieratic Texts from the Collection*, edited by Kim Ryholt, 9–38, Pl. 1–4A. The Carlsberg Papyri 7. Copenhagen: Museum Tusculanum Press.

Lieven, Alexandra von. 2006b. "Fragmente einer Liturgie für Sokar-Osiris (P. Vindob. Aeg. 12001)." In *"Von reichlich ägyptischem Verstande": Festschrift für Waltraud Guglielmi*, edited by Hans-Werner Fischer-Elfert and Karola Zibelius-Chen, 75–85. Philippika 11. Wiesbaden: Harrassowitz.

Lieven, Alexandra von. 2007. *The Carlsberg Papyri 8: Grundriss des Laufes der Sterne. Das sogenannte Nutbuch*. 2 vols. Carsten Niebuhr Institute Publications 31. Copenhagen: Museum Tusculanum Press.

Lieven, Alexandra von. 2010. "The Soul of the Sun Permeates the Whole World. Sun Cult and Religious Astronomy in Ancient Egypt." In *Sun Worship in the Civilisations of the World*, edited by Adalbert Gail and Jaroslav Vacek, 4: 29–60. Pandanus 10. Prague: Seminar of Indian Studies, Institute of South and Central Asia, Faculty of Arts, Charles University.

Lieven, Alexandra von. 2013. "Review of Daniela Wuensch and Klaus P. Sommer (Eds.), Ludwig Borchardt, Die Altägyptische Zeitmessung." *British Journal of the History of Science* 46: 716–718.

Lieven, Alexandra von. 2016. "The Movement of Time. News from the 'Clockmaker' Amenemhet." In *Rich and Great: Studies in Honour of Anthony J. Spalinger on the Occasion of His 70th Feast of Thoth*, edited by Renata Landgráfová and Jana Mynářová, 207–31. Prague: Charles University in Prague.

Lieven, Alexandra von. 2017. "Divine Figurations of Time in Ancient Egypt." In *The Construction of Time in Antiquity*, edited by Lutz Doering and Jonathan Ben-Dov, 97–123. Cambridge: Cambridge University Press.

Lieven, Alexandra von. 2018. “From Crocodile to Dragon. History and Transformations of the Dodekaoros.” In *The Interactions of Ancient Science*, edited by David Brown, 124–37. Bremen: Ute Hempen.

Locher, Kurt. 1998. “Middle Kingdom Astronomical Coffin Lids: Extension of the Corpus from 12 to 17 Specimens since Neugebauer and Parker.” In *Proceedings of the Seventh International Congress of Egyptologists, Cambridge, 3–9 September 1995*, edited by Christopher Eyre, 697–701. Orientalia Lovaniensia Analecta 82. Leuven: Peeters.

Lodomez, Guy. 2007. “Les fragments de clepsydre de dynastie des Argéades.” *Chronique d'Egypte* 82: 57–76.

Luft, Ulrich. 1987. “Der Tagesbeginn in Ägypten.” *Altorientalische Forschungen* 14: 3–11.

Mackensen, Ludolf von. 1978. “Neue Ergebnisse zur ägyptischen Zeitmessung.” *Alte Uhren* 1: 13–18.

Mendel, Daniela. 2005. *Die Monatsgöttinnen in Tempeln und im privaten Kult*. Rites Égyptiens 11. Turnhout: Brepols.

Mengoli, Pierangelo. 1986. “La clessidra egizia del Museo Barracco.” *Vicino Oriente* 6: 193–217.

Mengoli, Pierangelo. 1989. “La clessidra di Karnak: L'orologio ad acqua di Amenophis III.” *Oriens antiquus* 28: 227–71.

Mills, Allan A. 1982. “Newton's Water Clocks and the Fluid Mechanics of Clepsydrae.” *Notes and Records of the Royal Society of London* 37 (1): 35–61.

Mills, Allan A., and Sarah L. Symons. 2000. “The Karnak Clepsydra.” *Bulletin of the Scientific Instrument Society* 66: 18–20.

Molinero Polo, Miguel Ángel. 2014. “A Bright Night Sky over Karakhamun: The Astronomical Ceiling of the Main Burial Chamber in TT223.” In *Tombs of the South Asasif Necropolis Thebes, Karakhamun (TT 223), and Karabasken (TT 391) in the Twenty-Fifth Dynasty*, edited by Elena Pischikova, 201–38. Cairo: The American University in Cairo Press.

Müller-Roth, Marcus. 2008. *Das Buch vom Tage*. Orbis Biblicus et Orientalis 236. Göttingen: Vandenhoeck & Ruprecht.

Neugebauer, Otto, and Richard A. Parker. 1960. *Egyptian Astronomical Texts Vol. 1*. Providence, RI: Brown University Press.

Neugebauer, Otto, and Richard A. Parker. 1969. *Egyptian Astronomical Texts Vol. 3*. Providence, RI: Brown University Press.

Osing, Jürgen. 1998. *The Carlsberg Papyri 2: Hieratische Papyri aus Tebtunis I*. 2 vols. Carsten Niebuhr Institute Publications 17. Copenhagen: Museum Tusculanum Press.

Parker, Richard A. 1950. *The Calendars of Ancient Egypt*. Studies in Ancient Oriental Civilization 26. Chicago: University of Chicago Press.

Pfeiffer, Stefan. 2004. *Das Dekret von Kanopos (238 v. Chr.). Kommentar und historische Auswertung eines dreisprachigen Synodaldekrets der ägyptischen Priester zu Ehren Ptolemaios' III. und seiner Familie*. Munich: K.G. Saur.

Piankoff, Alexandre. 1942. *Le livre du jour et de la nuit*. Bibliothèque d'Étude 13. Cairo: Institut Français d'Archéologie Orientale.

Pogo, Alexander. 1936. "Egyptian Water-Clocks." *Isis* 25: 403–25.

Pries, Andreas H. 2009. *Das nächtliche Stundenritual zum Schutz des Königs und verwandte Kompositionen: der Papyrus Kairo 58027 und die Textvarianten in den Geburtshäusern von Dendara und Edfu*. Studien zur Archäologie und Geschichte Altägyptens 27. Heidelberg: Heidelberger Orientverlag.

Pries, Andreas H. 2010. "Nächtliche Stundengötter und vielköpfige Genien zum Schutz des Herrschers: zwei Bildzaubereien aus dem alten Ägypten." In *Bild und Ritual: visuelle Kulturen in historischer Perspektive*, edited by Claus Ambos, Petra Rösch, Bernd Schneidmüller, and Stefan Weinfurter, 45–54. Darmstadt: Wissenschaftliche Buchgesellschaft.

Pries, Andreas H. 2011. *Die Stundenwachen im Osiriskult: eine Studie zur Tradition und späten Rezeption von Ritualen im Alten Ägypten*. 2 vols. Studien zur spätägyptischen Religion 2. Wiesbaden: Harrassowitz.

Quack, Joachim F. 2002a. "Beiträge zu den ägyptischen Dekanen und ihrer Rezeption in der griechisch-römischen Welt." Habilitation thesis, Freie Universität Berlin.

Quack, Joachim F. 2002b. "Zeitrechnung im Alten Ägypten." In *Vom Herrscher zur Dynastie: zum Wesen kontinuierlicher Zeitrechnung in Antike und Gegenwart*, edited by Harry Falk, 27–67. Bremen: Hempen Verlag.

Quack, Joachim F. 2003. "Zum ägyptischen Ritual im Iseum Campense in Rom." In *Rituale in der Vorgeschichte, Antike und Gegenwart*, edited by Carola Metzner-Nebelsick et al., 57–66. Rahden: Verlag M. Leidorf.

Quack, Joachim F. 2006. "Eine Papyruskopie des Textes der Votivellen (Papyrus Carlsberg 419)." In *Hieratic Texts from the Collection*, edited by Kim Ryholt, 39–52, Pl. 5, 5A. The Carlsberg Papyri 7. Copenhagen: Museum Tusculanum Press.

Quack, Joachim F. 2012. "Anrufungen an Osiris als nächtlichen Sonnengott im Rahmen eines Königsrituals (pBerlin 23026)." In *Forschung in der Papyrussammlung: Eine Festgabe für das Neue Museum*, edited by V. Lepper, 165–187. Ägyptische und orientalische Papyri und Handschriften des Ägyptischen Museums und Papyrussammlung. Berlin 1. Berlin: Akademie Verlag.

Quack, Joachim F. 2013. "Zeit, Krise und Bewältigung: Ägyptische Zeiteinheiten, ihre Schutzgötter und deren bildliche Umsetzung." In *Das Bild der Jahreszeiten im Wandel der Kulturen und Zeiten*, edited by Thierry Greub and Dieter Boschung, 73–98. Morphomata 7. Munich: Fink.

Quack, Joachim F. Forthcoming. "The Early History of the 12-Hour Unit in Ancient Egypt."

Quirke, Stephen. 2001. *The Cult of Ra: Sun-Worship in Ancient Egypt*. London: Thames & Hudson.

Ritner, Robert. 2016. "Oriental Institute Museum Notes 16: Two Egyptian Clepsydrae (OIM E16875 and A7125)." *Journal of Near Eastern Studies* 75 (2): 361–89.

Roulin, Gilles. 1996. *Le Livre de la Nuit: une composition égyptienne de l'au-delà*. 2 vols. Orbis Biblicus et Orientalis 147. Fribourg: Universitätsverlag.

Schlüter, Katrin. 2017. *Die Kultstellen im Tierfriedhof von Tuna el-Gebel in frühptolemäischer Zeit: Der Gang C-B und die Kammer C-B-2*. 1st ed. Vaterstetten: Patrick Brose.

Schomberg, Anette. 2010. "Die Uhr des Archimedes—Zur Rezeption antiker Technik in arabischen Quellen." In *Wasserwelten. Badekultur und Technik*, edited by Mamoun Fansa and Karen Aydin, 184–97. Schriftenreihe des Landesmuseums für Natur und Mensch 77. Mainz: von Zabern.

Schomberg, Anette. 2017. "'To Amaze the World'—A Contribution to the Shape and Meaning of the Water Clock in Antiquity, Athens, Greece, 29–30 March 2015." In *Cura Aquarum in Greece. Proceedings of the 16th International Conference on the History of Water Management and Hydraulic Engineering in the Mediterranean Region*, edited by Kai Wellbrock, 301–39. Clausthal-Zellerfeld: Papierflieger Verlag.

Schomberg, Anette. 2017. *Berlin Waterclock Project*. Berlin: Edition Topoi, [online database] accessed June 5, 2019, http://repository.edition-topoi.org/collection/BWCP.

Schomberg, Anette. Forthcoming. *Die Vermessung der Zeit. Eine diachrone Untersuchung zu den ägyptischen Wasseruhren*. Göttingen: Göttinger Miszellen—Occasional Studies.

Sezgin, Fuat, ed. 2002. *Compendium on the Theory and Practice of the Mechanical Arts—Al-Jāmiʿ Bain Al-ʿilm Wa-l-ʿamal an-Nāfiʿ Fī Ṣināʿat Al-Ḥiyal, by Ibn Ar-Razzāz Al-Jazarī*. Frankfurt am Main: Institute for the History of Arabic-Islamic Science.

Simón Calero, Julián. 2008. *The Genesis of Fluid Mechanics, 1640–1780*. Studies in History and Philosophy of Science 22. Dordrecht: Springer.

Smith, Mark. 2017. *Following Osiris: Perspectives on the Osirian Afterlife from Four Millennia*. Oxford: Oxford University Press.

Stutzinger, Dagmar. 2001. *Eine römische Wasserauslaufuhr*. Berlin: Kulturstiftung der Länder.

Thissen, Heinz Josef. 2001. *Des Niloten Horapollon Hieroglyphenbuch, I: Text und Übersetzung*. Archiv für Papyrusforschung und verwandte Gebiete 6. Munich; Leipzig: K.G. Saur.

Turner, Anthony J. 1984. *The Time Museum I. Time Measuring Instruments*. Rockford, IL: The Time Museum.

Warburton, David. 2007. *The Egyptian Amduat: The Book of the Hidden Chamber*. Edited by Erik Hornung and Theodor Abt. Zurich: Living Human Heritage.

Winter, Eva. 2013. *Zeitzeichen: Zur Entwicklung und Verwendung antiker Zeitmesser*. 2 vols. Berlin: de Gruyter.

Wuensch, Daniela, and Klaus Sommer. 2013. *Ludwig Borchardt, Die altägyptische Zeitmessung. Reprint mit einer Einleitung von Daniela Wuensch: Was die alten Ägypter über Uhren und Zeitmessung wussten*. Göttingen: Termessos Verlag.

Yellin, Janice. 1984. "An Astronomical Text from Begrawiyeh South." In *Meroitische Forschungen 1980 Akten der 4. internationalen Tagung für meroitische Forschungen Vom 24. Bis 29. November 1980 in Berlin*, edited by Fritz Hintze, 577–82. Berlin: Akademie-Verlag.

Young, Suzanne. 1939. "An Athenian Clepsydra." *Hesperia* 8: 274–84.

Zeidler, Jürgen. 1999. *Pfortenbuchstudien. Teil I: Textkritik und Textgeschichte des Pfortenbuches. Teil II: Kritische Edition des Pfortenbuches nach den Versionen des Neuen Reiches*. 2 vols. Göttinger Orientforschungen IV. Reihe: Ägypten 36. Wiesbaden: Harrassowitz.

CHAPTER 3

Short Time in Mesopotamia

John Steele

At the beginning of the fifth tablet of the Babylonian creation epic *Enūma Eliš*, a widely copied literary work probably composed in Babylonia in the latter part of the second millennium BCE,[1] the god Marduk, having defeated Tiamat and formed the heavens and the earth from her corpse, creates the Sun, the Moon, and the stars and sets out temporal order in the universe:

> *ú-ba-áš-šim man-za-za an* DINGIR.DINGIR GAL.MEŠ MUL.MEŠ *tam-šil-šu-nu lu-ma-ši uš-zi-iz ú-ad-di* MU-AṆ-NA *mi-iṣ-ra-ta ú-aṣ-ṣir* 12 ITU.MEŠ MUL.MEŠ ⸢*šu-lu*⸣-[*šá-a*] *uš-zi-iz iš-tu* u_4*-mi ša* MU-AN-NA *uṣ-ṣ*[*i-r*]*u ú-ṣu-ra-ti ú-šar-šíd man-za-az* ^{d}*né-bé-ri ana ud-du-u rik-si-šú-un a-na la e-peš an-ni la e-gu-ú ma-na-ma man-za-az* ^{d}*en-líl u* ^{d}*é-a ú-kin it-ti-šú ip-te-ma* KÁ.GAL.MEŠ *ina ṣi-li ki-lal-la-an ši-ga-ru ú-dan-ni-na šu-me-la u im-na in aka-bat-ti-šá-ma iš-ta-kan e-la-a-ti* ^{d}*nanna-ru uš-te-pa-a mu-šá iq-ti-pa ú-ad-di-šum-ma šu-uk-nat mu-ši a-na ud-du-ú* u_4*-mu ar-ḫi-šam la na-par-ka-a ina a-ge ú-ṣir i-na reš* ITU*-ma na-pa-ḫi e-*[*l*]*i ma-a-ti qar-ni na-ba-a-ta a-na ud-du-ú za-ka-ri* u_4*-mu i-na* UD-7-KÁM *a-ga-a* [*maš*]*-la* [*š*]*á-pat-tu lu-ú šú-tam-ḫu-rat mi-ši*[*l ar-ḫi*]*-šam i-*[*n*]*u-ma* ^{d}*šamaš i-na i-šid* AN*-e ina-*[*aṭ-ṭa-l*]*u-ka ina* [*s*]*i-*[*i*]*m-ti šu-tak-ṣi-ba-am-ma bi-ni ar-ka-niš* [*bu-ub-bu-l*]*um a-na ḫar-ra-an* ^{d}*šamaš šu-taq-rib-ma šá* [x (x) UD-3]0-KÁM *lu šu-tam-ḫu-rat* ^{d}*šamaš lu šá-na-at*
>
> *Enūma Eliš* 5 1–22

He fashioned heavenly stations for the great gods, and set up constellations, the patterns of the stars. He appointed the year, marked off the divisions, and set up three stars each for the twelve months. After he had organized the year, he established the heavenly station of Nēberu to fix the stars' intervals. That none should transgress or be slothful he fixed the heavenly stations of Enlil and Ea with it. Gates he opened on both sides, and put strong bolts at the left and the right. He placed the heights (of heaven) in her (Tiāmat's) belly, he created Nannar, entrusting to him the

1 Lambert 2013.

 | DOI:10.1163/9789004416291_005

> night. He appointed him as the jewel of the night to fix the days, and month by month without ceasing he elevated him with a crown, (saying,) "Shine over the land at the beginning of the month, resplendent with horns to fix the calling of days. On the seventh day the crown will be half size, on the fifteenth day, halfway through each month, stand in opposition. When Šamaš [sees] you on the horizon, diminish in all the proper stages and shine backwards. On the 29th day, draw near to the path of Šamaš, [...] the 30th day, stand in conjunction and rival Šamaš."[2]
>
> LAMBERT 2013, 99

In this passage, which at its most basic level is about creating an ordered universe, Marduk sets up cycles of the appearance of the Sun and the Moon so that time can be fitted into a regular pattern governed by the calendar. The month is divided into thirty days, beginning with the appearance of the new Moon crescent on the first day of the month and progressing through the phases of the Moon, with full Moon set as the fifteenth day in the middle of the month and the Moon and Sun being in conjunction on the thirtieth day. Of course, in reality, the cycle of the Moon does not always take thirty days, but rather varies in an irregular sequence between twenty-nine and thirty days. The calendar described in this passage, therefore, is an idealized version of the calendar in which months always last thirty days and there are always twelve months in the year, making a year last for a total of 360 days. The actual Babylonian calendar, however, operated with lunar months of either twenty-nine or thirty days, and years of either twelve or thirteen months.[3]

Unfortunately, the preserved sources for the fifth tablet of *Enūma Eliš* are all badly damaged after this point, leaving large lacunae in the text. Nevertheless, from what little remains, it seems that Marduk's ordering of time goes no further than the creation of the year, the month, and the day. Line 46 contains a damaged reference to the *ma-aṣ-ra mu-ši u i*[*m-mi*] ("watches of the night and the d[ay]"), which I suspect refers simply to the division between day and night, but the context is lost.

The story of Marduk's creation of temporal order presented in *Enūma Eliš* echoes earlier ideas. For example, an early second millennium BCE Sumerian dedication inscription of Kudur-Mabug, ruler of Larsa, describes the Moon god Nanna as:

2 Unless otherwise noted, translations are the author's own.

3 Steele 2011.

> dingir-zi u$_4$-gi$_6$-bi ḫé-ḫé/ḫe-ḫe iti gi-en-gi-en mu silim-ma
>
> RIME 4 220 4–6

> The reliable god, who interchanges day and night, who establishes the month, and keeps the year intact.
>
> LAMBERT 2013, 172

A reworking of the *Enūma Eliš* story with different gods in the role of creator appears in a passage in tablet 22 of the Nineveh version of the celestial divination series *Enūma Anu Enlil*, which reads:

> *e-nu* ^{d}*a-nu* ^{d}*en-líl u* ^{d}*é-a* DINGIR.MEŠ GAL.MEŠ [A]N-*e u* KI-*ta ib-nu-ú ú-ad-du-ú gis-kim-ma ú-kin-nu na-an-za-za* [*ú-š*]*ar-ši-du gi-is-gal-*[*la*] DINGIR.MEŠ *mu-ši-tim ú-*[*ad-du*]-⸢*ú*⸣ *ú-za-i-zu ḫar-ra-ni* MUL.MEŠ *tam-ši-l*[*i-šú-nu uṣ-ṣ*]*i-ru lu-ma-a-*[*ši*] *mūša* U$_4$-⸢ZAL⸣ *ú-za-*⸢*'i*⸣-[*zu im-du-d*]*u* ITU *u* MU *ib-nu-*[*u*] *ana* 30 *u* 20 SU[M$^{?}$] x [... *purussê* AN]-*e u* KI-*tim* TAR-*s*[*u*]
>
> *Enūma Anu Enlil* 22

> When Anu, Enlil, and Ea, the great gods, created heaven and underworld, distinguished them, established stations, founded positions (for the stars), appointed the gods of the night, divided the courses, drew the constellations, the patterns of the stars, divided night from daylight, [measured] the month and formed the year, for Moon and Sun [...] they made decrease for heaven and underworld.
>
> LAMBERT 2013, 177

Like the account in *Enūma Eliš*, no mention is made of the division of time into units smaller than a day or night.

Day and night *were*, of course, divided into intervals of short time in two ways: 1) tripartite divisions of the day and night into three seasonally varying watches and 2) the fixed units of time *bēru* and UŠ, corresponding to two hours and four minutes respectively. But these parts of the day do not seem to have been created by the gods along with the calendar when the regular order of the universe was set up, and it is tempting to see a distinction between intervals of time which were created by the gods—the day, the month, and the year—and short time intervals which were created by humans.

The discrepancy between the significant attention paid to long time in Babylonian literature and thinking with the comparatively little consideration of the notion of short time in these same sources is mirrored by the attention paid to these two different forms of time by Assyriologists and historians of

Babylonian science. Numerous works have been published which examine the calendar and its ritual and administrative functions, year counts and enumerations of the length of reign of kings, conceptions of deep and near history, and other notions of long time in Mesopotamia,[4] but very few publications have examined short time, and those that have, have mostly focused on the design and/or accuracy of water clocks.[5] The principle exception is a study by Streck of the terminology used in Akkadian texts to refer to different times of the day or night. Streck's study, however, focuses on general terms such as "dawn," "morning," "time of lying down (siesta)," and "coolness of the day," rather than measurements of times or periods within the day.[6]

Before going further, it will be helpful to introduce some terminology. In his rather unfortunately named *Primitive Time-Reckoning*, Nilsson introduced a distinction between *time-reckoning*, which he defined as the continuous counting of the time-units, from *time-indications*, which he, in turn, defined as being "related to concrete phenomena of the heavens and of Nature."[7] Nilsson further explained: "Since these indications depend upon the concrete phenomenon, their duration fluctuates with the latter, or rather the duration does not stand out by itself but the phenomenon as such is exclusively regarded ... we also find that the phenomena to which the time-indications are related are of fluctuating and very unequal duration. Since the duration is indeterminate and fluctuating, and the time-indications are not limited one by the other but overlap and leave gaps, they cannot be numerically grouped together." Nilsson assumed that time-indication preceded the development of time-reckoning. In drawing up these two categories of time, Nilsson was considering the differences between continuous calendars such as the Gregorian calendar or, indeed, ancient Greek luni-solar calendars, and practices found around the world which link events, usually agricultural events, to observable natural phenomena. To give an example not discussed by Nilsson, the Hopi of the southern United States observed the setting of the Sun along the mountainous western horizon and linked the planting and harvesting of crops to when the Sun set between certain crevices and peaks on the horizon. The setting of the Sun in certain locations acted as an indicator of important times. But the intervals between these important moments were not subdivided into days or other units, and the important moments are not spread out at regular intervals either along the horizon or according to a fixed time system.[8] Nilsson's basic

4 E.g., Englund 1988; Sallaberger 1993; Cohen 1993; Livingstone 1999; Feliu, Llop, Mittet Albà, and Sanmartín 2013.

5 E.g., Brown, Fermor, and Walker 1999–2000, Fermor and Steele 2000.

6 Streck 2017.

7 Nilsson 1920, 9.

8 Zeilik 1985; McCluskey 1990.

distinction between time-reckoning and time-indications has been refined by later authors and, although his assumption that time-indications precede time-reckoning has unsurprisingly been shown to be wrong—the Hopi, for example, operated a lunar calendar alongside the solar time-indications just discussed[9]—it is a distinction that remains useful in how we think about conceptions and uses of time.

Nilsson and most other authors have generally used the terms time-reckoning and time-indications when discussing calendars and long time. But these concepts can equally well be applied to short time. Short time can be measured and quantified using time units, but can also be indicated by natural or human-created phenomena. Often the two exist alongside one another. Think, for example, of a school classroom with a clock on the wall reckoning time in uniform intervals, but where a bell is rung to signal the beginning and end of a class. The bell serves to indicate when an event should or does take place, but it might not be rung at regular intervals—the break between classes may be different at different times of day—and so it is hard to reckon time by the bell. Natural phenomena are still used as time indicators today. For example, many parks and gardens close "at dusk" or "at nightfall," times that change throughout the year, and which may also vary due to factors such as cloud cover, whether the Sun sets behind buildings or a flat horizon, and the subjective (and perhaps biased, when it is cold) opinion of the groundskeeper.

The aim of this study is to explore the different ways that people in ancient Mesopotamia divided the day into short time, how these divisions were used in different contexts, and how they were measured. Two parallel systems of short-time reckoning co-existed, one which used fixed-length time units, the other seasonally-varying time units. Van der Waerden has suggested that the former were "astronomical" whereas the latter were "popular" units of time,[10] although all of the evidence he adduces in support of this distinction comes from texts of the astral sciences. As we will see, however, an examination of a wider range of evidence only partially confirms van der Waerden's suggestion. In addition to these two systems of short-time reckoning, certain astronomical phenomena, in particular the culmination of certain stars, were used as short-time indicators.

9 Zeilik 1986.

10 van der Waerden 1951, 27.

1 Systems of Short-Time Reckoning: the Division of the Day into Units of Short Time

Two systems of short-time reckoning were used in Mesopotamia: seasonally-varying time units and fixed-length time units. Seasonally-varying time units divide daytime (the period from sunrise to sunset) and night (the period from sunset to sunrise) into parts which, as the name suggests, vary in length over the course of the year and between day and night. Fixed-length time units, by contrast, are not defined by the length of day and night, but instead remain constant throughout the year and do not differ when used during the day or the night.

1.1 *Seasonally-Varying Time Units*

Both daytime and night were divided into three "watches" (Akkadian *maṣṣartu*, usually written with the logogram EN.NUN), whose length varies throughout the year and, except at the equinoxes, between daytime and night. The word *maṣṣartu* literally means "observation, guard" and, among other things, the word is used to refer to military watches and to the practice of astronomical observation. Its use in time reckoning is almost certainly connected to periods of sentry duty. The three watches during the day are usually designated in Akkadian as *ina šērēti* ("in the morning"); *muṣlālu*, usually written AN.NE ("[around] noon" or "noontime"), and *ina līlâti*, usually written *ina* KIN.SIG ("in the afternoon"). The night watches are *barārītu*, usually written EN.NUN.AN.USAN or USAN ("evening watch"); *qablītu*, usually written EN.NUN.MURUB$_4$.BA or MURUB$_4$ ("middle watch"); and *namāritu*, usually written EN.NUN.ZALÁG.GA or ZALÁG ("dawn watch").

It is generally assumed by modern scholars that the three watches on any given day or night were of equal length. In other words, a night watch equaled one-third of the length of the night, and a day watch equaled one-third of the length of daytime. There is surprisingly little direct evidence in support of this conclusion, however. In fact, I know of only two passages, one in a Late Babylonian astronomical text and the other in a text containing a commentary on celestial omens, which indicate that the three watches are of the same length. The first passage is found in a text concerning what in Greek astronomy would be called the rising times of the ecliptic.[11] The text sets out a scheme that relates the rising times over the eastern horizon of every one-twelfth of a sign of the zodiac with positions at or behind certain stars which culminate during the corresponding period. The details of the scheme do not concern

11 Steele 2017.

us here, but the scheme is preceded by a short introduction that includes the following passage:

> 6 KASKAL u_4-*mu ana* 3 ⸢ḪA.LA.MEŠ? KAS$_5$⸣ *za-az-ma* ḪA.LA *reš-tú* ⸢2⸣ DANNA *še-rim* UD.28!.KAM 2-*tú* ḪA.LA 2 DANNA [AN.NE] ⸢UD.29.KAM⸣ 3-*tú* ḪA.LA 2 DANNA EN.USAN UD.30.KAM
>
> BN 34713 Rev. I 11–13

> 6 *bēru* is the day (which) into 3 portions ... is divided?: The first portion of 2 *bēru* is the morning (and corresponds to) the 28th day. The 2nd portion of 2 *bēru* is [the noontime] (and corresponds to) the 29th day. The 3rd portion of 2 *bēru* is the afternoon (and corresponds to) the 30th day.
>
> STEELE 2017: 51

A *bēru* is a fixed-length time unit corresponding to 2 hours (see section 1.2 below). The passage concerns the date of the spring equinox on which daylight lasts for 6 *bēru*, or 12 hours. The 6 *bēru* are divided into the three daytime watches of morning, noontime, and afternoon, all of which last for 2 *bēru*.

The second passage is found in a commentary on *Enūma Anu Enlil* 20 (al-Rawi and George 1991–92, 60 n. 24):

> 3 *ma-na* EN.NUN GE$_6$ *ša* [iti]BÁR 1 *ma-na-a-a a-na* 3 EN.NUN SUM-*in*
>
> K 3145 17–18

> 3 *minas* is the watch of the night in Month I; 1 *mina* is for each of the 3 watches.

A *mina* is a unit of weight and, in this context, corresponds to the amount of water collected from a water clock and is directly proportional to the passage of time (see section 1.2 below). In this passage, therefore, night is taken to last for a time period corresponding to 3 *minas* of water and is divided into three equal-length watches, each of which lasts for the time period corresponding to 1 *mina* of water.

Both of these passages, however, deal with the special case of the month of the spring equinox, when daytime and night are taken to be of the same length. We cannot be certain that the same principle of equal division into three parts held throughout the year. In any event, it is likely that, in practice, the division of daytime and night into three parts was usually neither intended to be nor in actuality mathematically precise. Instead, in most contexts the three watches probably represented only broad ranges of time that were only approximately the same length and were not defined or measured precisely, akin to the way that English speakers today may often refer to the period before lunch as the "morning," even if lunch is not eaten until well after 12 noon.

A second seasonally-varying time unit was the "seasonal hour," equal to one-quarter of a watch or one-twelfth of the length of a day (daytime seasonal hour) or one-twelfth of the length of a night (nighttime seasonal hour). As I will discuss in section 3.8, only a very few texts contain references to seasonal hours.

In neither the watch nor the seasonal hour system can the unit be divided into smaller subunits: there is no unit which corresponds to a particular fraction of a watch or a seasonal hour, and despite the fact that a seasonal hour is simply one-quarter of a watch, the two units were never combined (e.g., we have no references to the second seasonal hour of the middle watch; this moment would only ever be defined as being within the middle watch with no further precision, or as in the sixth seasonal hour). Nor are fractions of a watch ever recorded.

The watch and the seasonal hour systems both divide the day and night into fixed periods of time that correspond to specific periods of the day or night. For example, the "morning" watch runs, at least in principle, from sunrise to the moment when one-third of the time from sunrise to sunset has passed. Moments in time can only be said to occur within a specific watch or seasonal hour, and the interval between two moments in time can only be referred to relative to those specific divisions of the day or night. Thus, in the watch system the duration of an event such as a lunar eclipse can only be given by stating that it began in a certain watch and ended in a certain watch (which may be the same watch)—it is not possible to give a precise duration measured in parts of a watch.

1.2 *Fixed-Length Time Units*

The system of fixed-length time units was founded on the *bēru* (Sumerian: DANNA) The *bēru* is a unit used for both time and distance. Temporally, a *bēru* was equal to one-twelfth of a complete day (i.e., the period comprising nighttime and daytime), which is equivalent to 2 of our hours. The division of the day into twelve parts was almost certainly made through an analogy with the division of the year into twelve months. When used as a unit of distance, a *bēru* corresponds to about 7 miles and is attested as early as about 2400 BCE.[12] It would not be unreasonable to walk 7 miles in 2 hours, and so it seems highly

12 Neugebauer 1941, 16.

likely that the *bēru* of distance corresponds to the distance walked in one *bēru* of time. Metrological texts state that the *bēru* as a measure of distance was divided into 30 UŠ, which was itself divided into 60 NINDA. The same division was used for the *bēru* as a unit of time: the *bēru*, which corresponds to 2 hours, was divided into 30 UŠ, each of which corresponds to 4 minutes and was itself divided into 60 NINDA, each of which corresponds to 4 seconds.[13]

In some early theoretical texts, time intervals are given as weights using the unit *mina* and its subunit *šiqil* (60 *šiqil* = 1 *mina*). These weights can be interpreted as the weight of water flowing from an idealized water clock in the equivalent interval of time. Weight is taken to be directly proportional to the passage of time, and so the *mina* acts as a proxy for a time unit. In most texts that use weights to indicate time intervals, 1 *mina* is taken to be equivalent to 2 *bēru*.[14]

In contrast to the seasonally-varying watches and seasonal hours which define specific periods of time during the day or night, measurements in *bēru* and UŠ are always intervals of time, rather than specific moments. Thus, the *bēru* were not used as a continuous temporal framework akin to our notion of hours running from midnight to midnight. Instead, times stated in the *bēru* system are always relative to another moment, usually sunrise or sunset. Thus the *bēru* system is not related to systems of equinoctial hours (for example, a time could not be given as 2 *bēru* o'clock, but only as 2 *bēru* before or after sunrise or sunset). However, because *bēru* refer to intervals of time, measurements can also be made of the durations of events themselves, independent of any reference to sunrise or sunset. Thus, the fixed-length unit system of time reckoning is both more precise in the measurements made using it and more flexible in what kind of time intervals can be measured than the seasonally-varying time unit system.

2 Short-Time Indication: the *ziqpu* Stars

I define short-time indication here as marking the moment when one event occurred, should occur, or will occur by reference to the simultaneous occurrence of another event. This definition distinguishes short-time indication from references to the sequential occurrence of events, which can be presented in a form of narrative: one event taking place after another event, without a precise statement of how much time has passed between the two. Short-time

13 Brown 2000; Steele forthcoming a.

14 Brown, Fermor, and Walker 1999–2000.

indicators are usually natural events which occur on a cyclical basis each day, without human input. Examples of such natural events used as time indicators include the moments of sunrise or sunset, the moment when the Moon rises or sets, or the moment when a star rises, sets, or reaches a particular place in the sky.

The most widely used system of short-time indication referred to in cuneiform texts, aside from the moments of sunrise and sunset, is the culmination of stars at the meridian. Over the course of the night, the stars appear to rotate around the north celestial pole, some rising and setting over the course of this rotation and others remaining visible all night long. While performing this rotation, a star will reach its highest point as it crosses the north-south meridian. This action of reaching its highest point is known as its culmination. The Akkadian terminology for the highest point reached by a star (or anything else) is *ziqpu*.

References to the culmination of stars to indicate the moment when an event takes place usually name one of a fairly stable group of 25 stars or small asterisms, although a few other stars are mentioned very occasionally.[15] Collectively, these stars were known as *ziqpu* stars. Several lists of the 25 *ziqpu* stars are known, many of which also include statements of the distance between successive *ziqpu* stars given in *bēru* and UŠ. In modern astronomical terms, these differences can be understood either as differences in the right ascension of the two stars or as intervals in time between the two stars crossing the meridian. A few reports of astronomical observations indicate the time of the observation by stating that a point a certain distance in front of or behind a *ziqpu* star culminated.

3 References to Short Time in Cuneiform Texts

Historians of Babylonian science encounter an abundance of references to short-time reckoning in texts of the astral sciences. The picture is very different outside of the astral sciences, however. Very few texts in other genres refer to either short-time reckoning or short-time indication. This apparent lack of interest in short time stands in marked contrast to the extensive interest in longer time intervals found in a wide variety of texts. Historical texts, for example, regularly talk about the length of the reign of a king, or the number of days someone was away on a journey, and give precise dates for historical events. But when during a day something happened is very rarely mentioned.

15 Steele 2014.

Contracts, letters, treatises, etc., very often record the date on which they were written, mention future or past dates, and stipulate terms of days and months for contracted activity, but almost never refer to specific moments during a day or night.

In the following sections, I examine a range of genres of cuneiform texts in order to investigate how frequently short time is mentioned, in what form it appears (short-time reckoning using watches or *bēru*, or short-time indication), and in what context. I begin by looking at texts which are closest to everyday life, namely letters and economic documents, before turning to a range of different types of scholarly texts including historical texts, literature, medicine, and ritual texts, and finally to texts in the astral sciences. Neither the range of genres covered nor the corpus of texts within each genre that I examine is exhaustive. Rather, I have tried to pick representative samples that illustrate the broad trends in the way that short time appears in cuneiform texts.

3.1 *Letters*

A large number of letters are preserved from throughout the region of ancient Mesopotamia, dating from the late third to the first millennium BCE, and attested in both Sumerian and Akkadian (in a wide variety of dialects). These letters range from royal correspondence—between kings and between a king and his government, military, and scholarly advisors—to private letters, either between family members or between people engaged in business. Only a tiny percentage of these letters, however, include any reference to short time. For example, in a recent publication[16] of private letters from the Late Babylonian Period, only 2 out of 242 letters include statements concerning short time.

The first letter was sent by a man named Bazuzu to his brother Nidintu. After greeting his brother using the standard formula for Babylonian letters, Bazuzu reminds his brother of a conversation that they had on an earlier occasion:

> *šá* UD.5.K[AM] *ina bé-e-ri* *u*$_{4}$*-mu ina* ⸢É⸣ [*šá*] ^{I}*su-qa*$^{!}$*-*⸢*a-a*⸣ A-*šú* ^{id}[…] *a-na* ŠEŠ-*ia aq-bu-*[*ú*] *um-ma* 1 GUR Z[Ú].LUM.MA ⸢*a*⸣-[*na*] ^{I}SUM-⸢*nu*⸣*-ni i-di-*⸢*in ù*⸣ 1 ⸢GUR⸣ ZÚ.LUM.MA *a-*[*na*] ^{I}*ba-ni-ia* ^{lú}*úda-a-a-lu i-*⸢*din*⸣
>
> HACKL, JURSA, AND SCHMIDL 2014: No. 146 Obv. 4–8

> It should be noted that on the 5th day at (one) *bēru* of day(time) in the house of Sūqāja, son of […], I said to my brother, "Give one *kurru* of dates to Iddinūnu and one *kurru* of dates to Bānia, the night watchman."

16 Hackl, Jursa, and Schmidl 2014.

Bazuzu continues by asking his brother to write to confirm that he will quickly deliver these dates to Iddinūnu. The precise time given in the report of the earlier conversation may reflect a quasi-legal context for the letter, in which Bazuzu wants to confirm that he is not responsible for his brother's failure to deliver the dates.

The second letter concerns the repair of a boat and includes the statement that the workmen should come *a-di mu-ṣal-*⌈*la*⌉*-lu*$_4$ "before noontime."[17]

The large body of official correspondence preserved from the Neo-Assyrian period provides similar evidence for a lack of concern with short time in letters. Of the several thousand letters sent between the king's scholars, advisors, military and governmental offices, and members of the temples and the royal court, only a handful include references to short time. Most of these letters are sent to the king by his scholars and discuss the ominous interpretation of astronomical events. The reports of the observations and the quotations of omens given in these letters usually use the watch system of seasonal short-time measurement, following the model of celestial omen texts (see section 3.6 below). The few other letters in this corpus to contain references to short time deal with unusual situations. For example, a letter sent by Urad-Nanaya to king Esarhaddon concerns the treatment of the crown prince for an illness:

> *šul-mu ad-dan-niš a-na* DUMU.LUGAL *dul-lu ša a-na* GABA! *né-pu-šu-u-ni ni-din-u-ni* 5/6 DANNA u$_4$*-mu it-ta-lak iḫ-ti-ri-di uk-ti-il i-da-te it-tu-šib a-ke-e ta*!*-ri-*[*i*]*ṣ*
>
> SAA 10 322 Obv. 7–14

> The crown prince is doing very well. The treatment which we carried out and gave to the chest lasted for 5/6th of a *bēru* of day; he retained consciousness? and afterwards sat up (in his bed). How proper he is!
>
> PARPOLA 1993, 260, with modifications by the present author

A powerful storm is the subject of a letter sent by an unknown scribe to King Sargon II:

> *ina* GE$_6$ *ša* UD.4.KAM *šá-a-*⌈*ru*⌉ *dan-nu šá a-dan-niš i-*⌈*zi*⌉*-*[*qa*] TÚG*-maš-kan-*MEŠ *gab-bu mi-*[*ḫu-u*] *i-ba-áš-ši ú-ta-*⌈*si*⌉*-*[*ḫi*] UN.MEŠ *ip-tal-ḫu a-dan-*⌈*niš*⌉ ANŠE.KUR.MEŠ *ina* ŠÀ*-bi a-ḫa-*[*iš*] *it-ta-ad-bu-ku ina*

17 Hackl, Jursa, and Schmidl 2014: No. 146 Obv. 4.

⸢UGU⸣ MÚL *kip-pi-te ú-*⸢*sar*⸣-[*ri*] *ina* UGU MUL *taš-ka-*[*a-ti*] ⸢*it*⸣*-tu-ú-aḫ* [... ŠÀ-bu] *ša* LUGAL EN-[*ia lu* DÙG].

SAA 5 249 Obv 6'–16'

On the night of the 4th day an extremely strong wind was bl[owing]. The sto[rm] was so (strong) it tor[e off] all the tents; people got panicked, horses piled together making a heap. It started at the culmination of the Circle Star and subsided at the culmination of the Triplet Star. [All is well]; the king, my lord, [can be at ease.].

LANFRANCHI AND PARPOLA 1990, 178, with modifications by the present author

Here, the beginning and end of the storm are correlated with the culmination of two *ziqpu* stars that act as short-time indicators.

3.2 *Historical Texts*

The precise dating of events is a prominent feature of chronicles, annals, and other historical texts. Precise short-time intervals are rarely recorded, however. One exception is a Babylonian chronicle covering the period from middle of the eighth to the middle of the seventh century BCE. The chronicle reports the death of the Elamite king Ḫumban-ḫaltaš I in 681 BCE as follows:

ina ^{iti}DU$_6$ UD.23.KÁM *ḫum-ba-*[*ḫal*]*-da-šú šàr* ELAM *ina* AN.NE *ma-ḫi-iṣ-ma ina* ŠÚ-*e* ^{d}[UTU U]G$_7$*-ut*

BM 92502 = Grayson 1975 No. 1 III 30–31

Month VII, the 23rd day, Ḫumban-[ḫaltaš], king of Elam, became ill at noontime (i.e., the middle watch of the day) and died at [sun]set.

The royal annals of Sennacherib include a passage referring to a battle which the king ends at 2 *bēru* after sunset:

GIŠ.GIGIR.MEŠ *a-di* ANŠE.KUR.RA.MEŠ-*ši-na ša i-na qit-ru-ub* MÈ *dan-ni ra-ki-bu-šin de-ku-ma ù ši-na muš-šu-ra-ma ra-ma-nu-uš-šin it-ta-na-al-la-ka mit-ḫa-riš ú-ter-ra a-di* 2 DANNA GE$_6$ *il-li-ku da-ak-šú-nu ap-ru-us*

RINAP 3/1 22 VI 19b–23

I brought back all together the chariots along with their horses, whose drivers had been killed in the thick of (that) mighty battle and which had themselves been released so that they galloped about on their own. When 2 *bēru* of night had passed, I stopped their slaughter.

GRAYSON AND NOVOTNY 2012, 184, with modifications by the present author

The two examples are exceptional, however. Almost no other historical texts include statements of the time during the day or night at which an event took place.

3.3 *Literary Texts*

Although literary texts contain many qualitative references to "morning," "evening," and so forth, there are very few references to quantified short time in literary texts. Unlike historical texts, which are firmly grounded in a chronological framework at the day level, cuneiform literary works, where they are interested in time at all, are concerned with the sequence of events rather than locating them within a calendrical framework. Short time only rarely appears within literary texts, and when it does, it is almost always to serve a specific narrative purpose. For example, the ninth tablet of the Standard Babylonian version of Gilgameš tells of Gilgameš's journey along the "Path of the Sun" through the darkness as he seeks Ūta-napišti to learn the secret of immortality. The text is unfortunately badly damaged at this part of the tale, but its repetitive nature allows the basic outline of the description of the journey to be understood. Gilgameš travels through the darkness for 12 *bēru*. Quantities given in *bēru* can of course be understood as distances rather than as intervals of time, but I think it is almost certain, as already argued by George (2003), that time is meant here. 12 *bēru* of time corresponds to a day and a night, whereas 12 *bēru* of distance is a fairly arbitrary distance to travel. Thus, Gilgameš travels for a complete day and night. Rather than simply saying that Gilgameš's journey lasted for "a day and a night," the division into 12 parts emphasizes the difficulty of the journey through darkness. It is a long path, and after a *bēru* has passed it seems like the journey has barely begun: the same drudging march must be continued another eleven times. Furthermore, by recounting this journey in *bēru* rather than in watches which, except at the equinoxes, vary in length between day and night, the monotony of the journey is further emphasized. In addition, day and night watches are in a sense meaningless on a journey through darkness where there is no day and no night.

Elsewhere in Gilgameš, the night watches play a role: in part of an Old Babylonian version of the epic telling the story of Gilgameš and Enkidu's journey to the cedar forest, we are told that Gilgameš lay down to rest and was woken from a dream during the middle watch of the night.

> ^{d}GIŠ *sa-ki-ip ni-il šu-na-tam mu-ši-ia-tum ub-l*[*a*$^{?}$*-š*]*u*$^{?}$ *i-na qá-ab-li-tim ši-*⸢*it*⸣*-ta-šu ú-ga-al-li-is-sú it-bé i-ta-wa-a a-na ib-ri-šu ib-ri* ⸢*a*⸣*-ta-mar šu-ut-tam am-mi-*⸢*nim*⸣ *la te-ed-ki-a-ni ma-di-*⸢*iš pa-al-ḫa*⸣*-*[*at*]
>
> OB Schøyen$_2$ Gilgameš Obv. 1–4

> Gilgameš was lying down at rest; the night brought a dream. In the middle watch of the night he awoke with a start, he arose to talk to his friend: "My friend, I have seen a dream! Why did you not rouse me? It was very frightening!"
>
> GEORGE 2003, 233

After three more days' travel, Gilgameš is forced to sleep again:

> *ma-la-*⌜*ak*⌝ UD.1.⌜KAM *ši*⌝*-na ù ša-la-*⌜*ši-im*⌝ *šu-nu iṭ-ḫu-*<*ú*> *a-na ma-ti-*⌜*ib-la*⌝ *i-li-ma* ᵈG[I]Š *a-na ṣe-er* KUR *it-ta-na-ap-*⌜*la*⌝*-ás ka-li-šu-nu* ⌜*ḫur*⌝*-sa-*MI *i-na ki-im-ṣi-šu ú-um-mi-dam zu-qá-*[*a*]*s-sú ši-it-tum ra-ḫi-a-at ni-ši im-qù-*⌜*us*⌝*-sú i-na qá-ab-li-tim ši-it-ta-šu ú-ga-*⌜*al*⌝*-li-is-sú it-bé i-ta-wa-am a-na ib-ri-*[*š*]*u ib-ri a-ta-mar śa-ni-tam e-li šu-ut-tim* ⌜*ša*⌝ *a-mu-*⌜*ru*⌝ *pa-ni-tim pa-al-*⌜*ḫa-at*⌝
>
> OB Schøyen$_2$ Gilgameš Obv. 25–33

> A journey of one whole day, two, and a third, they drew near to the land of Ebla. Gilgameš climbed up to the top of a hill, he looked around at all the mountains, he rested his chin on his knees, the sleep that spills over people fell on him. In the middle watch of the night he awoke with a start, he arose to talk to his friend: "My friend, I have seen another! It was more frightening than the previous dream I saw."
>
> GEORGE 2003, 233

The fact that Gilgameš dreams and wakes up in the middle watch of the night may perhaps serve as an indication of the impact of the dream on Gilgameš himself. In addition to its possible mantic significance, Gilgameš is woken up in the middle of the night, before he is well rested after his long journey.

3.4 *Ritual Texts*

There are many texts describing ritual procedures from Assyria and Babylonia from the first millennium BCE. These texts often place great importance on performing rituals on the correct days. Indeed, several rituals are multi-day affairs, and the texts set out exactly what should be done on which day. Within a day, however, many ritual texts simply imply the order in which things are done—first an incantation is recited, then cultic foods are prepared, then the cultic statue is moved from one place to another, for example—rather than stating when these different events are meant to happen. The following is part of a ritual text from Neo-Assyrian Assur describing a multi-day ritual:

DIŠ ^{iti}GU$_{4}$ UD.13.KAM *e-maš* ^{d}15 *ina* ŠÀ-bi URU *ina* É *a-ki-te ip-pu-šú* ^{d}*aš+šur* ^{d}*nin-líl* ^{d}MAŠ.MAŠ ^{d}IM *ina a-ki-te il-la-ku* ^{giš}BANŠUR.MEŠ *i-rak-kas* UDU.SIZKUR.MEŠ *i-na-saḫ sal-qu e-pal*

BM 121206 VI 23'–26'

¶ On the 13th of Month II the *emaš* of Ištar make within the town in the *akītu*-house; Aššur, Ninlil, Ninurta, and Adad go to the *akītu*-feast; he arranges the tables, slaughters the sheep, offers cooked meat.

VAN DRIEL 1969, 89, with modifications by the present author

A clearer temporal order of activities is found in another part of the text, which concerns the 4th day of Month I:

mu-šú šá UD.4.KAM *šá* ^{iti}BÁR MUL *e-du šá* KUN-*šú ziq-pu* 5-*su* ^{dug}*qa-pu-u-tu* GAR-*at* A.MEŠ *e*-x-*šú*

---

bir-ti MUL e$_{4}$-*ru*$_{6}$ *bir-ti* ^{d}*na-sul-lu maḫ-ru-u* 6-*su qa-pu-tú* GAR-*at* LUGAL *ina* É *aš+šur i-ta-rab*

---

ina UGU MUL *na-dul-lu* EGIR$^{?}$-*u* ^{dug}*qa*<*-pu-tu*> *a-na gam-mu-ri* GAR LUGAL *ina* É ^{d}*nin-líl e-ta-rab*

BM 121206 VIV 20'–27'

When on the night of the 4th of Month I, the Single Star of His (the Lion's) Tail culminates, 1/5 *qapūtu*-pot is placed, water …

Between Eru and the Front Yoke 1/6 *qapūtu* is placed: the king has entered the temple of Aššur.

At the culmination of the Rear Yoke a *qapūtu* to use is fully placed; the king has entered the temple of Ninlil

VAN DRIEL 1969, 91 with modifications by the present author

In this passage, the ritual activities happen not only in a specified sequence but at specified moments. These moments are not defined by intervals of short time but rather by the culmination of *ziqpu* stars. The *ziqpu* stars act not as a system of time reckoning but as time indicators: visual triggers of when something should be done.

The tablet AO 6460 from Uruk, which concerns a nocturnal festival in the city's Rēš temple, contains further references to the time at which things are supposed to be done. The tablet is from an archive of scholarly tablets dating to the late third and early second century BCE from the Rēš temple itself. Although no colophon is preserved on the tablet, which is broken at the end, tablets from this archive were written by a small group of scribes, most of whom wrote tablets in a variety of genres including rituals, divination, and astronomy. This particular ritual is interesting because it takes place at night on the roof of a building in the temple and includes offerings to the Sun, Moon, and planets—one of the few references we have to astral religion in Babylonia.[18] The text contains several references to the time at which parts of the ritual should take place. For example:

> *ina* EN.NUN.USAN *ina ú-ru* BÁRA.MAḪ *ziq-qur-ra*[*t*] *šá* É *re-eš ki-ma šá* MUL [d]1 GAL-*ú šá* AN-*e it-tap-ḫa an-tu*$_4$ GAL-*tu*$_4$ *šá* AN-*e ina* MUL MAR. GÍD.DA *it-tap-ḫa a-na tam-šil zi-i-mu bu-un-né-e* MUL *šá-ma-mi* [d]*a-nù* LUGAL *it-ta-ṣa-a ṣa-lam ba-nu-ú*
>
> AO 6460 Obv. 14–17

> In the first watch of the night on the roof of the (topmost) sanctuary of the temple tower of the Rēš-temple, when the star Great Anu of Heaven rises (and the star) Great Antu of Heaven rises in the Wagon-constellation, (he will recite) "Rivaling the glow of the most beautiful of the stars of the sky" (and) "Anu, the king, has risen, the beautiful constellation."
>
> LINSSEN 2004, 247

This passage is interesting in that it combines both short-time reckoning using seasonally-varying units—the ritual activities take place during the first watch of the night—with the use of the rising of particular stars which are used as short-time indicators, similar to the use of the culmination of stars in the Neo-Assyrian ritual just discussed. Even more interestingly, later in the text we find a statement of another short-time interval, this time given using fixed-length units of time:

> UD.17.KAM 10 UŠ *u*$_4$-*mu* EGIR KUR-*ḫa* [d]UTU KÁ *ina pa-ni* [d]1 *u an-tu*$_4$ BAD-*te-ma ba-a-a-a*[*t*] DUḪ-*ár*
>
> AO 6460 Rev. 28–29

18 Krul 2018.

(On) the 17th day, 10 UŠ of day after sunrise, in front of Anu and Antu the gate will be opened, and the vigil will be ended.

LINSSEN 2004, 247, with modifications by the present author

Thus, the text uses both the seasonally-varying watch system and the fixed-length *bēru* system of short-time reckoning, as well as the rising of stars as short-time indicators. A further point to note is that, to my knowledge, this is the only example outside of the astral sciences where the time unit UŠ appears; all other non-astral science texts in which fixed-length units of time appear use only the unit *bēru*, not its subdivision (common fractions (1/2lf, 1/3rd, 2/3rd, and 5/6th) of a *bēru* occasionally appear but never intervals which would require the use of the smaller unit). I will return to the significance of this point later.

3.5 *Medical Texts*

Cuneiform medical texts may be divided into two main groups: texts concerned with the diagnosis of a patient and/or the prognosis for recovery, and texts concerned with the treatment of the patient. References to short time are comparatively rare in medical texts. They occur almost exclusively in the first group of texts and concern the time when the patient was affected by symptoms of the illness. For example, in the extensive composition named by modern authors the Diagnostic and Prognostic Series (DPS),[19] we find entries such as the following:

DIŠ *ina šá-šal-li-šú šá* 15 SÌG-*iṣ u ú-rap-pad* ŠU DINGIR URU-*šú ina* EN.NUN.ZALÁG.G[A SÌG]-⌜*iṣ*⌝ G[AM]

DIŠ *ina ina šá-šal-li-šú šá* 2,30 SÌG-*iṣ u ú-rap-pad* ŠU DINGIR-*šú ina* EN.NUN.MURUB$_4$-*t*[*i* SÌG]-*iṣ* GAM

DPS 12 86–87

If he was injured on his "right" upper back and he wanders about, "hand" of the god of his city. (If) he was injured in the dawn watch, he will die.

If he was injured on his "left" upper back (and) he wanders about, "hand" of his god. (If) he was injured in the middle watch, he will die.

SCURLOCK 2014, 100

These entries provide both a diagnosis of the illness—being struck down by a particular god (most illnesses were considered to be caused by the action of a god "touching" a person)—and a prognostication of whether the patient

19 Scurlock 2014, 13–272.

will recover. In the case of the first illness, if the patient was affected during the dawn (i.e., third) watch of the night, then he will die. In the second case, if he was affected during the middle watch, he will die. What happens if he was struck with these illnesses at other times is not stated, presumably indicating that in those situations he will recover.

All of the references to short time in the Diagnostic and Prognostic Series and in other medical texts that I have consulted are given using the seasonally-varying time unit of the watch. To my knowledge, the fixed-length *bēru* system of time reckoning never appears in medical texts.

3.6 *Omen Texts*

A significant proportion of scholarly cuneiform tablets contain collections of omens. Several large omen compendia are known containing omens relating to a wide range of topics including celestial phenomena, terrestrial phenomena seen in a city, the birth of malformed animals, liver divination, and the behavior of animals. In addition, many commentary texts were written which aim to explicate the omen texts.

As might be expected, short time appears most frequently in texts containing celestial omens. A wide range of celestial events were considered ominous, including the appearance of the Moon (especially at new Moon), the appearance of the Sun, lunar and solar eclipses, the appearance and motion of the planets, and the visibility and appearance of stars, comets, and meteors.

Eclipse omens frequently make reference to the watch in which an eclipse occurs. For example, the 17th tablet of the omen series *Enūma Anu Enlil* contains omens drawn from the month in which a lunar eclipse occurs and the watch of the night in which it begins. Here is a short extract:

> DIŠ *ina* ^{iti}GU$_4$ AN.KU$_{10}$ EN.NUN.AN.USÁN GAR LUGAL BA.BE-*ma* DUMU LUGAL.MES *ana* GIŠ.AŠ.TE AD-*šú-nu* SÁ.SÁ NU TUŠ.MEŠ
>
> DIŠ AN.KU$_{10}$ EN.NUN.MURUB$_4$.BA GAR EBUR KUR NU SI.SÁ KI.LAM TUR
>
> DIŠ AN.KU$_{10}$ EN.NUN.U$_4$.ZAL GAR ZI-*ut* LUGAL IM.GI *šá a-a-bi-šú i-kaš-ša-du u* ⸢É x x x⸣
>
> *Enūma Anu Enlil* 17 II 1–3

If an eclipse occurs in Month II in the evening watch, the king will die and the sons of the king will vie for the throne of their father but will not sit (on it).

If an eclipse occurs in the middle watch, the harvest of the land will not thrive; the economy will dwindle.

> If an eclipse occurs in the morning watch, attack of a rebel king who will conquer his enemies and the house ...
>
> ROCHBERG-HALTON 1988, 127

The text series *Šumma sin ina tāmartišu*, a commentary to *Enūma Anu Enlil*, contains in its fourth tablet a scheme that associates the watch in which an eclipse takes place with the astrological outcomes and countries that are affected by the eclipse. The text reads:

> DIŠ AN.KU_{10} EN.NUN.AN.USÁN *ana* ⌜NAM⌝.[UG_7.MEŠ : DIŠ AN.KU_{10}] ⌜EN⌝.NUN.$MURUB_4$.BA *ana* KI.LAM TUR.RA : DIŠ AN.KU_{10} EN.NUN.UD.ZAL.LE *ana* GIG ⌜AN⌝.[TI.LA]
>
> *Šumma sin ina tāmartišu* 4 5–6

> [An eclipse (during) the evening watch is] for deaths.
> An eclipse (during) the middle watch is for diminishing business.
> [An eclipse (during) the morn]ing watch is for sick (people) becoming well.
>
> WAINER 2016, 78

> ⌜DIŠ EN⌝.NUN.AN.USÁN KUR URI.KI : EN.NUN.$MURUB_4$.BA KUR SU.⌜BIR_4⌝.[KI ... DIŠ] ⌜EN⌝.NUN.UD.ZAL.LE [KUR.NIM.MA.KI]
>
> *Šumma sin ina tāmartišu* 4 17–18

> (An eclipse during) the evening watch is for the land of Akkad.
> (An eclipse during) the middle watch is for the land of Subartu.
> (An eclipse during) the morning watch is for [the land of Elam].
>
> WAINER 2016, 79

Outside of celestial divination there are only a very small number of omens that include reference to the time at which an ominous event occurred. In these cases, time is designated by the watch system. For example, in tablet 19 of the terrestrial omen series *Šumma Alu*, a series of omens appear where the time at which a ghost cries out determines the interpretation:

> DIŠ *ina* É NA GIDIM *ina* EN.NUN.$_{D}$USAN KA.KA-*si* LÚ BU *ulu l-tab-bar*
> DIŠ *ina* É NA GIDIM *ina* EN.NUN.$MURUB_4$.BA KA.KA-*si* KUR-*ád* ÁŠ : TIL u_4-*me*
> DIŠ *ina* É NA GIDIM *ina* EN.NUN.UD.ZAL.LI KA.KA-*si* ARḪUŠ DINGIR *ana* LÚ GÁL-*ši*
>
> *Šumma Alu* 19 57–59

If a ghost is persistently crying out in a man's house during the evening watch, that man will not grow old.

If a ghost is persistently crying out in a man's house during the middle watch—attaining of a wish (or alternatively): end of days.

If a ghost is persistently crying out in a man's house during the dawn watch, a god's mercy will be with a man.

FREEDMAN 1998, 281, with modifications by the present author

3.7 *Astronomical Texts*

In contrast to the use of seasonally-varying watches in astrological texts, astronomical texts rarely refer to the watches except in the broad sense of indicating observations made early in the night or those made before sunrise in the morning. Instead they generally use either the fixed-length time unit *bēru* and its subunits UŠ and NINDA or weight measurements that act as proxies for time units. For example, the early astronomical compendium known by its incipit as MUL.APIN (Hunger and Steele 2019), a very popular and influential text preserved in many copies from throughout Assyria and Babylonia, contains, alongside various star lists, intercalation schemes, and descriptions of the subdivisions of the synodic time of the planets, a simple mathematical scheme for the length of night and the duration of visibility of the Moon. The first few lines of the scheme read as follows:

DIŠ *ina* ^{iti}BÁR UD.1.KAM 3 MA.NA 10 GÍN EN.NUN GE_6 12 UŠ 40 NINDA ŠÚ *šá sin*
DIŠ *ina* ^{iti}BÁR UD.15.KAM 3 MA.NA EN.NUN GE_6 12 UŠ KUR *šá sin*
DIŠ *ina* $^{iti}GU_4$ UD.1.KAM 2 5/6 MA.NA EN.NUN GE_6 11 UŠ 20 NINDA ŠÚ *šá sin*
DIŠ *ina* $^{iti}GU_4$ UD.15.KAM 2 2/3 MA.NA EN.NUN GE_6 10 UŠ 40 NINDA KUR *šá sin*
DIŠ *ina* $^{iti}SIG_4$ UD.1.KAM 2 1/2 MA.NA EN.NUN GE_6 10 UŠ ŠÚ *šá sin*
DIŠ *ina* $^{iti}SIG_4$ UD.15.KAM 2 1/3 MA.NA EN.NUN GE_6 9 UŠ 20 NINDA KUR *šá sin*

MUL.APIN II ii 43–46

¶ On the 1st day of Month I, 3 *mina* 10 *šiqil* is the watch of the night; 12 UŠ 40 NINDA is the setting of the Moon.

¶ On the 15th day of Month I, 3 *mina* is the watch of the night, 12 UŠ is the rising of the Moon.

> ¶ On the 1st day of Month II, 2 5/6 *mina* is the watch of the night, 11 UŠ 20 NINDA is the setting of the Moon.
> ¶ On the 15th day of Month II, 2 2/3 *mina* is the watch of the night, 10 UŠ 40 NINDA is the rising of the Moon.
> ¶ On the 1st day of Month III, 2 1/2 *mina* is the watch of the night, 10 UŠ is the setting of the Moon.
> ¶ On the 15th day Month III, 2 1/3 *mina* is the watch of the night, 9 UŠ 20 NINDA is the rising of the Moon.
>
> HUNGER AND STEELE 2019, 155

This passage presents two simple zigzag schemes, which vary in steps of constant differences every 15 days between maximum and minimum values at the two solstices. The zigzag function for the length of night varies between a maximum of 4 *mina* at winter solstice to a minimum of 2 *mina* at summer solstice. *Mina* is a unit of weight and refers to the weight of water in a water clock. By comparing the scheme with other schemes for the length of night which use the units *bēru* and UŠ, it is clear that there is a direct relationship between the weight of water and the passage of time, with 1 *mina* corresponding to 2 *bēru*.

The scheme for the duration of visibility of the Moon on the 1st of the month and its invisibility on the 15th varies between a maximum of 16 UŠ at winter solstice to a minimum of 8 UŠ at summer solstice. The two zigzag functions are related in that the values for the visibility/invisibility of the Moon are one-fifteenth of those for the length of night, once the night-lengths have been converted from weight measures into time measures. This relationship, which is explained in a subsequent passage in the text, demonstrates the interchangeable nature of fixed-length time units.

The Astronomical Diaries and related texts, which date to the later first millennium BCE, contain records of systematic observations of celestial phenomena.[20] More than two thousand such texts are known, providing us with a clear picture of the practice of making celestial observations in Babylonia in the late period. Short time appears in three ways in these texts. First, observations of the passages of the Moon or a planet past either another planet or a reference star are customarily recorded as having been made "in the evening" or "in the morning." These terms may be referring to two of the three night watches, but equally may be used simply in the broader sense of indicating that the observation was made after sunset or before sunrise. Secondly, the times of astronomical phenomena that occur at a specific moment, such as eclipses, are

20 Hunger 1999.

usually reported as a time interval from sunrise or sunset. These time intervals are recorded in the *bēru* system of fixed-length time units. Finally, the durations of certain phenomena, including the lengths of the phases of an eclipse and the interval between the Sun and Moon crossing the horizon on certain specific occasions, are also recorded in the *bēru* system. Let us consider the last two cases in some detail.

The tablet BM 38462 contains a compilation of reports of lunar eclipse observations and predictions dating to the reign of Nebuchadnezzar II in the early 6th century BCE. The reports are fairly typical for this time. Each report of an observed eclipse begins with the date and the time at which the eclipse began. For example:

> MU-25 IZI 1 1/2 KAS GE_6 G[IN]
>
> BM 38462 = HUNGER 2001 No. 6 Rev. III 3'

> Year 25. Month V, 1 1/2 *bēru* after suns[et].
>
> HUNGER 2001, 29

The time is recorded with the unit *bēru*. In this case, the measurement is 1 1/2 *bēru*. Other examples give times in whole *bēru* and/or the common fractions of a *bēru*, 1/2, 2/3, and 5/6. The subunit UŠ is only used either in the case of 1/3 *bēru*—which is often given as 10 UŠ because Late Babylonian scribes tended to avoid using the sign for 1/3 since it could too easily be confused with the sign for 2/3—or when there is no common fraction sign available, such as in the following example:

> [MU-4 B]AR ⌜13 $MURUB_4$⌝ 3 KAS 5 UŠ GE_6 GIN [...] 3 *u* 2 SAR 3 *re-bít* [...] *ina* 2 ZÁLAG 2 GIN
>
> BM 38462 = HUNGER 2011 No. 6 Obv. I 8–10

> [Year 4 Mon]th I, the 13th, middle watch, 3 *bēru* 5 UŠ after sunset, it began in the west and north. Three quarters [was covered]. It cleared in the north. The north wind blew.
>
> HUNGER 2001, 27

What is significant here is that, in records such as these, the fundamental unit of short-time is the *bēru*. The way that the time of eclipses and other events is recorded changed dramatically in the early fifth century BCE,[21] as the following example from a later compilation text shows:

---

21 Rochberg-Halton 1989, 148.

GAN GE_6 15 AN-[KU_{10} *sin*] BE-*ti* DIR NUP PAP *ina* 1,4 GE_6 G[IN]

BM 37088+37652 = HUNGER 2001 No. 10 II 11'–13'

Month IX, night of the 15th, [lunar] eclipse, total; clouds, I did not watch. At 1,4 after sunset.

HUNGER 2001, 41, with modifications by the present author

In this record, the time of the eclipse is given by a sexagesimal number rather than as a quantity with its accompanying unit. In Mesopotamia, quantities—that is, numbers that refer to real, measurable, things—rarely need to use the sexagesimal system because there is usually a bigger unit that can be substituted. For example, in the KÙŠ or cubit system of measuring distances, where there are 24 SI or fingers in a cubit, we would not normally find measurements such as 3,6 fingers. Instead, this would be written as 7 cubits 18 fingers. It is surprising, therefore, to find a number such as 1,4 as a measurement. The context shows that the implied unit for this number is UŠ, and so this measurement is 1,4 or 64 UŠ, which equals 2 *bēru* 4 UŠ. The reason for this change in how time was recorded is not certain. For any scribe who knew anything about astronomy, there is no real ambiguity in writing the time using sexagesimal numbers without specifying the unit, and this method does save on space. Removing the metrological system might have made it easier to analyze or compare the times of different eclipses: it is easier to compare two numbers with the same unit rather than with different units (e.g., it is easier to see that the difference between 76 and 32 is 44 than that the difference between 2 *bēru* 16 UŠ and 1 *bēru* 22 UŠ is 24 UŠ). It is worth noting, as well, that this change takes place around the same time as changes in the way that some other astronomical phenomena were observed and recorded, and in a period when methods of predicting astronomical phenomena developed considerably (Steele forthcoming b). Interestingly, in texts containing compilations of eclipse records that pre-date this change in the way that short-time intervals were recorded, but which were put together after this change, the records have been altered to use the later system of recording time intervals as sexagesimal numbers without the units being given. For example, the report of an eclipse in the 3rd year of Sennacherib (21 April 686 BCE) preserved in the compilation text BM 35115+, which was put together in the late fourth century BCE, records the time of the eclipse in the sexagesimal format:

[GU_4?] ⌜5⌝ ITU NI *ina bi* [S]I *u* KUR TAB 2/3 ḪAB [ŠÚ? MA]R? GIN [*ina*] 1,40 [GE_6 G]IN

BM 35115+ Obv. = HUNGER 2001 No. 3 II 2–6

> [Month II$^{?}$], (after) 5 months …, between [nor]th and east it began; two-thirds of the disk [was covered$^{?}$. The we]st$^{?}$ wind blew. [At] 1,40 [after Sun]set.
>
> HUNGER 2001, 9

A small number of observations of lunar eclipses complement the measured time of the eclipse with a report of the culmination of a *ziqpu* star at that moment. For example, an Astronomical Diary for year 86 of the Seleucid Era contains the following account of the lunar eclipse of 2 August 226 BCE:

> ⸢52⸣ GE$_{6}$ GIN ⸢SA$_{4}$ *šá*⸣ GABA-*šú ziq-pi sin* AN-KU$_{10}$ Á KUR *ki* TAB-*ú ina* ⸢17⸣ GE$_{6}$ *gab-šú* ŠÚ 10 UŠ GE$_{6}$ ÍR *ana* ZALÁG *ki-i* TAB-*ú ina* 15 GE$_{6}$ TA ÙLU *ana* SI ZALÁG *ina* GAR *né-ḫi ina* ZALÁG *ḫa-muṭ* 42 GAR ÍR *u* ZALÁG AN-KU$_{10}$ SA$_{5}^{?}$ AN-KU$_{10}$-*šú* SI ŠÁR GIN ⸢AN$^{?}$-KU$_{10}^{?}$-*šú*$^{?}$⸣ ^{d}UDU-IDIM.MEŠ *gab-bi* NU GUB.MEŠ 5 KÙŠ *ár* MÚL *ár šá* SUḪUR MÁŠ ⸢KAxME$^{?}$⸣
>
> Diary No. –255 Rev. 4–8

> At 52 after sunset, the Bright Star of its Chest culminated (and there was a) lunar eclipse; when it began on the east side, in 17 nighttime it covered completely; 10 UŠ nighttime maximum phase; when it began to clear, it cleared in 15 nighttime from south to north; in (its) onset it was slow, in (its) clearing fast; 42 onset, maximum phase, and clearing; its eclipse was red$^{?}$; (in) its eclipse, a gusty north wind blew; (in) its eclipse all the planets did not stand there; 5 cubits behind the Rear Star of the Goatfish it became eclipsed.
>
> SACHS AND HUNGER 1989, 141, with modifications by the present author

This account begins with the measured time of the eclipse given as a sexagesimal number without a unit, as is normal for an eclipse report written at this period. Following this measured time, we find a statement that at that moment one of the *ziqpu* stars culminated. The culmination of this star is recorded as taking place at the same time as the beginning of the eclipse, which itself took place at a measured time of 52 UŠ after sunset. Thus, the record contains both a measured time of the eclipse using the fixed-length system of time reckoning and a report of the simultaneous culmination of a star, acting as a time indicator.[22]

22 Although in principle it would be possible to use the culmination of a *ziqpu* star to determine the time of the event in UŠ, it is clear both from the phrasing of this and similar records and from analysis of the error pattern of the times derived from reports of *ziqpu*

The final group of measured short-time intervals found in the Astronomical Diaries are reports of the so-called "lunar six." The term "lunar six" refers to a series of measurements of the interval between the Sun and the Moon crossing the horizon undertaken on six specific occasions during the month: one, named NA, on the evening which marks the beginning of the month; four, ŠÚ, NA, ME, and GE_6, on the days around full Moon; and one, KUR, before sunrise on the last morning when the Moon is visible at the end of the month. Observations of lunar six measurements are preserved from the mid seventh to the first century BCE. The following example is typical:

> BAR 30 22 *a muš ana šámaš* NIM
> 14 13 ME DIR NU PAP
> 14 5,40 ŠÚ DIR NU PAP
> 15 2,30 GE_6
> 15 3,20 NA DIR NU P[AP]
> 27 16,10 NU PAP
>
> BM 40091 = HUNGER 2001 No. 40 Obv. 12'–17'

> Month I, (the 1st of which was identical with the) 30th (of the preceding month). 22,(0) (sunset to moonset); earthshine; measured; it was high to the Sun.
> The 14th, 13,(0) moonrise to sunset; clouds, not observed.
> The 14th, 5,40 moonset to sunrise; clouds, not observed.
> The 15th, 2,30 sunset to moonrise.
> The 15th, 3,20 sunrise to moonset; clouds, not observed.
> The 27th, 16,10 (moonrise to sunrise); not observed.
>
> HUNGER 2001, 111

From the earliest attested records of the lunar six, the intervals are given as sexagesimal numbers without units. From the astronomical context it is clear that the implied unit must be NINDA, 1/60th of an UŠ. One could alternatively read these numbers as, for example 16;10, meaning 16 and 10-sixtieths, with the unit UŠ, but I think it is more natural to take the smallest place value as indicating the implied unit.

3.8 *Horoscopes*

Twenty-eight tablets containing in total thirty individual horoscopes that present astronomical data relevant to predicting the life of an individual are

---

star culmination as compared to the recorded time measurements, that the time has not been derived from the *ziqpu* star observation—see Steele 2000, 66.

known. These horoscopes contain astronomical data for the day or the days surrounding the birth of an individual. Typically, horoscopes contain a statement of when the child was born. Of the thirty horoscopes, six contain simply a statement that the child was born on a particular day or night (preserved cases date to 288–199 BCE), nine a statement of the night watch during which the child was born (410–83 BCE; day watches are not attested in the preserved horoscopes), one a precise time of birth given using the fixed length *bēru* system measured before sunset (251 BCE), and nine are broken at this point of the text (410–81 BCE). The remaining five horoscopes, dating between 125–69 BCE give the time of birth using the term *simānu*. For example:

> [MU-1.ME.1,3]7-KAM [I]⸢*ar*⸣ LUGAL [iti]ŠU 30 [GE_6] 7 *ina* 6 *si-man* LÚ.TUR *a-lid ina si-ma-ni-*[*šú sin ina*] ⸢ALLA⸣ *šamáš ina* TIL ALLA MÚL.BABBAR *ina* GU *dele-bat ina* [MAŠ.MAŠ[?] GU_4.UD] *ina* ⸢ALLA[?]⸣ GENNA *u* AN *šá* ŠÚ-*ú* NU IGI-'*u*
>
> BM 41301 = ROCHBERG 1998 No. 22 Rev.[?] 9'–13'

> [Year 19]7, Ar(saces) was king, Month IV, (the 1st of which was identical with the) 30th (of the preceding month), [Night of the] 7th, at the 6th *siman*, the child was born. In [his] hour (of birth), [the Moon was in] Cancer, the Sun was in the end of Cancer, Jupiter was in Aquarius, Venus was in [Gemini[?], Mercury] was in Cancer[?], Saturn and Mars which had set were not visible.
>
> ROCHBERG 1998, 123, with modifications by the present author

The Akkadian word *simānu* means "time" or "time interval," and is well-known from other texts. Indeed, this broad meaning of "time" is found in earlier horoscopes where we have statements like "in his time" followed by a list of astronomical phenomena at the moment of birth. What makes the term special here is that it is used effectively as a unit, following a number. Literally, we could translate *ina* 6 *si-man* as "at the 6th time interval" in this example. The preserved cases that use *simānu* in this way never give a number greater than 12, and by comparing the lunar positions given in the horoscopes with computation, Rochberg-Halton has argued that these *simānu* numbers should be understood as seasonal hours.[23] Rochberg-Halton also pointed to two Neo-Assyrian texts that had been interpreted by earlier scholars as referring to seasonal hours. One of these is a small fragment of an ivory prism that

23 Rochberg-Halton 1989.

contains a series of numbers.[24] While the numbers on this ivory prism *could* be interpreted as seasonal hours,[25] this interpretation is far from certain. The other text is a strange text which is part treatise, part letter. The text was first edited by Pingree and Reiner who concluded that part of the text provided statements of the length of the seasonal hours at different times of the year.[26] However, Hunger has published a new edition of the text including a substantial additional fragment which has been joined to the tablet.[27] As Hunger convincingly shows, the new fragment rules out Pingree's interpretation of the text as being concerned with seasonal hours. Thus, the very late horoscopes seem to be the only texts that refer to seasonal hours. Given their late date—recall that the earliest known example dates to 125 BCE—it seems plausible to think that this is an import of a Greek concept into Babylonian practice.

Although the number of preserved horoscopes which use seasonal hours is small, it is noticeable that all refer to seasonal hours during the day; no nighttime seasonal hours are attested. By contrast, all of the preserved horoscopes which use watches to define the time of birth concern nighttime births.

4 Conclusion

From the examples discussed in the preceding sections, it is possible to draw some broad conclusions about the place of short time in Mesopotamian society and scholarship. The most obvious point to make is that, outside of the astral sciences, references to either short-time reckoning or short-time indication appear in only a very small proportion of texts. In particular, short time is mentioned in very few letters, economic, or historical texts. It is tempting, therefore, to conclude that short-time was of comparatively little importance in everyday life in Mesopotamia. In support of this conclusion it can be noted that we have no archaeological or textual evidence for public timekeeping in any Mesopotamian city: no public water clocks or sundials have been identified, no references to such objects appear in Babylonian literature, and no letters mention meeting at a time governed by such a device. References to short time appear somewhat more often, though still quite infrequently, in certain types of scholarly texts, especially omen texts, medical texts, and ritual texts.

24 Steele forthcoming c.
25 Smith 1969.
26 Pingree and Reiner 1974–77.
27 Hunger 2017.

It is only in texts of the astral science, however, that short time is mentioned regularly.

As explained in section 1, there are two main systems of short-time reckoning: the watch, which is a seasonally-varying time unit, and the *bēru* and its subunits UŠ and NINDA, which are fixed-length time units. The two systems seem to have been used for different purposes: for example, omen and medical texts use exclusively the seasonally-varying watch system, while astronomical texts use the fixed-length *bēru* system. Indeed, use of the watch system seems to be strongly correlated with a divinatory context: the medical texts which refer to periods of short time are written in the form of omens ("if a patient suffers from *x*, he/she will *y*"), and when the watches appear in literature or in letters it is often either in reference to medicine or to some form of divination: witness, for example, how in the epic of Gilgameš the fixed-length *bēru* system is used when describing the duration of Gilgameš's journey but the seasonally-varying watch system is used when describing Gilgameš's dreams. Dream omens were a popular form of Mesopotamian divination[28] and Gilgameš's dream clearly carries the connotation of a warning in the story.

The fixed-length *bēru* system of time reckoning appears to dominate the references to short time in astronomical texts. Interestingly, however, this system appears in two forms: as regularly written quantities with numbers accompanied by units and as sexagesimal numbers without units. In the observational texts of the first millennium we find both systems. Before the middle of the fifth century BCE, the time of the beginning and the duration of the phases of an eclipse were given as quantities with the unit *bēru* and, where necessary, the subunit UŠ. After this date, these time measurements are given in the sexagesimal form. The lunar six intervals were apparently always recorded as sexagesimal numbers without units, and were given to a higher precision than eclipse times (0;10 = 0;10 UŠ = 10 NINDA rather than 1 = 1 UŠ). The purpose of observing the lunar six time intervals has only become clear in the past few years. Brack-Bernsen has shown that the lunar six could be used to predict the length of the month through simple mathematical procedures.[29] She has also shown that the sum of the mid-month intervals acts as a direct proxy for the lunar velocity, and it is all but certain that these sums were used in the development of the System A and B lunar theories.[30] Both the lunar six intervals and the eclipse timings were clearly analyzed by Babylonian astronomers, and this may very well be the reason why they were recorded in the sexagesimal

28 Oppenheim 1956; Butler 1998.

29 Brack-Bersen 1999; 2002.

30 Brack-Bersen 1997.

format, rather than as quantities with units, as would be expected. The switch to the sexagesimal format also highlights the importance of smaller short-time units. Previously, and outside of astronomical texts, it was the *bēru* that was the fundamental unit of time. Indeed, outside of astronomical texts, I know of only a single example of a time given using the unit UŠ. Significantly, this text, the ritual text from Hellenistic Uruk discussed in section 3.4, was written by a scribe who also wrote astronomical texts and who seems to have brought across from the astronomical genre the practice of writing 1/3 *bēru* as 10 UŠ. Within astronomy, the *bēru* of 2 hours was clearly too imprecise to be useful and so the smaller units, the UŠ and the NINDA, replaced the *bēru* as the unit of choice.

It remains unclear exactly how short-time intervals were measured. It is usually assumed that most timings were made using water clocks or perhaps some form of sundial for daytime measurements. Lexical and other texts from the early second millennium BCE provide us with two names of water clocks: *dibdibbu*, a name which may well reflect the sound of dripping water, and *maltaktum*.[31] Two Old Babylonian mathematical problem texts refer to the water in a *dibdibbu*.[32] Several authors have tried to combine evidence from these mathematical texts with a statistical analysis of eclipse timings from the second half of the first millennium BCE in order to provide information on the accuracy, precision, and certain physical characteristics of the water clocks and to reconstruct the form of the water clock.[33] There are significant methodological problems with such approaches, however: the assumption that things described in mathematical problems bear any relation to real life must be taken with a large grain of salt, and combining evidence from over a millennium and a half requires the assumption that there was only ever one type of water clock used in Mesopotamia and its form remained unchanged over a very long time period. This latter assumption can be easily disproved even within the late Babylonian period when, for example, the precision of reports of eclipse and other timings is one sixth lower than that of the lunar six measurements, and there is a difference in the pattern of their errors, suggesting that two different water clocks were used even within the context of astronomical observation.[34] I think all that can be said is that water clocks were probably inflow devices with a constant-head tank which measured time in the fixed-length units *bēru*, UŠ, and NINDA, and which were used during both the day and the night. But

31 Brown, Fermor, and Walker 1999–2000.
32 Thureau-Dangin 1932.
33 Brown, Fermor, and Walker 1999–2000; Fermor and Steele 2000.
34 Huber 2000.

as to the size of the tank, the rate of the water flow, and the dimensions of the spout, we are completely in the dark.

At first sight, we are slightly better off in studying gnomons and sundials. MUL.APIN and other texts contain mathematical schemes for the variation in the length of shadow,[35] and we even have two late texts which describe the construction of some type of sundial device (LBAT 1494 and 1495). However, it is important not to fall prey to over-interpreting this material. The shadow length schemes all concern the length of the shadow rather than its direction. The MUL.APIN scheme presents the time at which the shadow reaches specified lengths given at intervals of one cubit, which is not practical for measuring time because once the shadow becomes longer than two or three cubits, the equivalent times given by the scheme cluster near sunrise and sunset. Several of the other schemes are concerned not with what time it is when the shadow reaches a given length, but with what length the shadow will be either at noon or at a specific time after sunrise. It is worth remarking here that in many cultures (for example, in ancient China), the gnomon was essentially a calendrical tool, not a tool for measuring short time. I suspect that the same is true for the Babylonian gnomon. It was used—if it was used at all—to determine the solstice and equinoxes, not to measure short time.

However, the texts that describe the construction of instruments may reflect a true sundial used for measuring short-time. These two texts, which I am currently studying with Christopher Walker and Irving Finkel, are unfortunately incomplete and very difficult to understand. They abound with terminology that is otherwise unattested or, at least, unattested in astronomical contexts. However, we believe that these texts describe the construction of several different instruments including one called the "day writer," which may have been a traditional gnomon used to investigate the Sun's motion, and one called "the times," which was used for measuring time. Thanks to a perceptive suggestion by Alexander Jones, it seems possible that this instrument may have been some kind of equinoctial sundial, but we are still working on reconstructing what the device looked like. It is even possible that this device measured seasonal hours—the term "the times" (*simānu*) is the same one that is used in the horoscopes to refer to them, and part of the text refers to division into 12 parts, although this could equally well refer to the twelve months of the year. As discussed in section 3.8, seasonal hours appear very late in Babylonia: they are first securely attested in the late second century BCE and were perhaps introduced from Greece. It seems possible, therefore, that the sundials described

35 Steele 2013.

on these tablets were also introduced to Babylonia from Greece in the second century BCE. If this is the case, while this would be very interesting from the point of view of the circulation of astronomical and technical knowledge, the texts unfortunately therefore tell us little about earlier Mesopotamian practices of short-time measurement.

References

Brack-Bernsen, Lis. 1997. *Zur Entstehung der babylonischen Mondtheorie: Beobachtung und theoretische Berechnung von Mondphasen*. Stuttgart: Franz Steiner Verlag.

Brack-Bernsen, Lis. 1999. "Goal-Year Tablets: Lunar Data and Predictions." In *Ancient Astronomy and Celestial Divination*, edited by Noel M. Swerdlow, 149–177. Cambridge, MA: The MIT Press.

Brack-Bernsen, Lis. 2002. "Predictions of Lunar Phenomena in Babylonian Astronomy." In *Under One Sky: Astronomy and Mathematics in the Ancient Near East*, edited by John M. Steele and Annette Imhausen, 5–19. Münster: Ugarit-Verlag.

Brown, David. 2000. "The Cuneiform Conception of Celestial Space and Time." *Cambridge Archaeological Journal* 10: 103–122.

Brown, David, John Fermor, and Christopher Walker. 1999–2000. "The Water Clock in Mesopotamia." *Archiv für Orientforschung* 47: 130–148.

Butler, Sally A.L. 1998. *Mesopotamian Conceptions of Dreams and Dream Rituals*. Münster: Ugarit-Verlag.

Cohen, Mark E. 1993. *The Cultic Calendars of the Ancient Near East*. Bethesda, MD: CDL Press.

Driel, Govert van. 1969. *The Cult of Aššur*. Assen: Koninklijke van Gorcum & Comp.

Englund, Robert K. 1988. "Administrative Timekeeping in Ancient Mesopotamia." *Journal of the Economic and Social History of the Orient* 31: 121–185.

Feliu, Lluís, Jaume Llop, Adelina Millet Albà, and Joaquín Sanmartín, eds. 2013. *Time and History in the Ancient Near East: Proceedings of the 56th Rencontre Assyriologique Internationale at Barcelona 26–30 July 2010*. Winona Lake, IN: Eisenbrauns.

Fermor, John, and John M. Steele. 2000. "The Design of Babylonian Waterclocks: Astronomical and Experimental Evidence." *Centaurus* 42: 210–222.

Freedman, Sally A.L. 1998. *If a City Is Set on a Height: The Akkadian Omen Series Šumma Alu Ina Mēlê Šakin Volume 1: Tablets 1–21*. Philadelphia: University of Pennsylvania Museum.

George, Andrew R. 2003. *The Babylonian Gilgamesh Epic*. Oxford: Oxford University Press.

Grayson, Albert K. 1975. *Assyrian and Babylonian Chronicles*. Locust Valley, NY: J.J. Augustin.

Grayson, Albert K., and Jamie Novotny. 2012. *The Royal Inscriptions of Sennacherib, King of Assyria (704–681 BC), Part 1*. Winona Lake, IN: Eisenbrauns.

Hackl, Johannes, Michael Jursa, and Martina Schmidl, eds. 2014. *Spätbabylonische Briefe. Band 1: Spätbabylonishe Privatbriefe*. Münster: Ugarit-Verlag.

Huber, Peter J. 2000. "Babylonian Short-Time Measurement: Lunar Sixes." *Centaurus* 42: 223–234.

Hunger, Hermann. 1999. "Non-Mathematical Astronomical Texts and Their Relationships." In *Ancient Astronomy and Celestial Divination*, edited by Noel M. Swerdlow, 77–96. Cambridge, MA: The MIT Press.

Hunger, Hermann. 2001. *Astronomical Diaries and Related Texts from Babylonia. Volume V: Lunar and Planetary Texts*. Vienna: Österreichishe Akademie der Wissenschaften.

Hunger, Hermann. 2017. "The So-Called Report on Seasonal Hours (K 2077+): A New Interpretation." In *Studies on the Ancient Exact Sciences in Honor of Lis Brack-Bernsen*, edited by John M. Steele and Mathieu Ossendrijver, 149–163. Berlin: Edition Topoi.

Hunger, Hermann, and John M. Steele. 2019. *The Babylonian Astronomical Compendium MUL.APIN*. Abingdon: Routledge.

Krul, Julia. 2018. *The Revival of the Anu Cult and the Nocturnal Fire Ceremony at Late Babylonian Uruk*. Leiden: Brill.

Lambert, Wilfred G. 2013. *Babylonian Creation Myths*. Winona Lake, IN: Eisenbrauns.

Lanfranchi, Giovanni B., and Simo Parpolo. 1990. *The Correspondence of Sargon II, Part II*. Helsinki: Helsinki University Press.

Linssen, Marc J.H. 2004. *The Cults of Uruk and Babylon: The Temple Ritual Texts as Evidence for Hellenistic Cult Practice*. Leiden: Brill.

Livingstone, Alasdair. 1999. "The Magic of Time." In *Mesopotamian Magic: Textual, Historical, and Interpretative Perspectives*, edited by Tzvi Abusch and Karel van der Toorn, 131–137. Groningen: Styx.

McCluskey, Stephen C. 1990. "Calendars and Symbolism: Functions of Observation in Hopi Astronomy." *Journal for the History of Astronomy* 21: S1–S16.

Neugebauer, Otto. 1941. "Some Fundamental Concepts in Ancient Astronomy." In *Studies in the History of Science*. Philadelphia: University of Pennsylvania.

Nilsson, Martin P. 1920. *Primitive Time-Reckoning: A Study in the Origins and First Development of the Art of Counting Time among the Primitive and Early Culture Peoples*. Lund: C.W.K. Gleerup.

Oppenheim, A. Leo. 1956. "The Interpretation of Dreams in the Ancient Near East. With a Translation of an Assyrian Dream-Book." *Transactions of the American Philosophical Society* 46: 179–373.

Parpola, Simo. 1993. *Letters from Assyrian and Babylonian Scholars*. Helsinki: Helsinki University Press.

Pingree, David, and Erica Reiner. 1974–77. "A Neo-Babylonian Report on Seasonal Hours." *Archiv Für Orientforschung* 25: 50–55.

al-Rawi, Farouk N.H. and Andrew R. George. 1991–92. "Enūma Anu Enlil XIV and Other Early Astronomical Tables." *Archiv Für Orientforschung* 38–39: 52–73.

Rochberg, Francesca. 1998. *Babylonian Horoscopes*. Philadelphia: American Philosophical Society.

Rochberg-Halton, Francesca. 1988. *Aspects of Babylonian Celestial Divination: The Lunar Eclipse Tables of Enūma Anu Enlil*. Horn: Berger & Söhne.

Rochberg-Halton, Francesca. 1989. "Babylonian Seasonal Hours." *Centaurus* 32: 146–170.

Sachs, Abraham J., and Hermann Hunger. 1988. *Astronomical Diaries and Related Texts from Babylonia. Volume I: Diaries from 652 B.C. to 262 B.C.* Vienna: Österreichishe Akademie der Wissenschaften.

Sachs, Abraham J., and Hermann Hunger. 1989. *Astronomical Diaries and Related Texts from Babylonia. Volume II: Diaries from 261 B.C. to 165 B.C.* Vienna: Österreichishe Akademie der Wissenschaften.

Sallaberger, Walther. 1993. *Der kultische Kalender der Ur III–Zeit*. Berlin: de Gruyter.

Scurlock, Jo Ann. 2014. *Sourcebook for Ancient Mesopotamian Medicine*. Atlanta, GA: Society for Biblical Literature Press.

Smith, Sidney. 1969. "Babylonian Time Reckoning." *Iraq* 31: 74–81.

Steele, John M. 2000. *Observations and Predictions of Eclipse Times by Early Astronomers*. Archimedes: New Studies in the History and Philosophy of Science and Technology 4. Dordrecht: Kluwer.

Steele, John M. 2011. "Making Sense of Time: Observational and Theoretical Calendars." In *The Oxford Handbook of Cuneiform Culture*, edited by Karen Radner, 470–85. Oxford: Oxford University Press.

Steele, John M. 2013. "Shadow-Length Schemes in Babylonian Astronomy." *SCIAMVS* 14: 3–39.

Steele, John M. 2014. "Late Babylonian Ziqpu-Star Lists: Written or Remembered Traditions of Knowledge." In *Traditions of Written Knowledge in Ancient Egypt and Mesopotamia*, edited by Daliah Bawanypeck and Annette Imhausen, 123–151. Münster: Ugarit-Verlag.

Steele, John M. 2017. *Rising Time Schemes in Babylonian Astronomy*. New York: Springer.

Steele, John M. Forthcoming a. "Description, Calculation, Measurement: Different Uses of Numbers and Quantities in Babylonian Astronomy." In *Mathematical Practices in the Context of the Astral Sciences*, edited by K. Chemla, M. Husson, A. Keller, and C. Proust.

Steele, John M. Forthcoming b. *Astronomy in Babylonia*.

Steele, John M. Forthcoming c. "The Ivory Prism from Nineveh (BM 123340)."

Streck, Michael P. 2017. "The Terminology for Times of the Day in Akkadian." In *At the Dawn of History: Ancient Near Eastern Studies in Honour of J.N. Postgate*, edited by Yağmur Heffron, Adam Stone, Martin Worthington, and J.N. Postgate, 583–609. Winona Lake, IN: Eisenbrauns.

Thureau-Dangin, François. 1932. "Le clepsydra chez les Babyloniens." *Revue d'Assyriologie* 29: 133–136.

Waerden, Bartel Leendert van der. 1951. "Babylonian Astronomy. III. The Earliest Astronomical Computations." *Journal of Near Eastern Studies* 10: 20–34.

Wainer, Zackary. 2016. "Traditions of Mesopotamian Celestial-Divinatory Schemes and the 4th Tablet of Šumma Sin Ina Tāmartišu." In *The Circulation of Astronomical Knowledge in the Ancient World*, edited by John M. Steele, 55–82. Leiden: Brill.

Zeilik, Michael. 1985. "The Ethnoastronomy of the Historic Pueblos, I: Calendrical Sun Watching." *Journal for the History of Astronomy* 16: S1–S24.

Zeilik, Michael. 1986. "The Ethnoastronomy of the Historic Pueblos, II: Moon Watching." *Journal for the History of Astronomy* 17: S1–S22.

CHAPTER 4

Greco-Roman Sundials: Precision and Displacement

Alexander Jones

In the Greco-Roman world, among the various means known to have existed for determining times of day with any degree of precision, stationary sundials (i.e., sundials installed in a fixed location and with fixed orientation) were by far the most common. The Topoi Excellence Cluster project on Ancient Sundials (henceforth BSDP) has estimated that between 550 and 600 Greco-Roman stationary and portable sundials are extant.[1] Of these, fewer than thirty are portable sundials, while the remains of water clocks—whether of the clepsydra variety that showed the hour as the level of water against a column of marks spaced according to the seasonal variation of day or night, or the more mechanically sophisticated variety that employed a display dial—are comparably scarce.[2] Admittedly, these figures in part reflect differences in the survival rates of ancient artifacts determined by their composition. Most portable sundials, and some clepsydras and components of mechanical water clocks, were made of metal (typically bronze), which was generally melted down and reused when the original object was no longer wanted. Stationary sundials were sculpted from blocks of stone, except for their metal gnomons (which are almost invariably missing now), and even in subsequent repurposing—e.g., as building blocks or filler—they often preserved at least part of their dial surfaces. Nevertheless, we may be confident that when a Greek or Roman of the Hellenistic or Roman period wanted to know the time of day, he or she would most probably have consulted a stationary sundial.

This chapter considers two distinct but related questions concerning the performance of Greco-Roman stationary sundials. The first question is, can the rarity in Greco-Roman textual sources of time specifications to a precision finer than whole hours be attributed to limitations in the ability of sundials to

1 Graßhoff et al. 2016. For the purposes of this paper, Egyptian time-keeping devices that represent pre-Ptolemaic traditions are excluded even if dating from the Greco-Roman period.

2 Schaldach 2016, 91 (inventory of portable sundials), 64–65 (clepsydras), and 81–83 (mechanical water clocks).

 | DOI:10.1163/9789004416291_006

display more refined times? In all the varieties of Greco-Roman time-keeping devices, the indicator of time's passage (shadow point, spot of sunlight, water level, moving pointer or revolving dial) had a continuous motion, in the course of which it successively crossed a series of marks or lines corresponding to the moments demarcating the beginnings and endings of seasonal hours. Intermediate marks do not occur. The time read off the device could be thought of as a discrete entity, either a block of time whose identity and character are defined primarily by its position in a sequence of such blocks ("during the *n*th hour"), or a moment subject to a certain tolerance ("at the *n*th hour" or "at the end of the *n*th hour"). However, the Greco-Roman definition of seasonal hours was based on a somewhat simplified model, according to which time was continuous and measurable by the uniform circular revolution of the Sun in the sky through the course of each day. This definition might have encouraged people to think of the output of a time-keeping device as a measurement of elapsed time, such that the hour marks or lines function as a scale of units along a continuum that also allows for intermediate time specifications expressed in hours and fractions of hours. Since we do not see this happening outside of scientific contexts, it is worth investigating what role sundial design might have had in preserving the whole seasonal hour as the smallest quantitative unit of time specification.

My second question is actually twofold. First, how often were sundials installed in localities for which they were not designed? Second, how would such displacements have been apparent to a user, and how would they have affected the accuracy of times displayed on them?

The consequences of geographical displacement of sundials has been treated before in the technical literature on sundials, but still are subject to misconceptions. The danger here is of supposing that, since the lengths of seasonal hours vary significantly with terrestrial latitude as well as with the stage of the year, a displaced sundial will uniformly stretch or diminish the displayed time intervals. What actually happens is more complicated, as some visual illustrations will make clear.

1 Precision of Time Specifications in Ancient Scientific Contexts

In the context of observational astronomy, refined time specifications can be found as early as the 7th century BCE in Babylonia.[3] These observed times, presumed to have been made using water clocks, are expressed in units called

3 See, e.g., Steele's contribution to the present volume.

UŠ, equivalent to 1/360 of a night and day or 4 minutes in our time metrology, and *bēru*, equivalent to 30 UŠ. Thus Babylonian records of lunar and solar eclipses typically included measurements of the time since sunset or sunrise of the beginning of the eclipse, as well as the time intervals taken for the eclipse, either as a whole or divided into stages such as onset, totality, and clearing. The earlier records, up to about the first quarter of the 6th century, typically gave a precision of 5 UŠ (i.e., 20 minutes), later refined to single UŠ, but typical errors remained quite large through the more than six centuries of Babylonian eclipse records, even for short time intervals.[4] Besides eclipse timings, time measurements in UŠ were recorded for a variety of astronomical timings, especially for intervals separating sunset or sunrise from the setting or rising of the Moon and planets when near conjunction or opposition.

Although the Babylonian unit UŠ was carried over into Greek astronomy under the name χρόνος (literally "time period," but usually translated "time degree"), the preserved Greek observational records that specify times with any precision at all employ either seasonal or equinoctial hours. The reports from before Ptolemy's time in Ptolemy's *Almagest* use seasonal hours, sometimes accompanied by a qualitative refinement such as "at the beginning," "at the end," "at the middle," or simply "during." A few give fractions in addition to the whole number of hours since sunset or sunrise: 1/2 (Timocharis, 283 BCE, ed. Heiberg 2.29), 1/3 (anonymous, 200 BCE, ed. Heiberg 1.345, and Hipparchus, 127 BCE, ed. Heiberg 1.374), 2/3 (anonymous, 200 BCE, ed. Heiberg 1.346, and Hipparchus, 128 BCE, ed. Heiberg 1.363). As transmitted by Ptolemy, the reports give no indication of how the times were determined.

Ptolemy's own observation reports (including one unattributed eclipse observation that is likely to be his) use equinoctial hours relative to noon or midnight. Ptolemy's times are usually whole numbers of equinoctial hours, but we also have some fractions, including 1/2 (139 CE, ed. Heiberg 2.283), 1/4 (139 CE, ed. Heiberg 1.362), 3/4 (133 CE, ed. Heiberg 1.314, and 138 CE, ed. Heiberg 2.306), 3/5 (the anonymous eclipse, 125 CE, ed. Heiberg 1.329), and 5/6 (135 CE, ed. Heiberg 1.408). Most of these more precise times, if not all, were ostensibly determined by calculation from the culminating degree of the equator on Ptolemy's observational armillary instrument. The fact that Ptolemy consistently reports this figure to a precision of single degrees—effectively 1 UŠ—even though the times always come out as whole equinoctial hours or hours with simple fractions, possibly indicates that Ptolemy, in actuality, calculated the culminating degrees from the times in hours.

4 Steele 2000, 57–66.

We have few Greek observation reports from the time following Ptolemy—still fewer with precise recorded times—and not all of these turn out to be genuine evidence of observational time precision. In his commentary on Book 6 of the *Almagest*, Theon of Alexandria provides worked examples of how to calculate the circumstances of a lunar eclipse using the *Almagest*'s tables, and those of a solar eclipse using both the *Almagest* and the Handy Tables, choosing for each kind an eclipse that he claims to have observed ἀσφαλέστατα ("most securely"), namely the total lunar eclipse of 364 CE, November 25/26, and the partial solar eclipse of 364 CE, June 16.[5] In both examples, the calculations are preceded by a set of times for the stages of the eclipse. In the case of the solar eclipse, these stages are unambiguously said to have been observed, whereas for the lunar eclipse, Theon gives conflicting indications about whether they were supposed to have been observed or calculated.[6] The times given for the beginning, middle, and end of the solar eclipse are respectively 2 5/6, 3 4/5, and 4 1/2 seasonal hours past noon, whereas the times for the beginning, beginning of totality, end of totality, and end of the lunar eclipse are respectively 14 9/10, 16 17/30, 17 4/15, and 18 3/5 equinoctial hours past noon. Unfortunately, whatever Theon meant his readers to assume about his observational activity, the exact agreement of these times with those that he obtains from the *Almagest* calculations has only one plausible explanation, that they were adjusted or fabricated to produce that very agreement.[7]

5 For the lunar eclipse, the text according to the earliest and most reliable manuscript, Laur. plut. 28.18, f. 231r, is ποιησάμεθα δὲ τὴν ψηφοφορίαν ἐπὶ τῆς ἀσφαλέστατα ἡμῖν τετηρημένης ἐνταῦθα ἐν Ἀλεξανδρείᾳ τῇ πρὸς Αἴγυπτον. ("We have made the computation for the [eclipse] that was observed most securely by us here in Alexandria in Egypt.") The 1538 Basel edition, p. 319, lacks ἡμῖν so that the observation is not expressly attributed to Theon.

6 For the solar eclipse times, Theon writes (Laur. plut. 28.18, f. 242r, 1538 ed. p. 332) ἀσφαλέστατα ἐτηρήσαμεν ("we observed most securely"). His lunar eclipse times (Laur. plut. 28.18, f. 231r, 1538 ed. p. 319), however, are governed by the verb ἐπιλογισάμεθα [*sic* for ἐπελογισάμεθα], "calculated." But at the end of his calculations from the *Almagest*, Theon states that the computed times obtained from the *Almagest* are ἀκολούθως τοῖς κατὰ τὴν τήρησιν γεγενημένοις ἡμῖν τῶν τοιούτων χρόνων ἐπιλογισμοῖς ("in agreement with the calculations made by us of such times on the occasion of the observation"), which must mean times computed *from observational data*.

7 Delambre took the agreement of the ostensibly observed times with the calculations as an indication that the reports were fabricated: "*cette conformité si singulière, qu'on n'obtiendrait aujourd'hui même que par le plus grand des hasards, pourrait faire soupçonner que cette prétendue éclipse est arrangée pour les Tables.*" (Delambre 1817, vol. 2, 591 and 594). Theon's solar eclipse has been much discussed in modern scholarship, mostly without awareness that the observation report is suspect; an exception is Mercier (2011, 407), who however mistakenly claims that Theon "never reported his observations as such."

Lastly, among the observation reports believed to have been compiled by the Neoplatonist Heliodorus is a report of an observation of the Moon occulting Saturn on February 21, 503 CE, made by the writer and ὁ φίλτατος ἀδελφός ("my dearest brother," presumably Ammonius), in which the time of the end of occultation is recorded as 5 3/4 seasonal hours (scil. past sunset) obtained ἀπὸ ἀστρολάβου ("from an astrolabe"), which probably means an armillary like Ptolemy's rather than a plane astrolabe, since the plane astrolabe would not function in any straightforward way as an instrument of time measurement at night.[8]

Thus we have good evidence for Greek astronomers making determinations of times of day and night to a precision of a fraction of an hour, though never approaching the 4-minute precision claimed in the Babylonian observation reports. Only Hipparchus's observations of the Moon's elongation from the Sun, which of course were diurnal, could in principle have been made using a sundial. Sundials are more plausible as a source of the birth times in diurnal horoscopes, and—notwithstanding the professions in the astrological literature of extreme time sensitivity, distinguishing horoscopes of people born even a fraction of an hour apart—the birth times in horoscopic documents are invariably given as whole numbers of seasonal hours, at most modified by qualitative expressions such as "at the beginning," but probably never with a quantitative fraction or subunit.[9] Time specifications in civil and administrative contexts too were always by whole numbers of seasonal hours.[10]

2 Reading Fractional Hours on Stationary Sundials

Given the empirical fact that the limit of time precision for non-specialists in the Greco-Roman world was effectively the whole seasonal hour, the question I wish to address now is the speculative one of whether the prevalent sundial technology would have allowed more refined time specifications if people had

8 Jones 2005, 80–83. Since the armillary yields time in time degrees, a time in seasonal hours would have had to be the result of conversion, not direct measurement on the instrument.

9 A possible exception is *P.Oxy.* XII 1476, which gives the time of birth as "10th hour of day completed" followed by "2 degrees": ἡμέρας ὥρ(α) ι/ πεπληρωμένη (*vac.*) μοι() β/. Neugebauer and van Hoesen (1959, 60–61) interpret this as meaning 2 time degrees (i.e., 8 minutes) past the end of the tenth seasonal hour. But such a mixture of time units—moreover, one seasonal, the other constant—makes no sense, and I suspect that the "2 degrees" is a misplaced addition or correction to the astronomical data of the horoscope, not part of the birth time.

10 Remijsen 2007.

wanted them. We will be concerned here not with scientific instrumentation or hypothetical high-precision sundials graduated for fractional hours, but with the kinds of sundial represented by surviving examples, though we can reasonably limit consideration to those that were more carefully executed.

Greco-Roman sundial design originated in a definition of the seasonal hour that was intended to ensure that the duration of all seasonal hours within a single day (or night) was equal. This presumption, that the twelve seasonal hours making up a day or night should be equal in duration, is what made the seasonal hour a unit of time (albeit a seasonally variable unit) and not merely an ordinally-counted partition of a larger interval, and the status of a seasonal hour as a metrical unit is in turn a prerequisite for such expressions as "a third of an hour" to be meaningful.

The definition in question was, naturally, in terms of astronomy. Between sunrise and sunset the Sun was assumed to traverse, at uniform speed, the arc above the horizon of a declination circle on the celestial sphere, so that the seasonal hours of day correspond to equal twelfths of that arc. In terms of the conventional geocentric cosmology, this was a simplification. The proper cosmological assumption was that the fundamental uniform revolution was the one performed by the sphere of the fixed stars, while the Sun superimposes on this uniform revolution its own slower, non-uniform, and oblique revolution around the poles of the ecliptic—in the language of Plato's *Timaeus*, the two revolutions are the motions of the "same" and the "different."[11] In sundial design, however, the Sun is treated as if, during the course of a single day, it moves at a constant rate of right ascension relative to the (uniformly revolving) celestial sphere, with unchanging declination, with a discrete change of declination happening between one day and the next; thus, the small effects of the obliquity of the ecliptic and of solar anomaly on the diurnal change in the Sun's right ascension and declination are disregarded.

We can think of the sky itself, the half of the celestial sphere that is above the horizon, as the prototype of a Greco-Roman sundial. That is, the declination arc that the Sun approximately traces from sunrise to sunset functions as a day curve, indicating the stage of the year, while the fraction of that arc traced by the Sun from sunrise to the present moment indicates the seasonal hour. The day curves inscribed on any mundane sundial should be the projections of the declination arcs, through a gnomon point, upon the sundial's surface, while the hour curve for hour n should be the locus of the projections of the points on all declination arcs corresponding to the end of the nth seasonal

11 See Sattler's contribution to this volume.

hour—or, in other words, the projection of the locus of points on all declination arcs that lie $n/12$ of the total arc from the eastern horizon.

Thus, the geometry of a sundial's grid of day and hour curves was determined by the kind of sundial surface and, for nonplanar surfaces, the location of the gnomon point. The more common types were the following:

(a) concave spherical surface; gnomon point at the sphere's center
(b) concave surface of a right cone with axis perpendicular to the equator; gnomon point along the axis
(c) planar surface parallel to the equator
(d) horizontal planar surface
(e) vertical plane (oriented in a cardinal or intermediate direction)
(f) downward-facing concave spherical surface; gnomon point (in this case, an eyehole allowing a ray of sunlight to fall on the shaded surface) on the sphere's surface

In types (a), (b), and (c) the declination arcs and their subdivisions project as similar and similarly subdivided circular arcs. The day curves in (d) and (e) are hyperbolas, except for the curve corresponding to the equinoxes (declination 0°), which is a straight line. In (f) the day curves are more complex teardrop or cardioidal curves generated as the intersections of the spherical surface with a cone, except that the curve corresponding to the equinoxes is a circle. All these curves were within the scope of Greek geometry as definable mathematical objects, so that a sundial designer who wished to do so could have constructed them on a stone surface according to accurate mathematical principles.[12]

If Greco-Roman timekeeping had employed the constant equinoctial hour (1/24 of a mean solar day) as the unit of civil time, the points on the celestial sphere corresponding to the boundaries of the equinoctial hours would all have lain along equally spaced great circles passing through the celestial north and south poles. However, even on the celestial sphere itself, the hour curves bounding seasonal hours, as defined above, would not have been tractable to a Greek geometer except as pointwise-constructed loci. Fortunately these loci turn out to diverge only very slightly from great circle arcs on the celestial sphere, so that sundial designers were able to approximate them on concave spherical sundials (type a) by great circle arcs and on planar sundials

12 Interest in the geometrical properties of conic sections is abundantly attested; for possible connections with sundial theory, see Neugebauer 1948 and Rinner 2017. Jones (2017) discusses how a Greek geometer might have treated the day curves of type (f), the so-called roofed spherical sundials.

(types c, d, and e) by straight lines.[13] Most sundial grids comprised just the three day curves corresponding to the summer solstice, the winter solstice, and both equinoxes, and the hour curves corresponding to the ends of the first through the eleventh seasonal hours. To complete such a grid, one usually had to determine the thirty-three points corresponding to the Sun's projected location at the end of each hour on either solstice and the equinoxes (or at least the twenty-two points on the solstitial curves), and then the day and hour curves could be inscribed as passing through the points lying on them. Some grids, however, had day curves not only for the solstices and equinoxes, but also for the other dates when the Sun entered each of the zodiacal signs.[14] For these denser grids, the designer had the option of determining all eleven hour-boundary points on each additional day curve, or merely inscribing the additional day curves across a set of hour curves that were determined just from the solstitial and equinoctial hour-boundary points.

As an illustration of the foregoing, Figure 4.1 shows a complete set of points at which intersect the hour curves and day curves of an ideally accurate horizontal planar sundial intended for latitude 41°, suitable for, say, Pompeii (actual latitude 40° 45′), with day curves for all dates of the Sun's entry into the zodiacal signs. The points were all individually determined by trigonometrical calculation equivalent to the kind of nomographic procedure employed in antiquity for this purpose and known as an *analemma*.[15] The central vertical straight line in the diagram corresponds to the meridian (with north at the top), and its intersection with the horizontal (east-west) straight line would be directly below the gnomon point. The hour curves, as delineated by sets of seven points fanning out from south to north, are visually indistinguishable from straight lines, while the day curves lie precisely on hyperbolic arcs.[16] A horizontal sundial obviously had to be mounted lower than eye-height, hence

13 For a mathematical analysis of the hour curves on the celestial sphere and their deviations from great circle arcs, see Drecker (1925, 12–20). The maximum deviation in hour-angle applying to a locality with latitude 40° is less than half a minute of arc.

14 In Greco-Roman sundial design it was assumed that the solstices and equinoxes coincided with the Sun's entry into Cancer, Libra, Capricorn, and Aries.

15 For a horizontal sundial, the direction and length of the shadow are determined respectively by the Sun's azimuth and altitude, or equivalently, the angles called *horizontalis* and *descensivus* in William of Moerbeke's Latin translation of Ptolemy's *Analemma*. See Drecker 1925, 4–11, and Luckey 1927. Ptolemy's work also provides methods of determining the angles numerically by trigonometry.

16 The straight lines representing the hour curves do not all intersect at a single point, reflecting the fact that the great circles approximating hour curves on the celestial sphere do not all pass through a common point.

there would not have been a great distance between viewer and sundial that might have impeded an accurate reading of the indicated time and season.

In Figure 4.2, the points from Figure 4.1 are superimposed on the grid of a horizontal sundial from Pompeii (Museo Archeologico Nazionale, Naples, inv. 2476). This grid is atypical in having hour curves composed of straight lines joining the corresponding hour points on each of the seven day curves, and these hour points show curious and apparently systematic deviations. The origin of the deviations is not clear, but they might reflect inaccuracies in an underlying analemma construction. On the other hand, the day curves are quite accurately and smoothly drawn.

Figure 4.3 superimposes the points from Figure 4.1 on the grid of another Pompeian horizontal sundial (Soprintendenza Speciale per i Beni Archeologici di Napoli e Pompei inv. 49725) that more conventionally has day curves only for the solstices and equinoxes, showing that this was a reasonably well-executed sundial, except for some inaccuracy in the extremities of the winter solstitial day curve. (For both sundials, the estimated centers of the gnomon's base, marked by a small circle in the diagrams, are displaced from the point vertically below the gnomon point, so the gnomons apparently were sloping.)

Neither of this pair of horizontal sundials is an instance of spectacularly high accuracy, but nevertheless the elements that matter for telling the time of day—namely, the hour curves—are close to their correct theoretical paths, with deviations that are only a small fraction of the spaces between the curves. In other words, at any time of year, when the tip of the gnomon's shadow fell upon an hour curve on either sundial, the true local time in seasonal hours would have been within a few minutes of the indicated time. Moreover, if two people separately consulted one or the other of the sundials, they could coordinate a meeting time with, again, an uncertainty of just a few minutes. Notwithstanding Seneca's gibe that agreement is easier to find among philosophers than among sundials (*Apoc.* 2.2), synchronization to a precision of single seasonal hours by means of decently-executed and properly-oriented sundials was entirely attainable.

The potential for reading off fractional hours on these sundials, however, is more problematic. The spaces between hour curves for the seasonal hours closest to noon are small, perhaps too small for a viewer to identify a displayed time as, say, on the half hour with much confidence. Further from noon, the spaces widen so rapidly—becoming infinite for the first and last hours of the day, when the Sun is in the plane of the horizon—that it becomes a matter of guesswork what point along a day curve would correspond to a time halfway or a third or two-thirds of the way between whole hours.

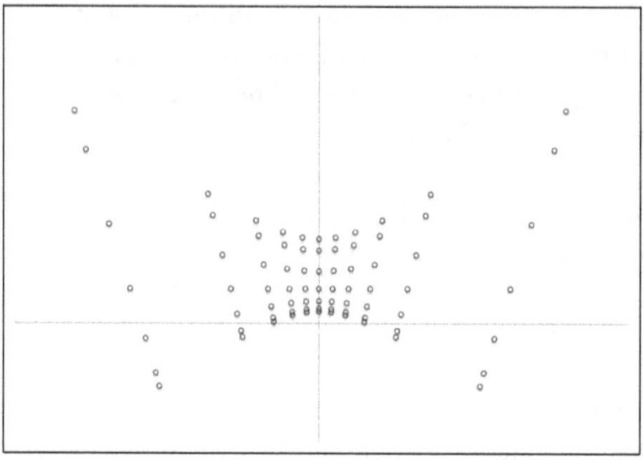

FIGURE 4.1
Seasonal hour-boundary points of a horizontal sundial for latitude 41°, computed for all solar declinations corresponding to the Sun's entry into a zodiacal sign. North is at the top. The uppermost set of points pertain to the winter solstice, and the lowermost to the summer solstice; seasonal hours run from left to right, starting with the end of the first hour of day.

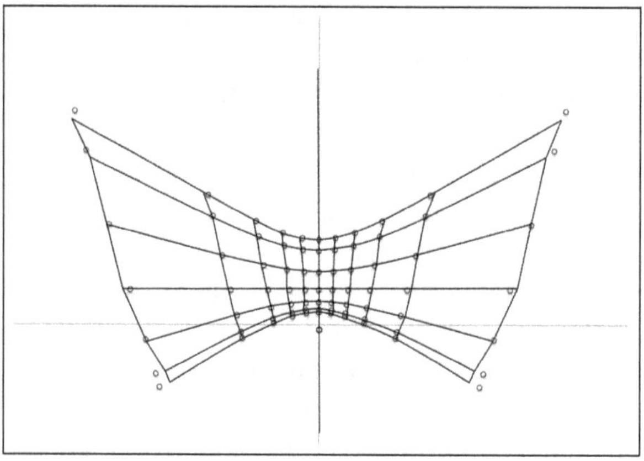

FIGURE 4.2
The hour-boundary points of Figure 4.1 superimposed on the grid of a horizontal sundial from Pompeii
MUSEO ARCHEOLOGICO NAZIONALE, NAPLES, INV. 2476

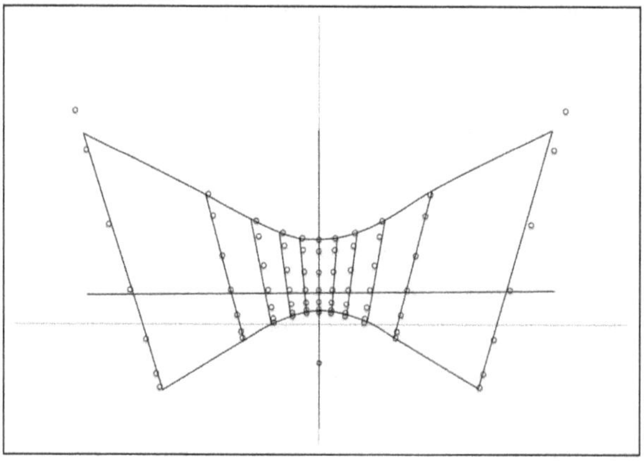

FIGURE 4.3
The hour-boundary points of Figure 4.1 superimposed on the grid of a horizontal sundial from Pompeii
SOPRINTENDENZA SPECIALE PER I BENI ARCHEOLOGICI DI NAPOLI E POMPEI INV. 49725

The same problem of visually interpolating fractional hours arises with all orientations of vertical planar sundials. For example, Figure 4.4 reproduces the grid of the east-facing sundial of the Tower of the Winds in Athens, which of course would have displayed times only before noon. Here, the closely-spaced hour curves are those closest to sunrise, near the top of the grid, whereas noon (with the Sun in the meridian plane) is at infinity, and the non-uniform subdivision of the seasonal hours immediately preceding noon would be difficult for a viewer to estimate correctly. Most viewers, if asked where the shadow indicates the time halfway through an hour on a planar sundial, would probably have pointed to the geometrical midpoint between the points for the whole hours, which is grossly incorrect for the majority of the displayed seasonal hours.

On sundials of type (f), known as "roofed spherical" sundials, the mathematics behind the grid of day and hour curves is rather complex, but an accurately-executed grid results in intervals between the hour-boundary points along the day curves that are comparatively uniform in size, as illustrated in Figure 4.5. In this case, treating each hour interval along a day curve as if the spot of sunlight traversed it with constant speed would result in a subdivision of the hours into fractions that would be entirely satisfactory for everyday purposes.

Lastly, we come to types (a), (b), and (c): the spherical, conical, and equatorial sundials. These types have in common the property (possessed, in fact, by any surface of revolution around the axis of the celestial sphere, when the gnomon point lies on the axis) that any arc of a declination circle on the celestial sphere projects through the gnomon point as a geometrically similar arc on the sundial. Consequently, fractional divisions of the hour intervals on these sundials are exactly proportional to the corresponding fractions of the seasonal hours. Presuming an accurately-drawn grid and that the sundial was mounted where one could see it reasonably well (e.g., not on top of a tall column), a viewer could easily have estimated halves, thirds, and quarters of hours within a few minutes' precision.

To sum up, four of the six major varieties of Greco-Roman sundial, if executed and aligned accurately and mounted where they could be seen sufficiently clearly, were capable of being read by a layperson to a precision of major fractions of a seasonal hour with sufficient accuracy for any plausible social purposes. Among these, the spherical and conical varieties by themselves account for more than half the known ancient sundials. The fact that Greco-Roman society never went beyond integer hour precision was not determined by limitations of easily available technology.

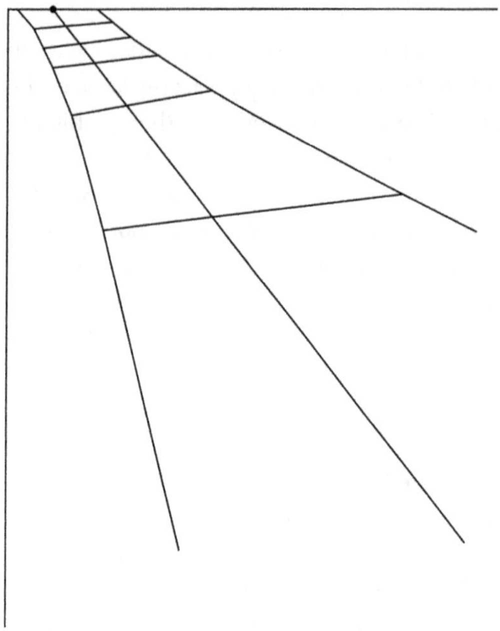

FIGURE 4.4
The grid of the east-facing sundial on the Tower of the Winds, Athens. The gnomon tip would have lain on a line perpendicular to the sundial face and passing through the point marked by a small circle. North is to the right, and the day curves from left to right are respectively for the summer solstice, the equinoxes, and the winter solstice. The hour curves correspond, from top to bottom, to the ends of the first through the fifth seasonal hour.

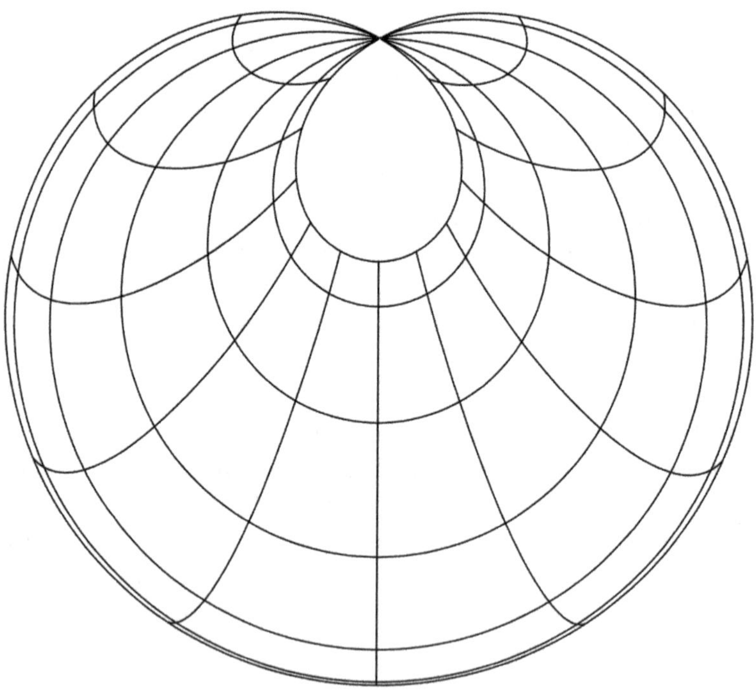

FIGURE 4.5 Grid for a roofed spherical sundial for latitude 48°, projected into the equatorial plane

3 Displaced Sundials

Discussions of geographically displaced sundials in antiquity almost always make reference to Pliny, *HN* 7.214:[17]

> M. Varro primum statutum in publico secundum Rostra in columna tradit bello Punico primo a M'. Valerio Messala cos. Catina capta in Sicilia, deportatum inde post XXX annos quam de Papiriano horologio traditur, anno urbis CCCCLXXXXI. nec congruebant ad horas eius lineae, paruerunt tamen ei annis undecentum, donec Q. Marcius Philippus, qui cum L. Paulo fuit censor, diligentius ordinatum iuxta posuit; idque munus inter censoria opera gratissime acceptum est.
>
> Marcus Varro reports that the first (sundial in Rome) was set up in public, on a column by the Rostra, during the first Punic War by the consul Manlius Valerius Messala after Catania in Sicily was taken, and that it was brought there 30 years later than is reported for the sundial associated with Papirius, in the year of the City 491 (i.e., 263 BCE).[18] Its lines did not agree with the hours, but nevertheless they obeyed (the sundial) for ninety-nine years, until Quintus Marcius Philippus, who was censor together with Lucius Paullus, installed next to it (another one that was) more carefully designed; this benefaction was received most favorably among the censor's accomplishments.

The passage comes within the context of an account (7.210–215) of three institutions that had come to be adopted universally and without explicit discussion (*gentium consensus tacitus*): use of the Ionian alphabet, shaving the beard, and observance of the division of days into hours (*in horarum observatione*). With respect to the hours, referring back to Book 2 (presumably 2.187) for their original "discovery" in Greece, Pliny focuses on their later adoption by the Romans, for which he cites two reported starting points: an assertion by Fabius Vestalis that the first sundial in Rome was erected by Lucius Papirius Cursor eleven years before the war with Pyrrhus (i.e., around 292 BCE), and Varro's possibly conflicting and more detailed report quoted above.[19]

---

17 A less detailed parallel account is Cens. 23.

18 The year number is corrupt in the manuscript tradition. I adopt Salmasius's emendation rather than Pighi's 490 which appears in some editions of Pliny.

19 It is not clear from his phrasing whether Pliny considers Varro's claim to have been that the sundial from Catania was the first one erected in Rome *at all*, or the first one erected in a public place. In the former case it is given as an alternate story to that of Vestalis, which Pliny complains lacked crucial details; in the latter case, the point would be that

What can we infer from Pliny's narrative? First, the sundial from Catania was placed on top of a column and was used to tell time for nearly a century. This tells us that the sundial was of a type that could be read from ground level when mounted high up—in other words, probably a spherical or conical sundial of type (a) or (b). The sundial's inscribed lines did not show the correct time, but we are not told how this error came to be known or when. Nor are we told that the error was a consequence of the sundial's having been made for a different locality; the remark that the sundial that supplanted it a century later was *diligentius ordinatum* ("more carefully designed") would seem to attribute the inaccuracies to sloppy execution rather than to geographical displacement. In 2.182 Pliny does state that the same time-keeping devices (*vasa horoscopa*) cannot be used everywhere because meridian-shadow-to-gnomon ratios vary from one locality to another, but it is by no means obvious that this was what he had in mind when he wrote the Book 7 passage.[20] Even in the Book 2 passage he fails to state that it is only the north-south component of a displacement that matters.

A naive expectation might be that a sundial made for use at Catania but installed at Rome would show the seasonal hours as if Rome lay on the same parallel of latitude as Catania. In fact, this could be accomplished, but only if the sundial was mounted in a special manner. Catania, at latitude 37° 30′, is situated about 4 1/2 ° south of Rome, latitude 41° 54′ (Figure 4.6).[21] Let us imagine that the sundial from Catania was, in the first instance, accurately constructed for Catania's latitude, and that it was remounted in Rome in an accurate north-south orientation, but tilted 4 1/2 ° from horizontal toward due south. In absolute terms, the sundial would be oriented exactly parallel to the orientation a sundial should have in a location on Catania's parallel, but due south of Rome, a place (actually in the sea) slightly south of Marsala that we may for

---

Papirius had dedicated his sundial in a sacred precinct, the temple of Quirinus. For a contrasting interpretation of this passage see Wolkenhauer 2011, 67–93.

20 The curious expression *vasa horoscopa* might mean all devices for determining times in seasonal hours, rather than specifically sundials; the supposition that it refers specifically to portable sundials is unjustified. Pliny's incompetence in matters of time-reckoning is glaring in 2.181, where he contends that a long-distance runner could travel the same distance east-to-west in significantly fewer hours than west-to-east because *cum sole iter erat* ("he travels with the Sun"), and moreover that a ship sailing westward can go further during even a winter day than during the nighttime because during the day it is *solem ipsum comitantes* ("travelling with the Sun").

21 The latitudes according to Ptolemy's *Geography*, respectively 37° 45′ and 41° 40′, may be taken as an indication of geographical data that might have been available in Pliny's time, if not in the third century BCE.

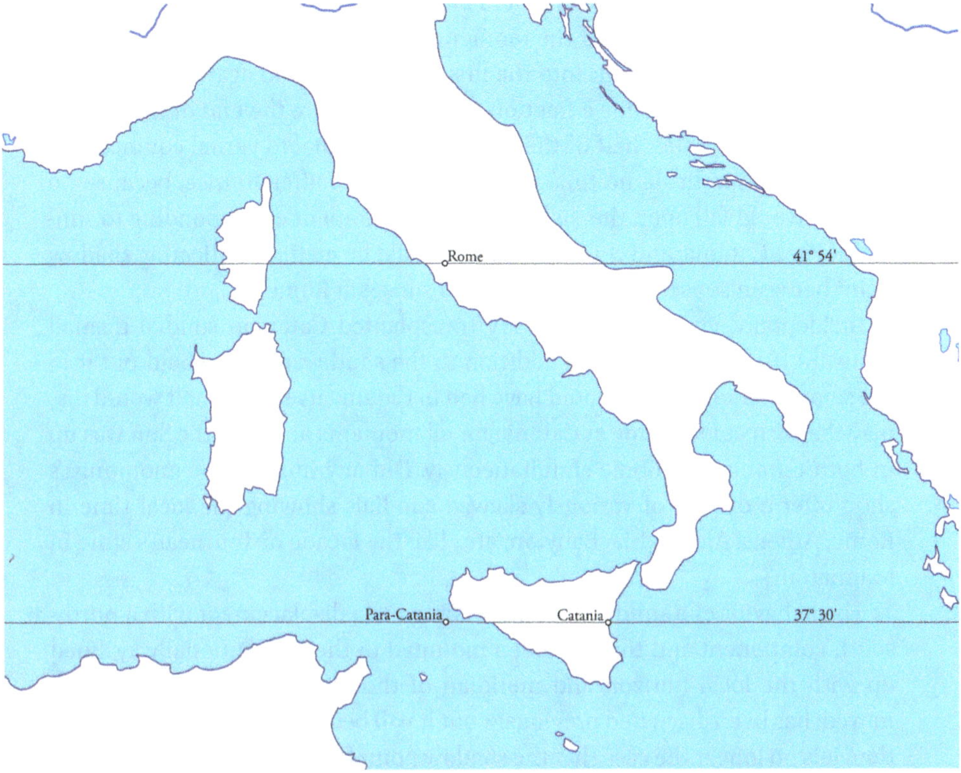

FIGURE 4.6 The central Mediterranean, with Rome, Catania, and the fictitious Para-Catania
HTTP://D-MAPS.COM/M/MEDITERRANEAN/MEDITMIN/MEDITMIN03.SVG

convenience name Para-Catania.[22] The result would be that the sundial would always show the correct local time for Para-Catania, as well as the correct stage of the year, so long as a ray of sunlight was able to fall upon the gnomon's tip and cast a shadow on the sundial's inscribed surface. In short, the shadow point always falls on exactly the same spot on the sundial as it would if the sundial were at Para-Catania. But except on an equinox, the sundial would show the correct time for Rome only at noon, whereas on the equinox it would show the correct Roman time all day.

During the half-year from the vernal equinox to the autumnal equinox, the interval of daytime from sunrise to sunset in Para-Catania, like in Catania itself, is longer than daytime in Rome, so that for the first and last bits of the

22 Para-Catania experiences the same variation in the lengths of days and nights through the year as Catania, but Para-Catania's local noon occurs about ten minutes later than Catania's because of their separation in longitude.

Para-Catanian day the sundial would show no time because the Sun would be below Rome's horizon. At the moment of sunrise, the sundial would show a time a fraction of an hour into the first hour of day, and again at sunset the sundial would show a time a fraction of an hour before the end of the twelfth hour. During the other half of the year, from autumnal to vernal equinox, the sundial would indicate no time until some interval after sunrise, because no shadow would fall upon the dial grid until the moment corresponding to sunrise at Para-Catania; and conversely there would be no time-indicating shadow point between sunset at Para-Catania and sunset at Rome.

Incidentally, giving our imaginary transplanted Catanian sundial a small eastward tilt (about 2 1/2 °) in addition to the southward one would put it in the same orientation as it would have had in Catania itself, so that it would display the correct local time at Catania for all moments at which the Sun was up in both Catania and Rome simultaneously. Did any imaginative gnomonist's shop offer a display of variously skewed sundials showing the local time in Rome, Athens, Alexandria, Babylon, etc., like the façade of Tourneau's store in Manhattan?

The behavior of a sundial that has undergone a displacement with a north-south component and that has been mounted in the conventional way, lined up with the local horizon and meridian of the place to which it has been moved, has been discussed previously, but it will bear revisiting.[23] In this situation, it is no longer the case that the shadow point falls on the same spot on the grid as it would have fallen in its original locality, nor is it ever in the same place relative to the day and hour curves as the shadow point on a sundial correctly constructed and mounted for the new location. Since our main concern is with how well it functions as a substitute for a sundial accurately calibrated for its place of exile, we can limit ourselves now to just this comparison—comparing, so to speak, an idealized version of the Catanian sundial installed near the Rostra by Valerius Messala to an idealized version of the sundial installed next to it by Marcius Philippus a century later.

The meridian hour curve marking the end of the sixth hour (i.e., local noon), is the projection of the celestial meridian, which is unaffected by north-south displacement on the Earth. On most sundial types, the meridian curve is also the line of symmetry of the grid.[24] Regardless of displacement, the shadow point will fall upon this line at noon throughout the year. Again, sunrise and

23 Gibbs 1976, 96; Hüttig 2000; Savoie 2001, 317–336; Hannah 2009, 134–136.

24 The principal exceptions are vertical sundials facing directions other than south or north. East-facing and west-facing vertical sundials have no meridian line; sundials facing intermediate directions ("declining" dials) have a meridian line but lack chiral symmetry.

sunset (i.e., the beginning of the first hour and the end of the twelfth) are moments when the Sun lies on the horizon and thus the shadow point lies on the projection of the horizon, whether or not the sundial is displaced. Thus, a displaced sundial of any type automatically displays the correct seasonal hour at these three moments on any day of the year.

To understand why these three times are always correctly displayed on the displaced sundial, as well as what happens with respect to the intermediate times during the morning and afternoon, it may be helpful to recall that the grid of any sundial accurately constructed and mounted for its intended location is a projection of a grid of day and hour curves on the celestial sphere. If we consider this celestial grid in the frame of reference of its poles and equator, the day curves, being the declination circles approximately traced diurnally by the Sun, are the same for all localities, but each locality on the terrestrial globe imposes its own system of hour curves, which include the local horizon (for the beginning of hour 1 and the end of hour 12) and the local meridian (for the end of hour 6). If we displace the sundial to another locality, we are effectively translating the associated celestial grid to a new position on the celestial sphere and pretending that in this position it is the appropriate grid for the sundial's new location.

The translation can be broken down into an east-west and a north-south component. A purely east-west translation is a rotation around the celestial poles. The celestial day curves slide along their declination circles, and the translated grid exactly coincides with the correct grid for the new position. Hence the same sundial will show correct seasonal hours and stages of the year in all localities having the same latitude. A purely north-south translation, on the other hand, is a rotation around a pair of poles lying on the equator, namely the rising and setting points of the Sun on the equinoxes. The translated day curves cease to be arcs of declination circles. The hour curves that were arcs of the local meridian and horizon are translated into arcs of the new local meridian and horizon, but the other translated hour curves will not coincide with the hour curves for the new locality.

Although the sundials of Pliny's anecdote were probably spherical or conical, the phenomena of transplantation can be illustrated more effectively by the grids of horizontal sundials, since using this type removes the awkwardness of representing a three-dimensional surface on the page, while adequately displaying details of sundial behavior through the entire day. Figure 4.7 shows a grid of a horizontal sundial computed for the latitude of Catania (fine black lines) superimposed on a grid computed for Rome's latitude (thick gray lines), on the assumption that the gnomon is identical in position and length for both grids. The grid for Rome represents the parts of the sundial surface that the

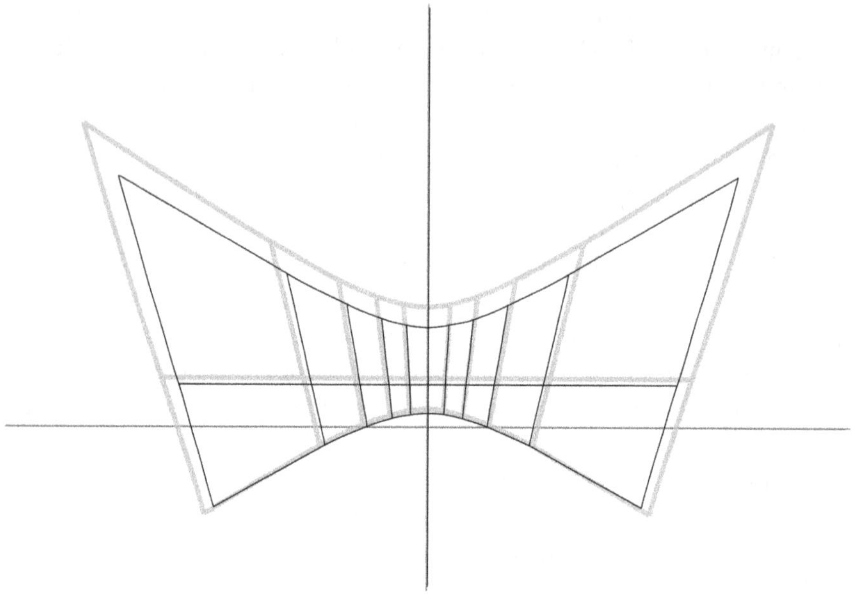

FIGURE 4.7 Grid of a horizontal sundial computed for the latitude of Catania (fine black lines) superimposed on a grid computed for Rome's latitude (thick gray lines).

shadow point can fall upon between the end of the first and the end of the eleventh hour, at any season of the year. Since the hour curves of the two grids are, for most of their length, nearly coincident, the Catanian sundial, if transplanted to Rome, will display seasonal hours that are in error by only a small fraction of an hour.[25] Around the winter solstice, the shadow point will trace paths close to the uppermost gray day curve, that is, noticeably outside the grid inscribed on the Catanian dial, though the point where the gnomon's shaft crosses the uppermost inscribed curve will give a reasonable approximation to the time. Around the summer solstice, the shadow point will never go beyond the lowermost gray day curve, which is slightly above the lowermost inscribed curve, though the horizontal sundial's projection compresses the summer day curves so much that the shortfall would scarcely be noticeable.

The seasonal errors of a displaced sundial could have been noticed simply by observing the disjunction between the inscribed day curves and the actual north-south range covered by the shadow through the year. Errors in time of day, by contrast, could only have been detected empirically if the displaced sundial was confronted with a more trusted sundial or water clock. Hannah

25 Gibbs (1976, 96 n. 25) calculates the greatest error as about 0.07 seasonal hours of the summer.

has suggested that when Pliny writes of the sundial from Catania *nec congruebant ad horas eius liniae*, he employs *hora* (as he does in a few other passages of the *Naturalis Historia*) in the sense of "time of year," thus "its lines did not agree with the seasons" (2009, 135). But the context of 7.214 is all about the Roman adoption of hours as subdivisions of the day, and *paruerunt* ... *ei* only makes sense as stating that the Romans trusted the sundial for marking the hours—they are unlikely to have adopted it as a calendar regulator. One would like to know what motivated Marcius Philippus to erect a new sundial a century later. Did people come to suspect that the sundial showed false times because it manifestly showed false seasons? Or did someone argue on theoretical, mathematical grounds that a sundial constructed for a latitude so far south of Rome must show false times?[26]

In any case, the ancient purchaser of an expensive, high-quality sundial could have taken comfort in the knowledge that, if for any reason he or she had to move to a distant place, the sundial would still work reasonably well as a time-keeping device, if not as a calendar.

4 Determining Intended Latitude and Evaluating Accuracy through Digital 3D Models

How common was it in antiquity for a sundial to be installed in a location whose latitude was significantly different from the one for which it was designed? This question cannot yet be answered satisfactorily on a broad base of data from throughout the Greco-Roman world. Ideally, this base ought to consist of sundials (1) whose findspots are known at least roughly (precise archeological context is not essential), (2) that were designed and executed with care and accuracy with respect to the geometry of their surfaces, day curves, and hour curves, and (3) that are sufficiently well preserved to allow a satisfactory analysis of their geometry.

In her *Greek and Roman Sundials*, Sharon Gibbs not only inventoried the 256 ancient sundials known to her but, for a large fraction of this corpus, reported estimates of the latitudes to which their grids correspond, based in large part on her own measurements.[27] Although the number of known Greco-Roman sundials has more than doubled since 1976—an expansion only partly

26 Wolkenhauer's contribution to this volume explores further motivations for erecting this second sundial alongside the first.

27 Gibbs 1976. One cannot help being astonished by the thoroughness with which Gibbs made personal inspection of the majority of the sundials reported in her monograph.

reflected in Eva Winter's recent survey of ancient time-reckoning devices[28]—with respect to measured data, Gibbs's book has not yet been systematically superseded, though it has been supplemented and corrected in parts by more recent studies of individual sundials and regional corpora. Moreover, the kinds of measurements from which she estimated the intended latitudes of sundials reflected what was practicable using surface measurements on objects that often were very far from completely preserved, and it is not always easy to tell how significant the results are.

The category known as roofed spherical sundials (type [f] above) illustrates the difficulties Gibbs faced. As noted above, this type projects a narrow beam of the Sun's rays upon a shaded spherical surface through an eyehole situated at the zenith of the sphere. The only geometrically simple curves on a roofed spherical sundial's grid are the equinoctial day curve, which is a complete circle passing through the eyehole and parallel to the equator, and the meridian hour curve, which is an arc of a great circle passing through the eyehole and bisecting the equinoctial circle. If the radius of the spherical surface is R, the radius of the equinoctial circle r, and the intended latitude of the sundial is φ, then:

$$(1)\ \cos \varphi = r/R$$

However, Gibbs could not measure R and r directly. To estimate R, she measured the arc of the meridian hour curve between the equinoctial circle and either of the solstitial day curves, from which she could derive the circumference and radius of the spherical surface by assuming a plausible value for the obliquity of the ecliptic. For r she measured the arc of the equinoctial circle between any pair of hour curves, since the geometry of the projection requires that these arcs should be 15° times the corresponding number of seasonal hours. Thus, each determination of φ could be affected by inaccuracies in the execution of the spherical surface, the solstitial day curves, and the hour curves, as well as some uncertainty about the assumed value for the obliquity of the ecliptic.

We may take as an example Pompeii Granario inv. 52789, a fragment of a roofed spherical sundial found at Pompeii (latitude 40° 45′, precise findspot unknown). When intact, the sundial took the form of an oversized drinking cup or *skyphos*. Somewhat more than half the grid survives, including much of the equinoctial and solstitial day curves and the hour curves for the ends of the fourth through the tenth hour. Gibbs gives two estimates of φ to 1° precision, derived from measurements (to 1 mm precision) of the meridian arcs between

---

28 Winter 2013. The technical information in this work is largely derived from earlier publications; because of a rather high rate of inconsistencies and errors, the user needs to check details systematically (see Schaldach 2015).

the equinoctial circle and the winter and summer solstitial curves, and one measurement of the equinoctial arc between the hour curves for the ends of the sixth and seventh hours: 50° using the winter meridian arc, and 53° using the summer meridian arc. These latitudes are not only much higher than that of Pompeii (and appropriate, say, for southern Great Britain), but also significantly different from each other. Similar, and even more extreme, scattering of latitude values occurs for several other roofed spherical sundials that Gibbs examined. One might suspect that this is due partly to imperfections in the mathematically complex solstitial and hour curves, whose errors become magnified in the calculation of r and R, and perhaps also to measurement errors.

Digital 3D models of ancient sundials, derived from either scanning or photogrammetry, offer the possibility of making—and repeating—measurements that would be difficult or impossible through the means that were available to Gibbs. The major repository of digital models is the BSDP, which has up to now made models of a substantial fraction of the known corpus of Greco-Roman sundials available for research. Most of these Berlin Project models are from laser scans, though some are photogrammetric. In the great majority of cases the models are sufficiently detailed and reliable to allow measurements from which the essential parameters of the sundials can be estimated. Of the thirty roofed spherical sundials that I am aware of, the BSDP currently has usable models of eight, and in addition, I have made photogrammetric models of three sundials of this type for which the BSDP currently lacks satisfactory models.

Digital models can be analyzed using specialized software that directly accesses the vertex coordinates of the mesh. However, it is possible to extract key parameters with more than satisfactory precision by applying fairly simple and easily accessible visual tools to the models. In the following analysis of the roofed spherical sundials, the principal software employed comprised MeshLab (open source software for editing 3D mesh models), Inkscape (open source vector graphics software, used here for fitting circles and straight lines to images and measuring lengths and angles), and Microsoft Excel (for calculating ideal sundial grids).

We can illustrate several tests for assessing the quality and estimating the intended latitude of roofed spherical sundials with the Pompeii fragment discussed above. Test 1 is of the quality of the spherical dial surface. After removing most of the surface mesh other than the dial surface (Figure 4.8), we select an arbitrary orientation of the surface and remove a series of thin, parallel planar slices (Figure 4.9). The perimeters of the remaining slices compare well with a set of concentric reference circles superimposed on the image (Figure 4.10), confirming that this is an accurate surface of revolution. Repeating this procedure using a different orientation for the slicing suffices to establish that the surface is accurately spherical.

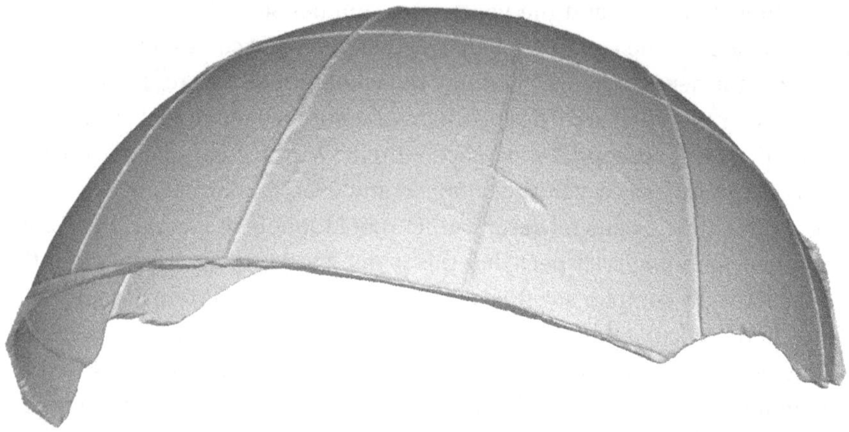

FIGURE 4.8 Orthographic image of the BSDP digital model of the dial surface of Pompeii Granario inv. 52789

FIGURE 4.9 Digital model of the dial surface with planar slices removed

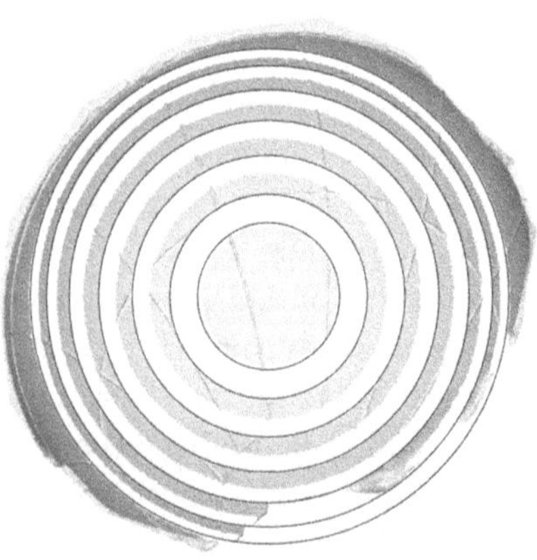

FIGURE 4.10 Orthogonal image of the sliced model in a plane parallel to the slices, with concentric reference circles in black

The remaining tests are methods of estimating the intended latitude of the sundial. For test 2, we cut away half the surface model of the complete sundial fragment along its meridian plane, and generate an orthographic image in the meridian plane (Figure 4.11). Fortunately the sundial's base survives, taking the form of a triangular projection behind the dial face. The angle between the base line and the plane of the equinoctial circle (which projects as expected as an almost exact straight line) is 52.5° to the nearest half degree (Figure 4.12). If the sundial was supposed to be mounted on a horizontal surface, this angle would be the local elevation of the equatorial plane, so that the intended latitude would be 37.5°, significantly south of Pompeii's actual latitude, 40° 45′.

If, on the other hand, the sundial was supposed to be mounted at a sufficient forward tilt of about 3° so that the equinoctial circle was parallel to the equatorial plane for Pompeii, there ought to be a corresponding deviation of the position of the eyehole from the point of the spherical surface that would be the zenith point if the sundial was installed horizontally. To check this, we fit a circle to the outline of the meridian arc and produce the projection of the equinoctial circle to meet it at the eyehole point. The angle subtended at the sphere's center between the eyehole point and the zenith point relative to the baseline is approximately 4°, near enough to the expected value to confirm that a forward tilt of the sundial was intended. We thus report the latitude resulting from Test 2 as 37.5° + 4°.

Test 3 uses the same image with the fitted meridian circle and projection of the equinoctial circle. Assuming that the eyehole was at the zenith of the bowl (relative to the local horizon), then from equation (1) we find φ is 40.5° to the nearest half degree.

Lastly, in Test 4 we generate an orthographic image of the dial surface in the plane of the equinoctial circle (that is, the equatorial plane) (Figure 4.13), upon which we superimpose equatorial projections of accurately-computed grids for different latitudes at 1° intervals, employing 23° 40′ as the obliquity of the ecliptic and assuming that the eyehole was at the zenith relative to the local horizon. The match for 41° (Figure 4.14) is quite impressive, with only very small deviations apparent along the winter solstitial curve and the hour curves, most noticeably where the hour curves approach the winter equinoctial curve. This was evidently a very well constructed sundial after all, and made for the latitude of Pompeii.[29] In this instance the principal cause of error in Gibbs's

29 It is true that the sundial would have functioned correctly if installed horizontally at a latitude about 37° 30′. But if it was made for that latitude and later displaced to Pompeii, one would have to explain why the eyehole was placed at just such an angle behind the zenith point so that the grid ended up being the normal one for the latitude of Pompeii.

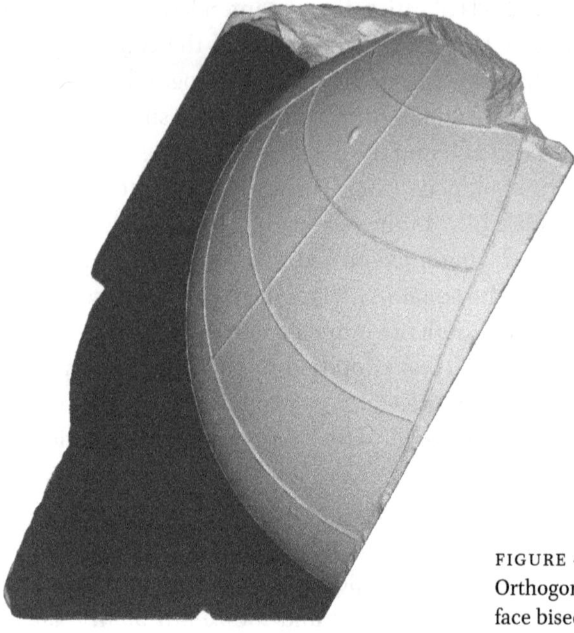

FIGURE 4.11
Orthogonal image of the sundial surface bisected along the meridian plane

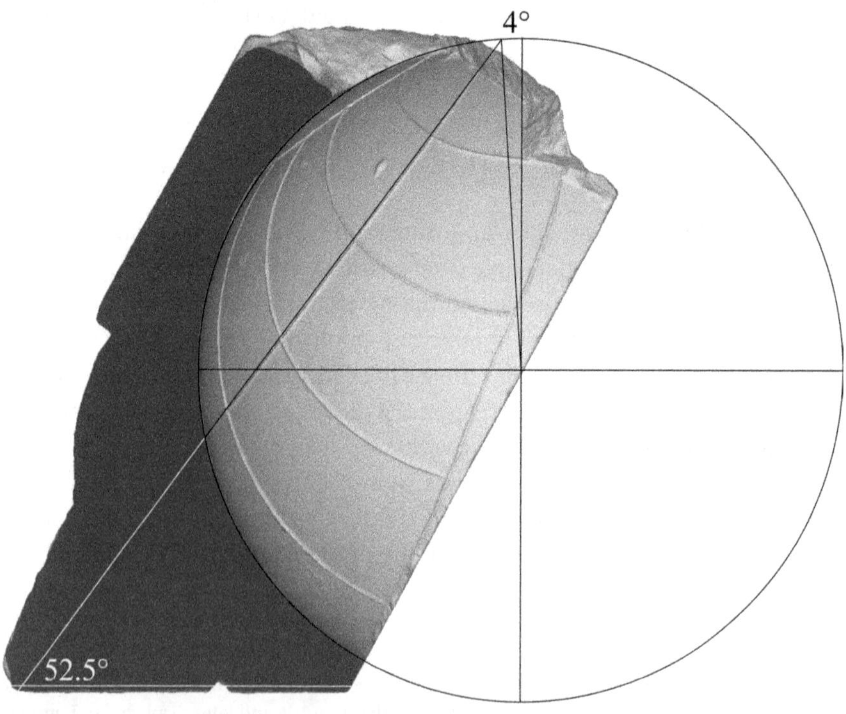

FIGURE 4.12 The orthogonal image from Figure 4.11 with lines traced for Tests 2 and 3

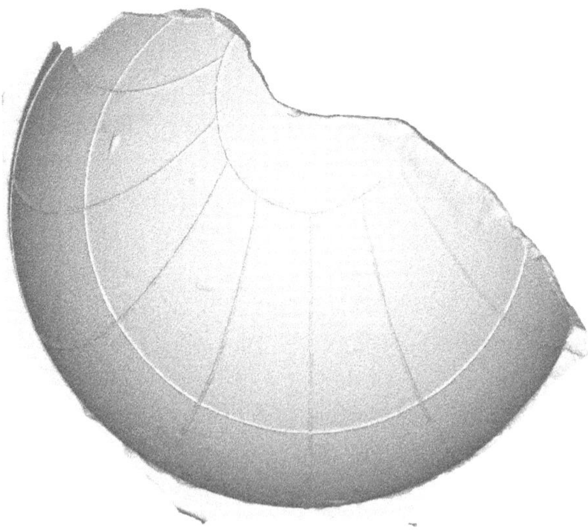

FIGURE 4.13
Orthogonal image of the dial surface in the equatorial plane

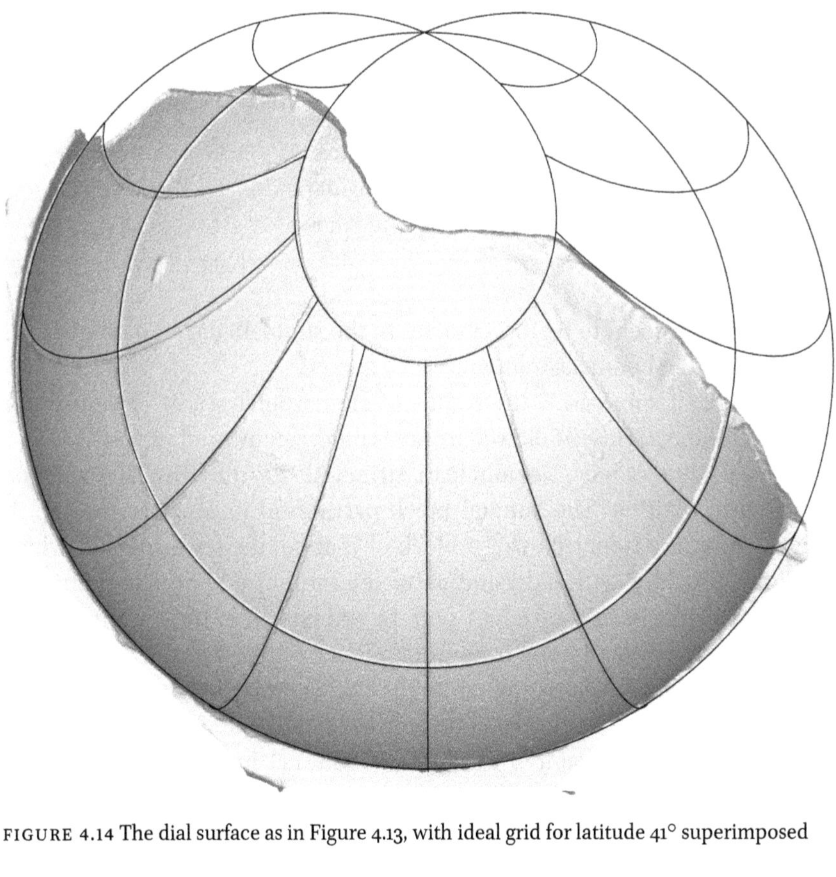

FIGURE 4.14 The dial surface as in Figure 4.13, with ideal grid for latitude 41° superimposed

TABLE 4.1 Analyses of eleven roofed spherical sundials

Gibbs	Winter	BSDP dialface	Present location	Findspot	Latitude of findspot
2001G	Altinum 3	1	Altino, Veneto	Altino	45° 35′
2018G	Fundort unbekannt 10	94	Bologna	unknown	44° 30′? (Bologna
2019G	Pompeji 15	95	Pompeii	Pompeii	40° 45′
2020	Baelo Claudia 1	96	Madrid	Baelo Claudia	36° 5′
2021G	—	97	Vatican	unknown	41° 54′? (Rome)
2023G	—	98	Berlin	Rome? (Acquired there)	41° 54′?
7001G	Tenos 1	280	Tinos	Tinos	37° 33′
—	Karthago 1	366	Paris	Carthage? (Reported)	36° 51′?
—	Ariminum 1	437	Verucchio	Verucchio?	43° 59′
—	Concordia Sagittaria 1	480	Concordia Sagittaria	Concordia Sagittaria	45° 45′
—	Ravenna 1	521	Ravenna	unknown	44° 25′? (Ravenna

results does not seem to be inaccuracies in the sundial's day and hour curves but some fault in her measurements.

In Table 4.1 I summarize the results of similar analyses of eleven roofed spherical sundials. This is followed by notes on the individual sundials.

1. **Museo Archeologico Nazionale di Altino, Altino, Italy, inv. AL 10.** Found near Quarto d'Altino. The sundial bowl, partly broken away at the top, is sculpted from the upper part of a block of marble, the lower part of which is a plinth provided with a decoration in the form of a boat. The plinth has several ostensibly horizontal edges, and the estimate of latitude from Test 2 is with respect to edges running from front to back of the plinth. The estimates from Tests 3 and 4 are close enough to the findspot's latitude to confirm that the sundial was made for this location; however, to function correctly it would have had to be installed at a forward tilt of about 12°. In fact the extrapolated

igital model	Test 1 (quality of spherical surface)	Test 2	Test 3	Test 4	Notes
SDP hotogrammetry	fair	34° + 10.5° = 44.5°	44.5°	44°	1
ones hotogrammetry	poor	N/A	44.5°	45°	2
SDP laser	good	37.5° + 4° = 41.5°	40.5°	41°	3
ones hotogrammetry	good	48° − 11.5° = 36.5°	37°	36°	4
SDP laser	good	43° + 0° = 43°	42°	N/A	5
SDP laser	fair	50° − 8° = 42°	41.5°	42°	6
ones hotogrammetry	good	37.5° + 1° = 38.5°	39°	37°	7
SDP hotogrammetry	good	44.5° − 2° = 42.5°	42°	42°	8
SDP hotogrammetry	fair	46° + 3° = 49°	46°	45°	9
SDP hotogrammetry	good	44° − 5° = 39°	41.5°	41°	10
SDP laser	fair	N/A	49	50°	11

highest point of the equinoctial day circle, which would have been where the eyehole was located, is about this angle to the rear of the extrapolated zenith of the spherical surface, assuming that the sundial was installed with no tilt.

Gibbs gives four rather scattered estimates of φ to 1° precision, derived from measurements (to 1 mm precision) of the meridian arcs between the equinoctial circle and the winter and summer solstitial curves, and measurements of the equinoctial arcs between the hour curves for the ends of the sixth, seventh, and eighth hours:

winter solstice, hours 6 and 7 $\varphi = 46°$
winter solstice, hours 7 and 8 $\varphi = 43°$
summer solstice, hours 6 and 7 $\varphi = 44°$
summer solstice, hours 7 and 8 $\varphi = 41°$

2. Museo Civico Archeologico, Bologna, without inventory number. The findspot is unknown. Parts of the sundial bowl are broken away along the top and sides. The digital model, which I reconstructed from BSDP photographs since the grid curves were not clear on the BSDP model, lacks any satisfactory indication of the intact sundial's vertical or horizontal directions. Despite the rather poor geometry of the bowl's surface, the latitude estimates are in good agreement with the latitude of Bologna. Using the meridian arc between the equinoctial circle and the summer solstitial curve with the arcs of the equinoctial circle between hours 1, 2, 3, and 4, Gibbs obtained somewhat scattered latitude estimates of 45°, 44° 13′, and 47° 30′.

3. Pompeii, Granario inv. 52789. (Analysis discussed in detail above.) The exact findspot is unknown but definitely Pompeii. The bowl is broken away along the top, sides, and part of the bottom edge.

4. Museo Arqueológico Nacional, Madrid, inv. 33.185. Found in the kitchen of a house at Baelo Claudia; it must have been moved from an outdoor location, perhaps for use as a table.[30] The sundial is essentially complete and in excellent preservation, though missing the metal plate that would have been perforated by the eyehole. The original publication of the sundial by Paris et al. reported the intended latitude as 41° 30′, measuring the elevation from horizontal of a line from the zenith of the bowl to the intersection of the meridian and equinoctial circles on the grid, on the assumption that the sundial would have been erected horizontally.[31] Raya Román, realizing that the eyehole must have been forward of the ostensible zenith point, applied the first step of our Test 2 and obtained 47° 30′, without considering the possibility that the sundial was meant to be installed at a tilt.[32]

5. Musei Vaticani inv. 53875. Findspot unknown. The majority of the bowl is broken away; what remains is most of the part between the equinoctial and summer solstitial day circles. Tests 2 and 3 are close enough to Rome's latitude to suggest that the sundial could have been made for Rome. Using the arc on the equinoctial circle between hours 6 and 7 and the meridian arc between the equinoctial circle and the summer solstitial curve, Gibbs estimated the latitude as 41° 13′.

6. Staatliche Museen zu Berlin inv. SK1049. Findspot unknown, but reportedly acquired in Rome.[33] The sundial is complete except for some breakage

30 Paris et al. 1923, 167.
31 Paris et al. 1923, 167 fig. 65.
32 Raya Román 1984, 111.
33 Schaldach n.d.

around where the lost eyehole plate would have been. Tests 2 (assuming an 8° backward tilt), 3, and 4 are all in good agreement with Rome's latitude. This sundial has received several previous analyses. Drecker found the intended latitude to be 41° 40′ from the ratio of directly-measured radii of the equinoctial circle and the sundial bowl.[34] Using the arc of the equinoctial circle between the 5th and 6th hours together with the (approximately equal) meridian arcs between the two solstitial curves and the equinoctial circle, Gibbs found 39°; this is the outlier among the estimates. Following Gibbs's method for the bowl's radius, but with direct measurement of the radius of the equinoctial circle, Schaldach gives 41.5° and, after more recent measurements, 41.67°.[35]

7. **Archaeological Museum of Tinos inv. A139.** Found near the altar of Poseidon at Kionia, Tinos. In addition to the roofed spherical bowl on the south side, the sundial has east-facing and west-facing vertical planar sundials, and a spherical bowl (type [a]) on its north side. The upper part of the roofed spherical bowl is broken away, except for some fragments that have been embedded in a (not particularly accurate) restoration. The latitudes from Tests 2, 3, and 4 are close enough to that of Tinos for the sundial to have been designed for that location. Using the arc of the equatorial circle between hours 4 and 5 together with the meridian arcs between the two solstitial curves and the equinoctial circle, Gibbs obtained latitude estimates of 37° (summer curve) and 20° (winter curve). Her measurement of the winter arc is suspect.

8. **Musée du Louvre inv. MNE 1178, *nº usuel* Ma 5074.** The sundial, in the form of an elaborately decorated *skyphos*, is complete and in excellent condition. Acquired by purchase, the findspot was reportedly Carthage.[36] The latitudes resulting from Tests 2 (assuming a 2° backwards tilt), 3, and 4 are consistent with each other and incompatible with the reported findspot, Carthage. Savoie and Lehoucq likewise estimated 41° 6′ from the ratio of radii of the equinoctial circle and the spherical bowl, with confirmation from measurements involving the other inscribed curves.[37] They conclude that the sundial was designed for the latitude of Rome but displaced to Carthage, remarking that the errors in the seasonal hours resulting from the displacement would have

34 Drecker 1925, 31.

35 Schaldach 2001, 107 (with a valuable discussion of the complications involved in estimating the latitude); Schaldach n.d.

36 According to Gagnaire (1998, 179), it was found in excavations of a Roman Villa at Carthage before the Second World War. Its earliest publicly accessible documentation seems to be its inclusion in an exhibition at The Hague in 1990; see Turner 1990, 60–61. The Louvre purchased it in 2000.

37 Savoie and Lehoucq 2001, 30.

been negligible. However, since the summer solstitial curve grazes the lower rim of the bowl (as was normal for roofed spherical sundials), at Carthage the spot of sunlight would have fallen not only outside the grid but entirely off the bowl's surface for an interval around the summer solstice. Rather, one may suspect that the findspot has not been accurately reported.

9. Museo Civico Archeologico, Verucchio. Findspot unknown, but presumably near Verucchio. The top and sides of the bowl are broken away. Because of the bowl's deviations from sphericity, its center cannot be determined with precision, so that the tilt component of Test 2 is not trustworthy. In any case Tests 2, 3, and 4 all suggest that the sundial was designed for a latitude slightly farther north than Verucchio. I am not aware of any previous estimate of the intended latitude.

10. Soprintendenza Archeologica del Veneto, Concordia Sagittaria. Found at Concordia Sagittaria. The bowl is completely preserved except for some breakage around the eyehole. Tests 2 (assuming tilt), 3, and 4 clearly indicate that the sundial was designed for a latitude several degrees south of Concordia Sagittaria. I am not aware of any previous estimate of the intended latitude of this sundial.

11. Museo Nazionale di Ravenna. Findspot unknown. The bowl is broken away at the top and sides; the model provides no satisfactory horizontal or vertical element for Test 1. Tests 2 and 3 give a latitude about 5° north of Ravenna (subject to modest uncertainty since the equinoctial circle is rather coarsely executed), so this may have been a displaced sundial if it was a local find, as seems likely. The same reservations concerning its usability in the summer would apply as for the "Carthage" sundial now in the Louvre. Mario Arnaldi, on the other hand, gives 45° as the latitude for which the sundial was made, without indicating how he obtained this figure.[38]

Of the eleven sundials examined above, five have securely-known findspots, and among these, four (Altino, Pompeii, Baelo Claudia, and Tinos) were constructed, with some care, for the latitudes of the localities where they were found. The sundial from Concordia Sagittaria, on the other hand, appears to have been made for somewhere significantly further south—say, between Rome and Naples. There are also four for which the default assumption would be that the sundial was found near its present location: those now at Bologna and the Vatican fit the presumed findspots, whereas those now at Verucchio and Ravenna seem to be displaced. Of the two with reported places of origin, the one now in Berlin may well have been made for Rome, where it was purchased if not actually found. The sundial now in Paris, however, though said

38 Arnaldi 1996, 14 n. 3.

to have been excavated at Carthage, clearly was made for somewhere around Rome's latitude—but, given the impossibility of confirming the claimed findspot or tracing the object's history during the half century following its supposed discovery, we should be circumspect about identifying this as a displaced sundial. Even so, three out of eleven is a surprisingly high rate of certain or probable displaced sundials. It would be interesting to find out whether extending the examination to a larger sample would yield similar results.

5 Conclusion

One way of describing the distinctive character of Greco-Roman sundials is that they were an embodiment of a fairly sophisticated mathematical theory in objects fashioned by craftspeople representing a wide range of skill, for practical use (as well as ornament) by lay people who were mostly ignorant of the underlying theory. The foregoing article highlights two strengths of the mathematical theory of sundials: the potential of certain popular types of sundial to facilitate reasonably accurate reading of the time of day to a precision significantly finer than whole seasonal hours, and the robustness of a well-executed system of hour curves with respect to even fairly large disparities between the latitude for which the sundial was designed and the latitude where it was installed. The former strength seems not to have been appreciated, at least outside the scientific community; the unit of the seasonal hour remained through antiquity the limit to which the capabilities of technology drew society in the direction of more refined time-management.

The frequency of demonstrable or probable instances of latitudinal displacement found in our—admittedly very small—sampling of roofed spherical sundials might suggest that the owners of these sundials were informed of how little effect terrestrial latitude has on the hour curves in the lateral (east-west) dimension. I suspect, however, that these displacements were not really symptomatic of mathematical sophistication on the part of the sundials' owners. In particular, the special characteristic of the roofed spherical type, that the time was indicated not by the shadow of a stick-like gnomon but by a small spot of sunlight, combined with the fact that the summer solstitial day curve in this type normally ran very close to the lip of the bowl for the middle part of the day, would have made this type useless for a large fraction of the summer days if displaced far south of its intended location: the narrow beam of light would miss the dial surface entirely. Probably the owners of the displaced sundials valued them more as prestige objects than as instruments for regulating daily life.

References

Arnaldi, Mario. 1996. "Due frammenti di orologio solare romano al museo nazionale di Ravenna." *Ravenna: Studi e ricerche* 3: 14–28.

Delambre, Jean B.J. 1817. *Histoire de l'astronomie Ancienne*. 2 vols. Paris: Courcier.

Drecker, Joseph. 1925. *Die Theorie der Sonnenuhren. Die Geschichte der Zeitmessung und der Uhren I E*. Berlin: de Gruyter.

Gagnaire, Paul. 1998. "Le Scaphé de Carthage." *L'astronomie* 112 (Juin–Juillet): 179–82.

Gibbs, Sharon L. 1976. *Greek and Roman Sundials*. New Haven, CT: Yale University Press.

Graßhoff, Gerd, Elisabeth Rinner, Karlheinz Schaldach, Bernhard Fritsch, and Liba Taub. 2016. *Ancient Sundials*. [Online database] accessed June 5, 2019, https://doi.org/10.17171/1-1.

Hannah, Robert. 2009. *Time in Antiquity*. New York: Routledge.

Heiberg, Johan L., ed. 1898. *Claudii Ptolemaei opera quae exstant omnia*. Vol. I. Leipzig: Teubner.

Hüttig, Manfred. 2000. "The Conical Sundial from Thyrrheion: Reconstruction and Error Analysis of a Displaced Antique Sundial." *Archive for History of Exact Sciences* 55 (2): 163–76.

Jones, Alexander. 2005. "Ptolemy's Canobic Inscription and Heliodorus' Observation Reports." *SCIAMVS* 6: 53–97.

Jones, Alexander. 2017. "The Roofed Spherical Sundial and the Greek Geometry of Curves." In *Studies in the Ancient Exact Sciences in Honour of Lis Brack-Bernsen*, edited by John Steele and Mathieu Ossendrijver, 183–203. Berlin: Edition Topoi.

Luckey, Paul. 1927. "Das Analemma des Ptolemäus." *Astronomische Nachrichten* 230 (5498): 18–46.

Mercier, Raymond. 2011. "Book Review, Ptolemy in Perspective, Edited by Alexander Jones." *Journal for the History of Astronomy* 42 (3): 405–8.

Neugebauer, Otto. 1948. "The Astronomical Origin of the Theory of Conic Sections." *Proceedings of the American Philosophical Society* 92 (3): 136–38.

Neugebauer, Otto, and Henry B. Van Hoesen. 1959. *Greek Horoscopes*. Philadelphia: American Philosophical Society.

Paris, Pierre, Jorge Eduardo Bonsor, Alfred Laumonier, Robert Ricard, and Cayetano de Mergelina. 1923. *Fouilles de Belo (Bolonia, Province de Cadix) (1917–1921). I. La ville et ses dépendances. Bibliothèque de l'école des hautes études hispaniques, Fascicule VI*. Bordeaux: Féret.

Raya Román, José María. 1984. "Reloj solar de Belo." *Boletin del Museo Arqueológico Nacional* 2 (1): 103–15.

Remijsen, Sofia. 2007. "The Postal Service and the Hour as a Unit of Time in Antiquity." *Historia* 56 (2): 127–40.

Rinner, Elizabeth. 2017. "Ancient Greek Sundials and the Theory of Conic Sections Revisited." In *Studies in the Ancient Exact Sciences in Honour of Lis Brack-Bernsen*, edited by John Steele and Mathieu Ossendrijver, 165–82. Berlin: Edition Topoi.

Savoie, Denis. 2001. *La Gnomonique. Nouvelle édition revue et augmentée.* Paris: Les Belles Lettres.

Savoie, Denis, and Roland Lehoucq. 2001. "Étude gnomonique d'un cadran solaire découvert à Carthage." *Revue d'archéométrie* 21: 25–34.

Schaldach, Karlheinz. n.d. "107898. Sonnenuhr. Berlin, Staatliche Museen, Antikensammlung Berlin." *Central Object database of the German Archaeological Institute (DAI) and the Archaeological Institute of the University of Cologne. Arachne.* [Online database entry] accessed June 5, 2019, http://arachne.uni-koeln.de/item/objekt/107898.

Schaldach, Karlheinz. 2001. *Römische Sonnenuhren. Eine Einführung in die antike Gnomonik.* 3rd ed. Frankfurt am Main: Harri Deutsch.

Schaldach, Karlheinz. 2015. "Book Review of Eva Winter, Zeitzeichen." *Gnomon* 87 (2): 144–49.

Schaldach, Karlheinz. 2016. "Measuring the Hours: Sundials, Water Clocks, and Portable Sundials." In *Time and Cosmos in Greco-Roman Antiquity*, edited by Alexander Jones, 63–93. Princeton: Princeton University Press.

Steele, John. 2000. *Observations and Predictions of Eclipse Times by Early Astronomers.* Archimedes: New Studies in the History and Philosophy of Science and Technology 4. Dordrecht: Kluwer.

Turner, Anthony J. 1990. *Time.* The Hague: Tijd voor Tijd Foundation.

Winter, Eva. 2013. *Zeitzeichen: Zur Entwicklung und Verwendung antiker Zeitmesser.* 2 vols. Berlin: de Gruyter.

Wolkenhauer, A. 2011. *Sonne und Mond, Kalender und Uhr: Studien zur Darstellung und poetischen Reflexion der Zeitordnung in der römischen Literatur.* Berlin: de Gruyter.

CHAPTER 5

Cosmology and Ideal Society: the Division of the Day into Hours in Plato's *Laws*

Barbara M. Sattler

In this chapter, I will show that Plato's last work, the *Laws*, provides us with the first use of the hour as a unit of division of the day in ancient Greek philosophy. I will argue that combining cosmology with the design of an ideal society in the way that we find in the *Laws* is precisely what leads to this conceptual advance. While the *Laws* may thereby also provide the first occurrence of understanding ὥρα in the sense of "hour" tout court, it is in any case the first occurrence in the field of philosophy as such where crucial changes in the conceptualization of time were taking place in the 4th century BCE.[1]

Plato's *Timaeus* is crucial for establishing the most encompassing account of time in Greek thinking so far and for providing a unified temporal framework that precedes any discussion about a unified calendar.[2] But in Plato's cosmology in the *Timaeus*, the day is introduced as the smallest unit of time measurement; in the context of natural philosophy, smaller units do not come into view for Plato. Smaller units dividing the day do, however, play a role for Plato in the discussion of a reform of society in the *Laws*, which is combined with an interest in the measurability of natural processes.

For the temporal units structuring the life of his ideal society, Plato does not rely solely on the common division of the day according to the beginning, end, and midpoint of the Sun in its motion, but also employs a notion that changes essentially in its meaning in the later 4th century BCE, the notion of ὥρα. Before that time, ὥρα commonly refers to the seasons, a special or fitting time to do something, or the springtime of life. The big change in the notion of ὥρα in Greek society that seems to have taken place toward the end of Plato's lifetime is the increasing understanding of ὥρα as what we call "hour," originally

---

1 I present a detailed account of these conceptual changes in my book manuscript *Ancient Notions of Time from Homer to Plato*.

2 Not only do we not find a unified calendar throughout the ancient Greek world before Hellenistic times, more importantly, we also do not find a unified notion or idea of time before Plato. The effects this lack of unification has on the experience of time are spelled out in Sattler (2017).

 | DOI:10.1163/9789004416291_007

seen as a twelfth part of the day.[3] And the first occurrence of an understanding of ὥρα as hour in Greek philosophy handed down to us can be found, or so I want to argue, in Plato's *Laws*—more exactly speaking, in his discussion of the length of council meetings.

1 The Development of Short Units of Time

Subdivisions of the day do not seem to have been important for the earliest Greek philosophers. The "day" itself, by contrast, is discussed occasionally in early Greek philosophy, but the term can cover at least three different things:

a. Day as phenomenon: the sequence of light and darkness in the outside world that we call "night" and "day" seems to be one of the astronomical phenomena that early Greek philosophers try to explain (so, for example, Heraclitus and Empedocles).[4]
b. Day as content: day as the processes or activities—in short, the content—of what is done during the time span in which it is light (or during the time span from one sunrise to the next).[5]
c. Day as a unit of measurement: in this sense, the day must have been important for the early Greek philosophers engaged in astronomy, but unfortunately we do not have much good evidence for it. Diogenes Laertius tells us that Thales is the first to divide the year into 365 days (DK11 A1), but this may just be part of the general astronomical persona that was later ascribed to Thales.[6] We see concerns about the day as a unit of measurement reflected in Heraclitus and again in Democritus.[7]

It seems that only if we focus on this last understanding of the day, as a measurement unit, do smaller units also come into sight.[8] If we look at smaller temporal structures, at what divides the day, it seems to me that we find three stages of development in the ancient Greek world:

3 Cf. the entry on ὥρα in the Liddell-Scott-Jones dictionary, meaning II, 2b, and in Montanari (2015), meaning 3 B.

4 See, for example, Heraclitus DK22 A1 and B6, and Empedocles DK31 A 30.

5 See, for example, Antiphon DK B49 and 62, and Kritias DK88 B6.

6 Similarly, scholars have been skeptical about whether Thales did indeed predict the solar eclipse ascribed to him in fragment DK11 A5 (Herodotus I.74.2).

7 See Heraclitus B 106 and Democritus DK68 B 14 (this fragment is also discussed below).

8 We may think that subdivisions of the day should also come into view with (b), day as content, but smaller divisions of the day thus understood would actually lead to divisions of the activities performed, rather than to more fine-grained temporal structures.

1. With respect to astronomical questions, units of time are important for the earliest natural philosophers, but the focus is mainly on long units, like years, seasons, and months. Hours do not seem to be important for the astronomical questions of the earliest philosophers.[9] The day is prominent in Greek literature outside natural philosophy, but without being seen as a neutral unit of measurement.
2. Days surface explicitly as neutral units of measurement in investigations of natural philosophy and astronomy in the 5th and 4th centuries BCE and become a central focus.
3. Once natural philosophy is brought together with philosophical concerns about the best legislation for a *polis* and about building up a new, ideal society, we also get the notion of an hour. Thus, the combination of natural philosophy with social concerns also brings with it some concern for regular sub-divisions of the day into hours.[10]

We should also bear in mind that some of these steps overlap in time and that they are specific to the Greek world—we do not find the same pattern with the Egyptians or Babylonians. Let us now look at these three steps in somewhat more detail.

1.1 *The Earliest Natural Philosophers*

The earliest Greek philosophers, the Presocratics, were, for the most part, pursuing natural philosophy, and astronomical and cosmological questions were of central concern. These included questions like: what does the cosmos look like? Where is the earth situated? Where are the Moon, the Sun, the five planets, and the fixed stars? How can we explain solar and lunar eclipses?[11] For these kinds of questions time was important, but usually longer time spans,

---

9 See, for example, Heraclitus DK22 A1 (Diogenes IX, 10–11) where we read about the coming into being of days and nights: ἡμέραν τε καὶ νύκτα γίνεσθαι καὶ μῆνας καὶ ὥρας ἐτείους καὶ ἐνιαυτοὺς ὑετούς τε καὶ πνεύματα καὶ τὰ τούτοις ὅμοια κατὰ τὰς διαφόρους ἀναθυμιάσεις. τὴν μὲν γὰρ λαμπρὰν ἀναθυμίασιν φλογωθεῖσαν ἐν τῷ κύκλῳ τοῦ ἡλίου ἡμέραν ποιεῖν, τὴν δὲ ἐναντίαν ἐπικρατήσασαν νύκτ' ἀποτελεῖν: καὶ ἐκ μὲν τοῦ λαμπροῦ τὸ θερμὸν αὐξόμενον θέρος ποιεῖν, ἐκ δὲ τοῦ σκοτεινοῦ τὸ ὑγρὸν πλεονάζον χειμῶνα ἀπεργάζεσθαι. ("Day and night, months, seasons and years, rains and winds and other similar phenomena are accounted for by the various exhalations. Thus the bright exhalation, set aflame in the hollow orb of the Sun, produces day, the opposite exhalation when it has got the mastery causes night; the increase of warmth due to the bright exhalation produces summer, whereas the preponderance of moisture due to the dark exhalation brings about winter," translation by Hicks [1972]). For seasons see also Thales A1 and Heraclitus B100.

10 For the possibility that a unit like an hour may previously have been used for merely astronomical purposes with Democritus, cf. below.

11 Cf., for example, Anaximander's and Philolaos's account of the cosmos and Dicks 1970.

larger units of time, like the solar and lunar year. Units of time smaller than the day do not seem to play a role in answering these questions in the very beginning.

The day is, of course, important as a temporal unit for "daily" life. But the day as we most prominently encounter it in early Greek writing, which is outside an astronomical and philosophical context, is not a neutral unit of measurement. Rather, each day displays intrinsic differences:[12] while the day is in some sense a measurement unit, it is affected by what is, should, or could be done during that particular day.[13]

The assumption that days are seen as having certain characters tied to different activities may be seen most obviously in Hesiod's *Works and Days*, the work that already prominently employs "days" in its title.[14] In the so-called "catalogue of days," towards the end of the poem, the character of different days of the month is the focus, and in accordance with this character, recommendations are given for what to do on this day and what to avoid on that day. For example, we find:

> δύω γε μὲν ἤματα μηνὸς
> ἔξοχ' ἀεξομένοιο βροτήσια ἔργα πένεσθαι,
> ἑνδεκάτη δὲ δυωδεκάτη τ' ἄμφω γε μὲν ἐσθλαί,
> ἠμὲν ὄις πείκειν ἠδ' εὔφρονα καρπὸν ἀμᾶσθαι·
>
> Two days of the waxing month are outstanding for toiling at a mortal's works, the eleventh and the twelfth. Both of them are fine, for shearing sheep and for gathering together the gladdening wheat.
>
> *Works and Days* lines 772–775, translation by MOST 2007

If we want to shear sheep or harvest wheat, we should choose the eleventh and twelfth day of the month, since these days in themselves are good for carrying

---

12 This understanding of "day" is close to the understanding of the English word "day" in sense 3 of the OED: "3. Chiefly with specifying adjective. A day characterized by the prevalent conditions or dominant mood experienced; esp. a day described in terms of the weather or its impact. Cf. *have a nice day*."

13 If we compare this understanding of days outside the astronomical and philosophical context with the three ways in which the day comes into view for the Presocratics, it seems that the first sense, "day as a phenomenon to be explained," is not discussed, while the second and third sense, "day as content" and "day as measurement unit," are merged.

14 However, the title does not seem to be attested before Lucian, cf. West 1978, 136. Modern scholars have often assumed that the catalogue of days is not by Hesiod (cf., for example, Nilsson [1911] and Solmsen [1963]), while West (1978, 346–350) thinks it is authentic. For our purposes it does not matter whether the catalogue is indeed by Hesiod or not.

out these activities. The twelfth month is also good for weaving, as we read in the following lines, 777–778, since then the spider also allegedly weaves its web.[15] Thus, in Hesiod we find the idea that different days of the month have different intrinsic qualities in the sense that they themselves are suitable or unsuitable for certain activities. Furthermore, it has been argued that in Homer a day is essentially seen as the unity connecting different experiences,[16] and it may also be identified with the fate experienced.[17]

Finally, we should bear in mind the important fact that days for the early Greeks were seen as essentially variable, in the sense that days are longer in the summer and shorter in the winter. Thus, in Pindar's *Olympian Ode* 2 we find the idea that it is only the blessed ones in the afterlife who will enjoy days and nights of equal length.[18] Equinoctial days are obviously not something *we* are blessed with every day; rather, such regularity is only for the selected few.

For philosophers interested in natural philosophy, and more specifically in measurement of the motions of the heavenly bodies, the variability of day-length is a rather unpleasant instability (I will come back to this). This is also something Plato complains about in his middle period, in the *Republic*. There, the motions of the heavenly bodies are understood to serve only as an insufficient example for "real astronomy," which is independent of all the irregularities observable in the sky. These irregularities show that the motions of the different heavenly bodies do not display a stable ratio to one another:

> Τῷ ὄντι δὴ ἀστρονομικόν, ἦν δ' ἐγώ, ὄντα οὐκ οἴει ταὐτὸν πείσεσθαι εἰς τὰς τῶν ἄστρων φορὰς ἀποβλέποντα; νομιεῖν μὲν ὡς οἷόν τε κάλλιστα τὰ τοιαῦτα ἔργα συστήσασθαι, οὕτω συνεστάναι τῷ τοῦ οὐρανοῦ δημιουργῷ αὐτόν τε καὶ τὰ ἐν αὐτῷ· τὴν δὲ νυκτὸς πρὸς ἡμέραν συμμετρίαν καὶ τούτων πρὸς μῆνα καὶ μηνὸς πρὸς ἐνιαυτὸν καὶ τῶν ἄλλων ἄστρων πρός τε ταῦτα καὶ πρὸς ἄλληλα, οὐκ ἄτοπον, οἴει, ἡγήσεται τὸν νομίζοντα γίγνεσθαί τε ταῦτα ἀεὶ ὡσαύτως καὶ οὐδαμῇ τὸν νομίζοντα γίγνεσθαί τε ταῦτα ἀεὶ ὡσαύτως καὶ οὐδαμῇ οὐδὲν παραλλάττειν, σῶμά τε ἔχοντα καὶ ὁρώμενα […]; Προβλήμασιν ἄρα, ἦν δ' ἐγώ, χρώμενοι ὥσπερ γεωμετρίαν οὕτω καὶ ἀστρονομίαν μέτιμεν, τὰ δ' ἐν τῷ οὐρανῷ ἐάσομεν, εἰ μέλλομεν ὄντως ἀστρονομίας μεταλαμβάνοντες χρήσιμον τὸ φύσει φρόνιμον ἐν τῇ ψυχῇ ἐξ ἀχρήστου ποιήσειν.

15 Cf. also Onians 1954, 411–415.

16 The day does not seem to have been employed for measuring time, but rather for providing unity for certain experiences; it is what Fränkel (1955) has called an "Erlebniseinheit."

17 Cf. Onians (1954, 413f.) who describes such a day as "a phase of fortune of greater or less duration."

18 *Olympian Ode* 2, lines 61ff.

> Then don't you think that a real astronomer will feel the same when he looks at the motions of the stars? He'll believe that the craftsman of the heavens arranges them and all that's in them in the finest way possible for such things. But as for the ratio of night to day, of days to a month, of a month to a year, or of the motions of the stars to any of them or to each other, don't you think that he'll consider it strange to believe they're always the same and never deviate anywhere at all or to try to in any sort of way grasp the truth about them, since they're connected to body and visible? [...] Then if, by really taking part in astronomy we're to make the naturally intelligent part of the soul useful instead of useless, let's study astronomy by means of problems, as we do geometry, and leave the things in the sky alone.
>
> 530a–c, translation by GRUBE, 1997

1.2 *The Day as a Neutral Unit of Measurement*

As a next step, we can see the explicit understanding of the day as a neutral measurement unit, which can then, in connection with night, also be seen as a regular unit. Already with Heraclitus, we find an explicit rejection of Hesiod's idea that different days of the month differ in their qualities:

> Ἡράκλειτος ἐπέπληξεν Ἡσιόδωι τὰς μὲν ἀγαθὰς ποιουμένωι, τὰς δὲ φαύλας, ὡς ἀγνοοῦντι φύσιν ἡμέρας ἁπάσης μίαν οὖσαν....
>
> Heraclitus rebuked Hesiod for making some days the good ones, others the bad ones, that he was ignorant that the nature of each day is one and the same....[19]
>
> PLUT. *Camill.* 19, DK22 B 106

Here we are told that the φύσις, the nature, essence, or character, of each day, is the same. The day is understood as a neutral stretch of time—a container, if you will—for different kinds of activities. There is nothing about the day's own character that would prejudge which activity is especially suited to be performed on this day—intrinsically the days are all exactly alike.[20] Similarly, Plato in his *Timaeus*, a work from his late period, does not take up the idea

19 Translations are my own, unless otherwise indicated.

20 We may see this fragment as an expression of conflict between a scientific understanding of the day that we find in Heraclitus and a social demand for predictability of what may happen on a day that Hesiod may be interested in.

of different days having different qualities, but rather introduces the day as a neutral unit for quantitative measurement. It is the smallest unit of time measurement dealt with:

> καὶ διακοσμῶν ἅμα οὐρανὸν ποιεῖ μένοντος αἰῶνος ἐν ἑνὶ κατ' ἀριθμὸν ἰοῦσαν αἰώνιον εἰκόνα, τοῦτον ὃν δὴ χρόνον ὠνομάκαμεν. ἡμέρας γὰρ καὶ νύκτας καὶ μῆνας καὶ ἐνιαυτούς, οὐκ ὄντας πρὶν οὐρανὸν γενέσθαι, τότε ἅμα ἐκείνῳ συνισταμένῳ τὴν γένεσιν αὐτῶν μηχανᾶται: ταῦτα δὲ πάντα μέρη χρόνου, καὶ τό τ' ἦν τό τ' ἔσται χρόνου γεγονότα εἴδη....
>
> By bringing order to the universe, he [the demiurge] makes an everlasting image of eternity remaining in unity, (an image) moving according to number; this[21] is, of course, what we have named "time." For before the heavens came to be, there were no days or nights, no months or years. But now, together with framing the heavens, he devised their coming to be. These all are parts of time, and *was* and *will be* are forms of time that have come to be....
>
> 37c6–e5, translation and italics by ZEYL 2000, with alterations

Days and nights are introduced here as parts of time and as coming about due to the fact that the demiurge, Plato's "creator god," orders the heavens. Thus "the day" is a cosmological unit, and there is no indication that one day would be in any qualitative sense different from the next; rather it is a neutral unit of measurement.[22]

While days are neutral units here, phenomenologically speaking, days and nights are of course not straightforwardly regular units, since the lengths of days and nights are not equal during the year.[23] We saw above that Plato complains about this fact in *Republic* VII, when he suggests that true astronomy should be done to some degree independently of the changing phenomena. Given the focus on measurability in the *Timaeus*, it seems surprising that the day is introduced as a unit separate from and not including night, as this seems to mean forgoing the possibility of understanding it as a regular unit; in the passage quoted above, "day" is named first, then "night." If we look a

21 I will follow the tradition in understanding "this" (τοῦτον) as referring back to the whole preceding clause. Brague (1983) has suggested understanding it as referring back only to ἀριθμὸν; I argue in detail against this reading in *Ancient Notions of Time from Homer to Plato*, chapter 6.

22 For further support for the neutrality of the day unit, see below.

23 Of course, this depends on where you live, but it certainly holds true if you live in Greece and around the Mediterranean Sea.

little bit further in the *Timaeus*, however, we see that Plato, in what follows, understands day and night together as one unit. This allows him to posit the regularity of the day-night unit as the basis for the intelligibility of empirical processes. In 39b-d we are told that in order for there to be a visible common measure for all the motions of the heavenly bodies, the demiurge lit the Sun and let it shine through the whole universe:

> νὺξ μὲν οὖν ἡμέρα τε γέγονεν οὕτως καὶ διὰ ταῦτα, ἡ τῆς μιᾶς καὶ φρονιμωτάτης κυκλήσεως περίοδος·
>
> Thus day and night came into being as the period of the single and most intelligent revolution.
>
> 39c1–3

Plato avoids the problems from the *Republic* by working with the day-night period as one single unit here—"day and night" are called "the period" (singular) which is always the same. This is what in Greek is called the νυχθήμερον;[24] the word νυχθήμερον itself, however, only appears in later sources.[25]

In Timaeus's account, the planets, Sun, and Moon are moving along two circles that cross and are arranged like an armillary sphere: they all move around the so-called circle of the Same, which is along the equator; in addition, each moves around in its own circle of the Different, which is around the ecliptic.[26]

The Sun's movement creating day and night is the "single and most intelligible one" in so far as the diurnal movement of the Sun follows only the orbit of the Same.[27] Hence, night and day only make obvious a movement that the

24 Cf. also Cornford 1997, 115.

25 Bolus, who flourished in the 3rd century BCE, in his *Physica et Mystica*, vol. 2, 41, lines 6 and 10, and 42, line 2, seems to be our earliest source for this word (while LSJ and Montanari [2015] give a passage from the letters of Paul [2 *Corinthians* 11.25] as the earliest occurrence); it can, of course, also be found in Ptolemy's *Almagest*.

26 For more details on this set-up, see Sattler (in preparation a).

27 Cf. the end of the sentence preceding the one just quoted in the main text: God lit the Sun ἵνα ὅτι μάλιστα εἰς ἅπαντα φαίνοι τὸν οὐρανὸν μετάσχοι τε ἀριθμοῦ τὰ ζῷα ὅσοις ἦν προσῆκον, μαθόντα παρὰ τῆς ταὐτοῦ καὶ ὁμοίου περιφορᾶς (39b5–c1) ("in order that he might fill the whole heaven with his shining and that all living things for whom it was meet might possess number, learning it from the *revolution of the Same and Uniform*," Cornford's translation, my italics). A day is simply the motion of the Sun around the circle of the Same, which is the same motion for all planets, and the only motion the fixed stars perform. By contrast, the yearly movement of the Sun manifests the one orbit of the Different specific to the Sun: ἐνιαυτὸς δὲ ὁπόταν ἥλιος τὸν ἑαυτοῦ περιέλθῃ κύκλον (39c5) ("a year has come to be when the Sun has completed its *own* circle").

planets, Sun, and Moon share, which can thus be called a "single" one, namely the movement along the circle of the Same, while these heavenly bodies differ in their movements along the orbit of the Different.[28]

Thus, the day unit has a prominent place in the set-up of the universe, but there is no discussion of a further division of the day into smaller units in the *Timaeus*. The demiurge does not seem to require any smaller units in order to make the cosmos resemble its intelligible model in the best possible way. And, for purposes of time measurement in the context of natural philosophy, smaller units do not yet seem to be relevant, as we also see with other natural philosophers of the time.

Where such smaller units do, however, play some role for Plato is in the discussion of setting up a society. While the divinely created time units are years, months, nights, and days, smaller units come into view once Plato turns also to human affairs. And it is not sufficient that the focus of the philosophical discussion is on the manner in which a new *polis* and society can be set up—we find this question already discussed in the *Republic*. Rather, it is only when such a discussion is intimately connected with concerns about astronomy that these smaller units come into view, as is the case with Plato's last work, the *Laws*.[29] We may think that the connection between society and astronomy in the *Laws* is not all that different from what we find in the *Republic*, as both works address these themes in the context of education. However, the big difference is that, in contrast to the *Republic*, in the *Laws* Plato asserts the intelligibility of natural processes, such as those relevant for astronomy (the motions of the heavenly bodies are understood as a combination of regular circular motions, and as having a ratio to each other).

In philosophy, the notion of the hour becomes important once the motions of the heavens and of society are combined in one investigation: only when natural philosophy is brought together with concerns about the ideal society do regular sub-divisions of the day become a focus of discussion, and we get a division into hours, or so I want to argue.

In the *Timaeus* we do, in fact, have a discussion of heavenly motions and of society together in the same dialogue. But this dialogue works *sub specie naturae*, so to speak—the discussion of ancient Athens and Atlantis at the very

28 Cf. 38c8–9 and also Cornford 1997, 115. The different movements of the planets along their particular circle of the Different produce additional units of time which are bigger than the night-and-day unit.

29 While the *Republic* talks about astronomy, as we saw in the quotation above, astronomy is only introduced as part of the mathematical curriculum and does not have any special status.

beginning of the work is solely an announcement of what is to come in the *Critias*. It is only in the *Laws* that we find this discussion *sub specie societatis*.

1.3 *The Connection between Astronomy and Social Concerns*

The *Laws* connect Plato's interest in a reform of society with his interest in demonstrating the measurability and thus intelligibility of natural processes. This can be seen from, among other things, the fact that his society, in order to be set up in accordance with the laws, has to have its own calendar. Given the state of the Athenian calendar at that time, Plato seems to assume a form of idealized calendar.[30] At least this is suggested if we look at the Athenian calendar from the 5th century onwards, which is the best-attested Greek calendar we have.[31] Very roughly put, the Athenians lived with at least two different calendars simultaneously, each used for different purposes: there was a calendar for the festivals, which consisted of 12 months based on the cycle of the Moon,[32] and a "political" calendar regulating economic and administrative life, which consisted of 10 months and was based on the motions of the Sun.[33] Originally, these two calendars did not necessarily begin or end on the same day, and it was difficult to keep the lunar calendar in accordance with the seasons. Thus, additional days or months would be inserted. While astronomers suggested different cycles for inserting additional months, it seems that the *poleis* did not adhere to a fixed pattern for these insertions.[34] Finally, in agricultural planning, points in time were fixed according to the risings of stars.[35]

In contrast to this intricate Athenian calendar, the calendar Plato sets up in the *Laws* seems to be an idealized calendar, where everything is done

30 Cf. Pritchett and Neugebauer 1947; Parke 1977; Hannah 2005.

31 Cf. Bickermann 1968, 34.

32 Even though the purpose of this calendar was mainly religious in the 5th century, it is usually called the "civil calendar" in the literature. Because some effort was made to connect it also with the solar year, it is sometimes also called a luni-solar calendar; for both points see Pritchett and Neugebauer (1947, 5), who, however, argue against understanding it as a luni-solar calendar.

33 This second calendar is called the "prytany calendar" or "senatorial calendar." See Pritchett and Neugebauer 1947, 35. See Meritt (1928, 123ff.) for the relationship between prytany (senatorial) and civil years.

34 Cf. Bickermann 1968, 28–30 and 35. For tinkering with the calendar, cf. Pritchett and Neugebauer 1947; Dunn 1999; Hannah 2005, 47-49, and my book manuscript on ancient notions of time, chapter 5. In Athens only the civil calendar was tampered with, while the prytany calendar seems to have been free from intercalation. See Meritt 1928, 71, and Bickermann 1968, 37.

35 As we find it, for example, in Hesiod's *Works and Days* 614–617; cf. also Hannah 2005, 25 and 46.

according to the seasons,[36] and where there is no suggestion that two different and unconnected ways of dividing up time need to be combined and thus may require tampering with the calendar. At least this is what Plato's brief sketch of the calendar to be used in the new society suggests. Plato mentions this calendar in the context of the discussion of the right education system in the new *polis*—the young must learn about literature, playing the lyre, arithmetic, and warfare:

> καίτοι τὰ μὲν περὶ τὸν πόλεμον ἃ δεῖ μανθάνειν τε αὐτοὺς καὶ μελετᾶν ἔχεις τῷ λόγῳ, τὰ δὲ περὶ τὰ γράμματα πρῶτον, καὶ δεύτερον λύρας πέρι καὶ λογισμῶν, ὧν ἔφαμεν δεῖν ὅσα τε πρὸς πόλεμον καὶ οἰκονομίαν καὶ τὴν κατὰ πόλιν διοίκησιν χρῆναι ἑκάστους λαβεῖν, καὶ πρὸς τὰ αὐτὰ ταῦτα ἔτι τὰ χρήσιμα τῶν ἐν ταῖς περιόδοις τῶν θείων, ἄστρων τε πέρι καὶ ἡλίου καὶ σελήνης, ὅσα διοικεῖν ἀναγκαῖόν ἐστιν περὶ ταῦτα πάσῃ πόλει—τίνων δὴ πέρι λέγομεν; ἡμερῶν τάξεως εἰς μηνῶν περιόδους καὶ μηνῶν εἰς ἕκαστον τὸν ἐνιαυτόν, ἵνα ὧραι καὶ θυσίαι καὶ ἑορταὶ τὰ προσήκοντ' ἀπολαμβάνουσαι ἑαυταῖς ἕκασται τῷ κατὰ φύσιν ἄγεσθαι, ζῶσαν τὴν πόλιν καὶ ἐγρηγορυῖαν παρεχόμεναι, θεοῖς μὲν τὰς τιμὰς ἀποδιδῶσιν, τοὺς δὲ ἀνθρώπους περὶ αὐτὰ μᾶλλον ἔμφρονας ἀπεργάζωνται—ταῦτα οὔπω σοι πάντα ἱκανῶς, ὦ φίλε, παρὰ τοῦ νομοθέτου διείρηται·
>
> Of course, we have told you what military skills they must practise and learn, but what about literature, playing the lyre, arithmetic? We stipulated that they must each understand enough of these subjects to fight a war and run a house and administer a state; for the same reasons they must acquire such knowledge about the heavenly bodies in their courses—sun, moon and stars—as will help them with the arrangements that every state is forced to make in this respect. You ask what arrangement we are referring to? We mean that the days must be grouped into months, and the months into years, in such a way that the seasons, along with their various sacrifices and festivals, may each receive proper recognition by being duly observed in their natural sequence. The result will be to keep the state active and alert, to render the gods due honour, and to make men better informed on these matters—all this, my friend, has not yet been adequately explained to you by the legislator.
>
> *Laws* 809c2–e1, translation by SAUNDERS 1997

36 Thus there is no danger that the calendar of the *Laws* would get out of sync with the seasons, one of the problems that the Athenian civil calendar was facing.

We see that the expected effects of such a calendar are quite remarkable: (1) it guarantees that the seasons as well as the festivals are properly recognized; (2) they are duly observed in their *natural* sequence (κατὰ φύσιν), which (3) keeps the state active,[37] (4) renders the gods due honor,[38] and (5) makes men better informed.[39]

This passage shows that discussions about calendars were influenced by astronomical, religious, and political concerns. And the ideal calendar of the *Laws* is meant to provide the ideal society with a religiously as well as astronomically correct and intelligible framework. But in the context of the temporal structures concerning a society, further divisions of the day into smaller units seem to be important.

2 Smaller Units of Time

When we now move on to the first occurrences of units of time in ancient Greece that are smaller than the day unit, we will start by looking briefly at different ways in which such smaller units were expressed. Subsequently, we will look at the different meanings the Greek word ὥρα originally possessed, before we concentrate on the first passage in Greek philosophy where it can be interpreted as meaning "hour." Finally, we will briefly look at the cultural and scientific background that allowed ὥρα to be understood as hour.

2.1 *Different Ways to Express Smaller Units*

We find different ways to express smaller temporal units in early Greek thinking, depending on whether a certain point in time or a certain duration needs to be expressed. From the very beginning of Greek literature, we find a division of the day according to the beginning, end, and midpoint of the Sun in its motion. Already in Homer, the sequence of night, dawn, midday, and so on,

37 England (1921, 288 [vol. II]) remarks at d5ff.: "Mathematics wakes up the sluggish mind; the necessity for the due observance of the calendar keeps the public mind alert in the same way."

38 Making clear the essentially religious dimension the calendar had for the ancient Greeks.

39 We are not told in which matters exactly human beings will be better informed, but presumably it will be on calendar issues and, related to that, on theological and astronomical questions. Schöpsdau (2011, 169ff.) remarks with respect to the calendar suggested in book VIII three unusual points in Plato's suggestion for a calendar: (1) that the festival calendar is modelled on the Egyptian solar year of 365 days; (2) that each month in this calendar is dedicated to one divinity and includes a festival for this divinity (the connection of 12 divinities with 12 months is not attested before Plato in Greece); (3) that one of the 12 divinities is Pluto, which is unique.

plays a crucial role in structuring the days narrated.[40] Here the focus is on a (more or less) specific point in the course of the day, that then also allows us to mark off different phases in the day. But the duration of these phases is somewhat unclear, and it would also vary according to the seasons.

Another way to divide the day is by breaking it down into parts (μέρη) or portions/watches (μοῖραι). Μοῖρα is found in Homer for the division of the night. For example, in *Iliad* X.252–253 we read: παροίχωκεν δὲ πλέων νὺξ τῶν δύο μοιράων, τριτάτη δ' ἔτι μοῖρα λέλειπται ("of the night, more than two watches have past, and the third is still left"). Μέρος appears first as referring to a temporal part in Herodotus.[41] While μέρος does not seem to have been used very frequently in the sense of a specific temporal part, we do also find it in the following passage in Plato's *Laws*:

> πεζῶν δὴ μόνον θήρευσίς τε καὶ ἄγρα λοιπὴ τοῖς παρ' ἡμῖν ἀθληταῖς, ὧν ἡ μὲν τῶν εὑδόντων αὖ κατὰ μέρη, νυκτερεία κληθεῖσα, ἀργῶν ἀνδρῶν, οὐκ ἀξία ἐπαίνου·
>
> That leaves only land animals for the athletes of our state to hunt and capture. Now sometimes this is done by what is called "night-hunting," when the participants, sluggard as they are, take it in turns to sleep κατὰ μέρη [for some parts]. This sort of hunting is not to be recommended.
>
> *Laws* 824a1–6, translation by SAUNDERS 1997 with alterations

In the context of distinguishing two kinds of hunting—night hunting, on the one hand, and hunting with nets and traps, on the other—this passage portrays night hunting as an activity where the hunters take turns on the watch. The people who are not on watch sleep κατὰ μέρη, which Susemihl (1862), for example, translates as "einige Stunden." So the notion of μέρη provides us with a way to divide the day or night into smaller units, which gives us some duration, but no points in time.

Finally, the day can be divided more specifically with the help of hours (ὧραι). Hours, first of all, give us certain determined sections, i.e., durations. With the help of the ordinal numbers, we can then use hours also in order to indicate a specific position in the course of the day. Accordingly, when we are told that some event takes place from the 7th to the 8th hour, we are given both

40 E.g., *Odyssey* VII, 288. Cf. also Taplin 1992, especially 14–26.

41 And this seems to be the only passage earlier than Plato which talks of μέρος not merely as the remaining part of the day, cf. below.

a temporal position in the day and a duration (it takes what we would call 60 minutes).

Talking about morning, noon, and so on gives us solely a position in time. And the way in which the notion of parts (μέρη) of the day is used seems to focus mainly on duration. By contrast, the notion of an hour works for positions in time as well as durations. In principle, such a combination could also be achieved if we combined talk about parts of time with ordinal numbers—"the first part of the day" could indicate a duration as well as a position. But it seems that both features were only brought together in Greek thinking with the notion of hours.[42]

While the Greeks use the term ὥρα in our earliest textual evidence—we find the word already in Homer—it is not originally used to indicate what we understand by "hour." And it is only at the very end of the fourth century BCE, so after the time of Plato, that the Greek understanding of ὥρα is changing in an essential way—or so the story is usually told.

2.2 *The Greek Understanding of ὥρα*

Before the fourth century, ὥρα commonly refers to (1) the seasons, (2) a special or fitting time to do something, (3) the springtime of life, or (4) any period fixed by the natural revolutions of the heavens. In all these senses we find the term frequently in pre-Platonic writers as well as in Plato.[43] The sense of ὥρα that is captured in English by the word "hour" and from which the English word ultimately derives, appears later, however, and is usually seen to reflect certain changes in society and science. The question is when these changes took place. Scholars have claimed that ὥρα has this meaning at the earliest from the very late 4th century onwards. For example, Bickermann (1968, 15) claims that we find ὥρα in the sense of hour only with Pytheas of Massalia as reported in Geminus (*Element. Astron* 6, 9) and in Aristotle (fragment 161).[44] I will show below that there is some astronomical and cultural background that clearly suggests the possibility of an earlier date. And I want to suggest that we have an earlier explicit occurrence of ὥρα in the sense of hour, and the first

42 For a more in-depth discussion of the connection between duration and points in time that the notion of an hour provides see Sattler (in preparation b).

43 For seasons see, for example, Homer, *Odyssey* XI, 294, and Plato, *Philebus* 30c2–7; for ὥρα as a special or fitting time to do something see, for example, Aristophanes, *Ecclesiazusae* 1163, and Plato *Republic* 462e4 or 474c4; for ὥρα as the springtime of life see, for example, Aeschylus, *Suppliant Women* 997, and Plato, *Symposium* 219c4; for ὥρα as a period fixed by natural revolutions see, for example, Euripides, *Alcestis* 449, and Plato, *Republic* 527d2–4.

44 Pape (1880) even claims that ὥρα is first used as meaning "hour" in Hipparchus in the 2nd century BCE.

occurrence in a clearly philosophical text, in Plato's *Laws*. In this text, Plato combines his ideas about how to set up the best possible society with an interest in astronomy and its effects on society.

In a passage that discusses the length of the meeting of a certain council, we find what I consider to be the first occurrence of ὥρα meaning "hour" in a philosophical text:

> ἐπίσκοποι δ' ἔστωσαν τούτων ἃς εἱλόμεθα γυναῖκες, πλείους εἴτ' ἐλάττους, τοῖς ἄρχουσιν ὁπόσας ἂν δοκῇ προστάττειν τε καὶ ὁπόταν, πρὸς τὸ τῆς Εἰλειθυίας ἱερὸν ἑκάστης ἡμέρας συλλεγόμεναι μέχρι τρίτου μέρους ὥρας, οἷ δὴ συλλεχθεῖσαι διαγγελλόντων ἀλλήλαις εἴ τίς τινα ὁρᾷ πρὸς ἄλλ' ἄττα βλέποντα ἄνδρα ἢ καὶ γυναῖκα τῶν παιδοποιουμένων ἢ πρὸς τὰ τεταγμένα ὑπὸ τῶν ἐν τοῖς γάμοις θυσιῶν τε καὶ ἱερῶν γενομένων.
>
> They [husband and wife] should be supervised by women whom we have chosen (several or only a few—the officials should appoint the number they think right, at times within their discretion). These women must assemble daily at the temple of Eileithuia[45] for μέχρι τρίτου μέρους ὥρας, and when they have convened, each must report to her colleagues any wife or husband of childbearing age she has seen who is concerned with anything but the duties imposed on him or her at the time of the sacrifices and rites of their marriage.
>
> 784a1–b1, translation by SAUNDERS 1997

In this passage, the Athenian Stranger suggests that a council, which oversees whether couples of childbearing age follow their duties, should meet every day for a certain time. In order to figure out the duration of their meeting, and thus in which sense Plato is talking about ὥρα here, we have to understand what the phrase μέχρι τρίτου μέρους ὥρας means. And for this we not only need to find a way to decide whether ὥρα here refers to hour or not, but also what μέχρι actually means; and, if ὥρα refers to hour, we may also want to know whether it refers to seasonal or standard hours.[46] As far as I can see, there are three

45 Eileithuia is the Greek goddess for childbirth and midwifery.

46 While astronomers used standard hours, which are more or less equivalent to our understanding of hours, for most other purposes seasonal hours were more common in ancient Greece; that is, the time when it was light was divided into 12 parts, and thus these parts would be shorter in winter, and longer in summer. While our explicit evidence for the distinction between the length of standard hours (ὧραι ἰσημεριναί) and seasonal hours (ὧραι καιρικαί) seems to be mainly derived from Ptolemy, the sundial findings of equinoctial

different readings of this phrase proposed in the literature, either implicitly or explicitly:

Reading 1: ὥρας has been deleted as a gloss, and the third part has then been seen as referring to ἡμέρας, mentioned beforehand. This is an explicit suggestion by Bury (1948), and Badham (1883) also brackets ὥρας. Bury understands "day" here as νυχθήμερον, that is, as 24 hours (rather than just as the time when it is light); thus a third of the day would be eight hours.

I do not see any good reason for this deletion, since it goes against all the manuscripts and does not seem to be necessary in order to make sense of the passage.[47]

Reading 2: ὥρας has been interpreted to mean "day" here, for example, by Jowett (1892) and Apelt (1916). ὥρας is then not deleted, but the understanding is the same as in 1, that we are dealing with the third part of the day.[48]

The problem I see with this reading is that without ἡμέρα qualifying ὥρα, or a special context, ὥρα, during this period, does not usually mean simply "day" (though with qualifications it can, of course, refer to a specific time of the day or the duration of a day).[49]

Reading 3: Finally, we can see ὥρας as one of the first expressions of "hour," so that the council meetings would last a third of an hour. England, in his 1921 edition of the *Laws*, understands it like this,[50] as do Taylor (1934) and Cousin (1831).

As far as I can see, contemporary debates hardly engage with this question. Saunders (1997) just translates it as a third of the day, and Schöpsdau (2011) in his commentary simply refers to Badham and Bury, and also understands it as "Dauer bis zu einem Drittel des Tages." Brisson and Pradeau (2006) at least make it clear that the reading of Badham and Bury has no support from the manuscripts (cf. above). They translate it as "le tiers d'une heure."

In order to decide between understanding the duration of the council meeting as a third of a day (as in readings 1 and 2) or a third of an hour (as in reading

---

sundials in Hermann, Sipsi, and Schaldach (2015) seem to suggest a much earlier awareness of it.

47 Also Brisson and Pradeau (2006) in their note 20 ad locum make the criticism that "la modification n'a pas d'autorité textuelle."

48 Montanari (2015) also seems to understand ὥρας as "day" since he translates the whole phrase as "until the third part of the day."

49 Cf. the entry "ὥρα" in LSJ, meaning A II, in Montanari (2015) meaning 3 A, and in Pape (1880), meaning 1.

50 England (1921, 639): "we may conclude that 20 minutes was the *minimum* time of attendance" (his italics). But we can of course have of a third of an hour without this yet being spelt out as what we would think of as 20 minutes (which is also not taking into account the possibility of seasonal hours).

3), we have to take into account that "day" can either refer to the time when it is light, from sunrise to sunset, as opposed to night; or to the period of time between one sunrise and the next, including night, in other words, to what we would call a 24-hour day. "Hour" could refer to standard or seasonal hours.[51] And μέχρι can either mean "up to" or "no less than": Bury and others understood it as the former, England and Taylor as the latter. And according to LSJ, μέχρι can also mean "nearly" in connection with numbers. So in principle, the phrase in question, μέχρι τρίτου μέρους ὥρας, could yield the following times for the length of this meeting:

i. if we assume ὥρα to mean "a day including night": (a) up to 8 hours, (b) no less than 8 hours, (c) nearly 8 hours;
ii. if we take ὥρα to mean "day as the time when it is light": (d) up to anything between 3 hours and 10 minutes and 4 hours and 56 minutes,[52] (e) no less than anything between 3 hours and 10 minutes and 4 hours and 56 minutes, (f) nearly anything between 3 hours and 10 minutes and 4 hours and 56 minutes;
iii. if we understand ὥρα to mean standard hour it should be (g) up to what we would call 20 minutes, (h) no less than 20 minutes, (i) nearly 20 minutes;
iv. and finally, if we understand ὥρα to mean seasonal hour, we get (j) up to anything between 16 and 25 minutes,[53] (k) no less than anything between 16 and 25 minutes, (l) nearly anything between 16 and 25 minutes.

We see that there are a lot of options between more than 8 hours and up to 16 minutes. It is also no surprise that those scholars who understand ὥρα here as "day" are inclined to understand μέχρι as meaning "up to" (so, for example, Bury), while those who understand it as "hour" rather seem to go for "no less than" (so, for example, England and Taylor). The main question is, however,

51 As mentioned above, our earliest testimonies for this division are from later times. But since standard hours would have been used by astronomers, while seasonal hours would have been used in more practical contexts, if Plato was talking about hours here, and if talking about hours was already a common practice at this time, it is not unreasonable to assume that Plato was aware of both kinds of hours.

52 I have taken today's minimal and maximal daylight times of Athens as a rough approximation—Athens has daylight for about 9 hours and 31 minutes around the winter solstice, and for about 14 hours and 48 minutes around the summer solstice. This also fits roughly with the ratio commonly assumed by ancient astronomers for Greece of 9 to 15; cf. Neugebauer 1972.

53 Again, I take here as a rough basis today's minimal and maximal daylight time of Athens for calculating a third of the minimum and maximum hour—around the winter solstice an hour could be as little as around 48 minutes, around the summer solstice as long as around 74 minutes.

whether we think that these meetings should last somewhere in the region of several hours, or more around the 20 minute mark.

Whether we understand it as a third of a day or of an hour also depends on what we think is realistic for a council meeting. Bury simply states that thinking of this meeting as taking up to 20 minutes is "absurd," without, however, giving any reason for the absurdity. Part of the motivation for my reading may be that I just find it too forbidding (too "absurd") to conceive of a *daily* council meeting that takes up to eight hours. But apart from the question how many hours of council meeting we think a person can endure every day, a more scholarly question is how we think this council operates. If the women constituting the council have a meeting every day that lasts up to (or around or no less than) eight hours, could these women do anything else? We know that their task is not simply to talk to the other women on the committee. Rather, their main responsibility is to interact with the different husbands and wives, in order to find out whether any of them is insufficiently focused on producing offspring and, if they are not fulfilling their duty, to exhort them or threaten them into better behavior (784c). If the women on the council cannot get a young couple to change its behavior, then these women also have to discuss the cases with the guardians of the laws (the νομοφύλακες).

Later on in the text we learn that the tasks of the women on the council are even more far-reaching: they also have to deal with cases of divorce and supervise the children's nurses. At 794b we read that one of the women on this council (there are twelve of them in total) is always elected for a year to supervise the nurses who watch over the games of the children from age three to six and to watch over the children themselves by going to the sanctuaries where the children play every day. In 929e9 ff. we hear that if a husband and a wife do not get on at all, then ten middle-aged men from the guardians of the law, and ten of the women who regulate marriages, so from the council we are talking about, should look into it. They should try to reconcile the couple, but if this is not possible, they should try to find another partner for each. A little bit further, in 930c, we also hear that if a husband dies and leaves a young widow, the women of the council will advise the relatives on what to do (presumably, on which new partner to secure). Finally, they are also involved in punishing grown-up children who do not show due respect and care for their parents (932b).

We see that the women on the council for overseeing marriages have quite a range of tasks, which eight-hour reporting meetings every day would certainly not permit. Even if we read μέχρι as "up to," and assume that on many days the meeting would not really last a full eight hours, it seems that these reporting committees would take up most of their members' time, and not leave enough time for their actual work. If we understand "day" as "the time when it is light," we still get meetings lasting three to five hours, which seems scarcely manageable every day. Thus, given the multitude of these women's tasks, of which the daily reporting meeting seems to be only one small part, it seems very likely to me that Plato is in fact suggesting that the meetings last for at least around one third of an hour.

We may think that this would make meetings too short—can we have a serious meeting in 20 minutes? We should not forget that this meeting is taking place *every day*, so all that needs to be talked about are possible changes from the previous day. Furthermore, it is simply a reporting meeting on the question of whether any couples of childbearing age do not fulfil their duties; they do not need to discuss any basic values or principles as these are settled. We saw that other meetings are invoked once measures need to be taken or serious cases need to be investigated. So, the meetings our passage talks about seem to be similar to today's so-called "stand-up meetings" in industries where groups of people work together. Meant as brief status updates every day, rather than as meetings for full discussions, they are not typically supposed to exceed 15 minutes (which is why people are not meant to sit down).[54]

Thus I think there is good reason to assume that Plato talks about a third of an hour in our passage. Accordingly, our passage should be seen as the first occurrence of ὥρα meaning "hour" in a Greek philosophical text. The cultural and scientific background for understanding ὥρα as hour seems to have been clearly in place during Plato's time, as I will argue below.

But let us first consider the possibility of earlier instances of understanding ὥρα as "hour" in the extant ancient Greek corpus. In medicine we find two passages in the earlier texts of the Hippocratic Corpus that seem to use ὥρα in the sense of hour.[55] They may date to the beginning of the 4th century.[56] And in astronomy, we find the following fragment of Democritus:

> καθ' ὃν ἡ μεγίστη τῶν ἡμερῶν ὡρῶν ἐστιν ἰσημερινῶν ιε....
>
> According to him [Democritus], the longest period of daylight is equivalent to 15 equinoctial hours....
>
> PTOLEMY, *Apparit. epileg.* 275, 1, in Heiberg's edition p. 67, lines 16–18, DK 68 B 14

54 Cf., for example, Stray, Sjøberg, and Dybå 2016.

55 *Epidemics* IV. 5, 150, 17–18 Littré and *On Internal Infections* VII, 238, 22–23 Littré.

56 Cf. Langholf 1973. It is notoriously difficult to date the texts from the Hippocratic Corpus. Craik (2015) thinks that Epidemics IV can be dated to around 400 BCE "though possibly incorporating material from different dates and different sources." And she thinks that a date early in the first half of the 4th century is probable for *On Internal Infections.*

If we think that this fragment actually gives Democritus's own words, then this would count as the earliest astronomical text, and perhaps the earliest text in general, in which ὥρα is understood as "hour."[57] However, this fragment is from a part of Ptolemy's *Phaseis*,[58] and not only does Ptolemy in no way indicate that he is in fact using Democritus's own words, it also seems common practice for Ptolemy to translate data from his predecessors into his own language and metrics.[59] So there does not seem to be too much reason to assume that Democritus did indeed use ὥρα in the sense of hours. But even if he did not use the word ὥρα and it is only Ptolemy who supplied the word, it seems that Democritus did talk about the relationship between the length of the day at the equinoxes and the length of the day at the summer solstice. In order to come up with such a numerical relation, not only must he have understood the day to be divided into parts, but he must also have worked with a standardized sub-unit of the day—whether or not he called it ὥρα.[60] There is, however, another wrinkle to this passage in Ptolemy: Ptolemy ascribes this claim not only to Democritus, but also to Caesar and Hipparchus (in contrast to the position taken by Callipus, Eudoxus, and others), and the passage quoted may thus actually refer to a later view, i.e., to that of Hipparchus in the second century.

If Democritus did not in fact use the word ὥρα, and the passages from the *Hippocratic Corpus* are later, then Plato's passage may be the earliest to use ὥρα in the sense of hour transmitted to us tout court. In any case, Plato's passage seems to be the first in a strictly philosophical text. But at whatever juncture the astronomers and doctors started using the notion of hours, the cultural practice of dividing the day into 12 hours was clearly in place long before Plato in the Near East. We find 12 day hours and 12 night hours established in Egypt probably from around 2000 BCE onwards,[61] and in Babylon we find the whole

57 We also find Eudoxus in one fragment (Lasserre F128, *Ars* 14. 11.) talking about hours, but this is probably not earlier than Plato's *Laws*.

58 A book that deals with the rising and setting of stars, weather signs, and so on, which can be considered to belong to the tradition of the ancient *parapegmata* (as may the cluster of Democritus's fragments grouped under B14 in DK).

59 So Henry Mendell in e-mail conversation, who also suggested that Democritus might have just said that the ratio of day to night is 5:3 or that the day is 5 parts, taking the night-and-day as 8 parts, or that the day is 15 parts, taking the night-and-day as 24 parts; cf. also Neugebauer (1972) for 15:9 as the standard ratio assumed for Greece in ancient times, and also the division of the year in the *Ars Eudoxi*.

60 Henry Mendell, however, doubts that Democritus had the notion of equinoctial hours or parts. They do appear in *P. Hibeh* I 27, which Neugebauer (1972) and Mendell date to around 300 BCE, and Mendell argues that the notion of equinoctial hours cannot precede it by many years.

61 Cf. Zitman 2010, 160–1. Scholars from Daressy (1900) onwards have assumed the Coffin of Mesehti (S1C) to be evidence that we can trace night hours back to 2100 BCE; this coffin is now taken to be roughly a century later, from the time of Mentuhotep II, cf. Kahl (2016,

day-night unit divided into 12 "double-hours." According to Herodotus II, 109, and thus quite some time before Plato, the Greeks learned the arrangement of dividing the day into twelve parts from the Babylonians.[62] In Aristophanes's *Ecclesiazusae* 651–652, which was performed around 392 BCE, we find a reference to the length of the shadow of the gnomon as indicating the progress of the day and as marking dinner time.[63] Here again we find that people in their daily lives break down the day into smaller units that resemble hours, and do not just deal with a division into morning, midday, and evening. Furthermore, we have sundials from the 4th century BCE that also suggest such divisions.[64] So it seems that the cultural practice had clearly been in place by the time of Plato, and there is no reason to think that Plato was unaware of it.

Now, importantly, neither the passage from Herodotus nor the one from Aristophanes speaks yet about hours. And while we find mention of a dining ὥρα in Herodotus and Thucydides, it is not clear that this necessarily refers to a dining *hour*: when Herodotus talks about the ὥρα for dinner (ὥρα δεῖπνον), for example, in I, 119, ὥρα may just mean the "right time" for a meal or the time for a certain activity, in the way that Herodotus uses ὥρα to talk about the time for bed in I, 10. When Thucydides talks about ὥρα ἀρίστου (probably lunch time) in VII, 81, 1, line 5, however, he clearly seems to indicate a point in time, since we are told that it is at the ὥρα ἀρίστου that the Syracusans and allies overtook the escaping Athenians. But we do not get ὥρα as indicating a specific duration.

We do, however, find hour-like divisions of the day before Plato, since we saw in Herodotus a division of the day into 12 μέρη, and in Aristophanes the

---

33–36) and Kahl, El-Hamrawi, and Verhoeven (2018). The oldest coffin positively attesting a division of the night into 12 hours seems currently to be that of Aashyt, from around 2050 BCE. I owe these references to Symons, Kahl, and a reviewer.

62 Against this, however, see Dicks (1970, 166). Given that the basic Babylonian unit of time is what we would understand as a double hour, *bēru*, Dicks thinks it to be more likely that the division into twelve units came from Egypt. For our purposes it does not matter whether the Greeks got the division from Egypt or Babylon; all that matters is that Herodotus depicts it as a common practice in Greece at the time of his writing, and therefore before the birth of Plato.

63 There were tables that showed the relation between the length of the shadow and the seasonal hour, cf. Neugebauer (1975, 736–740) who thinks that these Greek shadow tables probably start in the fourth or fifth century BCE. Cf. also Bickermann (1968, 14–15) and Schissel (1936) for shadow tables of a later date. Diogenes, II 1–2 (DK A1) claims that Anaximander was the first to erect a gnomon in Sparta—but even if true, this does not yet show that divisions into parts derived from the sundial are commonly used (though it would show that such divisions were of interest for natural philosophers very early on).

64 We have one from Olympia from the 4th century BCE according to Hermann, Sipsi, and Schaldach (2015).

practice of measuring the progress of the day with the help of sundials so as to indicate the dining time. But before we get ὥρα meaning "hour," we also get a step that may be seen as a transition from the understanding of ὥρα as any period fixed by the heavenly motions to hour. We find this expressed in a text by Plato himself, at the beginning of the *Phaedrus*:[65]

> Φαῖδρος
> εἰς καιρόν, ὡς ἔοικεν, ἀνυπόδητος ὢν ἔτυχον: σὺ μὲν γὰρ δὴ ἀεί. ῥᾷστον οὖν ἡμῖν κατὰ τὸ ὑδάτιον βρέχουσι τοὺς πόδας ἰέναι, καὶ οὐκ ἀηδές, ἄλλως τε καὶ τήνδε τὴν ὥραν τοῦ ἔτους τε καὶ τῆς ἡμέρας.
>
> Phaedrus: How lucky, then, that I am barefooted today—you, of course, are always so. The easiest thing to do is to walk right in the stream; this way we will also get our feet wet, which is very pleasant, especially at this hour and season.
>
> *Phaedrus* 229a3–6, translation by NEHAMAS AND WOODRUFF

Here we have the ὥρα of ἐτός and ἡμέρα—that is, the clearly marked subdivision of the year, which is the month or season, as well as of the day, which is the hour or something like it. While Nehamas and Woodruff (1997) as well as Cousin (1883), Schleiermacher (1990), and Brisson (2004) translate the "ὥρα of

65 We also get several passages talking about ὥρα τοῦ νυκτός or ὥρα τῆς ἡμέρας in Xenophon, an exact contemporary of Plato, for example at *Mem.* 4.7.4: ἐκέλευε δὲ καὶ ἀστρολογίας ἐμπείρους γίγνεσθαι, καὶ ταύτης μέντοι μέχρι τοῦ νυκτός τε ὥραν καὶ μηνὸς καὶ ἐνιαυτοῦ δύνασθαι γιγνώσκειν ἕνεκα πορείας τε καὶ πλοῦ καὶ φυλακῆς, καὶ ὅσα ἄλλα ἢ νυκτὸς ἢ μηνὸς ἢ ἐνιαυτοῦ πράττεται, πρὸς ταῦτ' ἔχειν τεκμηρίοις χρῆσθαι, τὰς ὥρας τῶν εἰρημένων διαγιγνώσκοντας: καὶ ταῦτα δὲ ῥᾴδια εἶναι μαθεῖν παρά τε νυκτοθηρῶν καὶ κυβερνητῶν καὶ ἄλλων πολλῶν οἷς ἐπιμελὲς ταῦτα εἰδέναι. ("Similarly he recommended them to make themselves familiar with astronomy, but only so far as to be able to find the time of night, month, and year, in order to use reliable evidence when planning a journey by land or sea, or setting the watch, and in all other affairs that are done in the night or month or year, by distinguishing the times and seasons aforesaid. This knowledge, again, was easily to be had from night hunters and pilots and others who made it their business to know such things.") And in *Mem.* 4.3.4: πάνυ γ', ἔφη, καὶ τοῦτο χάριτος ἄξιον. οὐκοῦν καὶ ἐπειδὴ ὁ μὲν ἥλιος φωτεινὸς ὢν τάς τε ὥρας τῆς ἡμέρας ἡμῖν καὶ τἆλλα πάντα σαφηνίζει, ἡ δὲ νὺξ διὰ τὸ σκοτεινὴ εἶναι ἀσαφεστέρα ἐστίν, ἄστρα ἐν τῇ νυκτὶ ἀνέφηναν, ἃ ἡμῖν τῆς νυκτὸς τὰς ὥρας ἐμφανίζει, καὶ διὰ τοῦτο πολλὰ ὧν δεόμεθα πράττομεν; ἔστι ταῦτα, ἔφη. ἀλλὰ μὴν ἥ γε σελήνη οὐ μόνον τῆς νυκτός, ἀλλὰ καὶ τοῦ μηνὸς τὰ μέρη φανερὰ ἡμῖν ποιεῖ. ("And since the night by reason of her darkness is dim, whereas the Sun by his brightness illuminates the hours of the day and all things else, have they not made stars to shine in the night, that mark the watches of night for us, and do we not thereby satisfy many of our needs?" "That is so." "Moreover, the Moon reveals to us not only the divisions of the night, but of the month too," translations by Marchant 2013). Xenophon also talks about the ὥρα for meals.

the day" as "hour," it could in principle also simply mean "a part," or "any period fixed by the natural revolutions of the heavens that divides the day." Given that ὥρα here does not only refer to the division of the day, but also to that of the year, it does not seem to mean hour in the primary sense here, but rather, more generally, "the natural sub-unit dividing a time span," whether this time-span is a year or a day.[66] So ὥρα seems to be understood in a general sense here as indicating a regular *sub-unit*.[67]

The treatment of ὥρα in the *Laws* differs from the one in the *Phaedrus* in two points: there is no talk about ὥρα ἡμέρας in the *Laws*, but only about ὥρα (though, because of the preceding ἑκάστης ἡμέρας, the word ἡμέρα may still sound in the ear when we read μέχρι τρίτου μέρους ὥρας); and in the *Laws* ὥρα indicates only the sub-unit specific for the day, the hour. In fact, having ἡμέρα not in the same phrase as ὥρα, but as a background from what just preceded, may have been one step for the development of an absolute usage of ὥρα as meaning the subunit of the day, the hour. In the *Laws*, ὥρα indicates a certain duration, not a specific point of time in the day. Ultimately, however, the notion of ὥρα will combine both aspects.

3 Conclusion

This chapter tried to reconstruct one stage in the development of the understanding of ὥρα as "hour." It attempted to show that the usage of ὥρα meaning "hour" is earlier than usually assumed, and that the first occurrence in a philosophical text is in Plato's *Laws*, where astronomy and concerns about the design of an ideal society are combined. We saw that in Plato's natural philosophy in the *Timaeus*, in order to make the cosmos as much like its intelligible model as possible, the demiurge introduces time and, with it, days, nights, months, and years, but not hours. Thus, to make the cosmos most good and intelligible, no hours are required. But once we investigate how human beings should best set up their society in such a cosmos, hours come into view.[68]

66 In contrast to the older understanding of ὥρα as any specific period fixed by the heavenly motions, ὥρα here refers specifically to sub-units of this period.

67 Cf. also "τοῦ γνώμονος ἡ σκιὰ ἐπιοῦσα ἐπὶ τὰς γραμμὰς σημαίνει τὰς ὥρας τοῦ ἐνιαυτοῦ καὶ τῆς ἡμέρας" IG12(8).240 (Samothrace).

68 I want to thank participants of the conference "Down to the Hour: Perspectives on Short Time in the Ancient Mediterranean," at the University of Chicago, February 2017, members of the Day Unit Group at the IIAS, Michael Della Rocca, and the editors of this volume for useful feedback on this paper. I am also very grateful to the IAS Durham, where I wrote a first draft of this paper while being a fellow.

References

Apelt, Otto, trans. 1916. *Platon. Gesetze*. Leipzig: Meiner.

Badham, Charles. 1883. "Platonis Legum liber sextus." *Mnemosyne* 11: 47–58.

Bickermann, Elias. 1968. *Chronology of the Ancient World*. London: Thames & Hudson.

Brague, Remi. 1982. *Du temps chez Platon et Aristote. Quatre études*. Paris: Presses Universitaires de France.

Brisson, Luc, trans. 2004. *Platon. Phèdre*. Paris: GF Flammarion.

Brisson, Luc and Jean-François Pradeau, trans. 2006. *Platon. Les lois: Livres VII à XII*. Paris: GF Flammarion.

Bury, Robert G., trans. 1926. *Plato. Laws*. Cambridge, MA: Harvard University Press.

Bury, Robert G. 1948. "Two Notes on the Laws." *The Classical Review* 62: 108–9.

Cornford, Francis M. 1997. *Plato's Cosmology. The Timaeus of Plato. Translated with a Running Commentary*. Indianapolis, IN: Hackett Publishing.

Cousin, Victor, trans. 1831. *Platon. Les lois*. Paris: Pichon et Didier Libraires.

Cousin, Victor, trans. 1849. *Platon. Phèdre*. Paris: P.-J. Rey.

Craik, Elizabeth M. 2015. *The 'Hippocratic' Corpus. Content and Context*. London and New York: Routledge.

Daressy, Georges. 1900. "Une ancienne liste des décans égyptiens." *Annales du service des antiquités de l'Égypte* 1: 79–90.

Dicks, David R. 1970. *Early Greek Astronomy to Aristotle*. London: Thames & Hudson.

Dunn, Francis. 1999. "Tampering with the Calendar." *Zeitschrift für Papyrologie und Epigraphik* 123: 213–31.

England, Edwin B. 1921. *The Laws of Plato. The Text Edited with Introduction and Notes*. New York: Arno Press.

Fränkel, Hermann. 1955. "Die Zeitauffassung in der frühgriechischen Literatur." In *Wege und Formen frühgriechischen Denkens. Literarische und philosophiegeschichtliche Studien,* edited by Franz Tietze, 1–22. Munich: Beck.

Grube, George M.A., trans. 1997. "Republic." In *Plato. Complete Works*, edited by John Cooper. Indianapolis, IN: Hackett Publishing.

Hannah, Robert. 2005. *Greek and Roman Calendars. Constructions of Time in the Classical World*. London: Bristol Classical Press.

Heiberg, Johan L., ed. 1898. *Claudii Ptolemaei opera quae exstant omnia*. Vol. I. Leipzig: Teubner.

Hermann, Klaus, Maria Sipsi, and Karlheinz Schaldach. 2015. "Frühe Arachnen – über die Anfänge der Zeitmessung in Griechenland." *Archäologischer Anzeiger*, 1. Halbband: 39–67.

Hicks, Robert D., trans. 1972. *Diogenes Laertius. Lives of Eminent Philosophers*. Cambridge, MA: Harvard University Press.

Jowett, Benjamin, trans. 1892. *The Dialogues of Plato*. Oxford: Clarendon Press.

Kahl, Jochem. 2016. "Zum Alter der Sargtext-Artefakte aus Assiut." In *Zwischen Philologie und Lexikographie des Ägyptisch-Koptischen. Akten der Leipziger Abschlußtagung des Akademieprojekts "Altägyptisches Wörterbuch", ASAW 84, 3*, edited by Peter Dils and Lutz Popko, 29–55. Leipzig: Sächsische Akademie der Wissenschaften zu Leipzig.

Kahl, Jochem, Mahmoud El-Hamrawi, and Ursula Verhoeven. 2018. "The Asyut Project: Thirteenth Season of Fieldwork (2017)." *Studien zur Altägyptischen Kultur* 44: 137–148.

Langholf, Volker. 1973. "'Ὥρα—Stunde. Zwei Belege aus dem Anfang des 4. Jh. v.Chr." *Hermes* 101 (3): 382–84.

Lasserre, François, trans. 1966. *Die Fragmente des Eudoxos von Knidos*. Berlin: de Gruyter.

Littré, Émile, ed. 1839-1861. *Oeuvres complètes d'Hippocrate*. 10 vols. Paris: Les Belles Lettres.

Marchant, Edgar C., trans. 2013. *Xenophon. Memorabilia; Oeconomicus*. Cambridge, MA: Harvard University Press.

Meritt, Benjamin D. 1928. *The Athenian Calendar in the Fifth Century*. Cambridge, MA: Harvard University Press.

Montanari, Franco. 2015. *The Brill Dictionary of Ancient Greek*. Edited by Madeleine Goh and Chad Schroeder. Leiden: Brill.

Most, Glenn, trans. 2007. *Hesiod. Theogony. Works and Days. Testimonia*. Cambridge, MA: Harvard University Press.

Nehamas, Alexander and Paul Woodruff, trans. 1997. "Phaedrus." In *Plato. Complete Works*, edited by John Cooper. Indianapolis, IN: Hackett Publishing.

Neugebauer, Otto. 1972. "On Some Aspects of Early Greek Astronomy." *Proceedings of the American Philosophical Society* 116 (3): 243–51.

Neugebauer, Otto. 1975. *A History of Ancient Mathematical Astronomy*, New York: Springer.

Nilsson, Martin P. 1911. "Die älteste griechische Zeitrechnung. Apollo und der Orient." *Archiv für Religionswissenschaft* 14: 423–48.

Onians, Richard B. 1954. *The Origins of European Thought: About the Body, the Mind, the Soul, the World, Time and Fate*. Cambridge: Cambridge University Press.

Pape, Wilhelm. 1880. *Griechisch-Deutsches Handwörterbuch*. Brunswick: Vieweg.

Parke, Herbert W. 1977. *Festivals of the Athenians*. London: Thames & Hudson.

Pritchett, William K., and Otto Neugebauer. 1947. *The Calendars of Athens*. Cambridge, MA: Harvard University Press.

Sattler, Barbara M. 2017. "How Natural Is a Unified Notion of Time? Temporal Experience in Early Greek Thought." In *The Routledge Handbook of the Philosophy of Temporal Experience*, edited by Ian Phillips, 19–29. New York: Routledge.

Sattler, Barbara M. In preparation a. "Thinking Makes the World Go Round—Intellection and Astronomy in Plato's *Timaeus*."
Sattler, Barbara M. In preparation b. "Duration versus point in time—Early Greek thought and the conceptual complexity of the notion of hour."
Saunders, Trevor, trans. 1997. "Laws." In *Plato. Complete Works*, edited by John Cooper. Indianapolis, IN: Hackett Publishing.
Schissel, Otmar. 1936. "Antike Stundentafeln." *Hermes* 71: 104–17.
Schleiermacher, Friedrich, trans. 1990. *Platon. Phaidros, Parmenides, Epistolai*. Edited by Dietrich Kurz. Darmstadt: Wissenschaftliche Buchgesellschaft.
Schöpsdau, Klaus, trans. 2011. *Platon. Nomoi (Gesetze), Buch VIII–XII*. Göttingen: Vandenhoeck & Ruprecht.
Solmsen, Friedrich. 1963. "The 'Days' of the Works and Days." *TAPA* 94: 293–320.
Stray, Viktoria, Dag I.K. Sjøberg, and Tore Dybå. 2016. "The daily stand-up meeting: A grounded theory study." *The Journal of Systems and Software* 114: 101–124.
Susemihl, Franz, trans. 1862. *Platon. Die Gesetze*. Stuttgart: Metzler.
Taplin, Oliver. 1992. *Homeric Soundings: The Shaping of the Iliad*. Oxford: Clarendon Press.
Taylor, Alfred E., trans. 1934. *The Laws of Plato*. London: Dent.
West, Martin L., ed. 1978. *Hesiod. Works and Days*. Oxford: Oxford University Press.
Zeyl, Donald, trans. 2000. *Plato. Timaeus*. Indianapolis, IN: Hackett Publishing.
Zitman, Marcel. 2010. *The Necropolis of Assiut: A Case Study of Local Egyptian Funerary Culture from the Old Kingdom to the End of the Middle Kingdom*. Orientalia Lovaniensia Analecta. Leuven: Peeters.

CHAPTER 6

Diurnal Selves in Ancient Rome

James Ker

In ancient Roman literary representations of social and ethical life we frequently encounter the day as a privileged temporal framework, within which Roman authors sought to define people's forms of life through descriptions of "diurnal patterns" (on this term, see below).[1] Examining these patterns and their internal subdivisions can attune us to the social and ethical significance assigned to this domain of "short time" in Roman cultural history—sometimes for polemical ends. In this chapter I will approach the day under a number of different aspects, ranging from the sub-parts into which the day is divided through to the longer span of recurring days—the everyday. In the present context I can only survey diurnal patterning in general terms, and I will move fairly swiftly to my case-study: a letter of Seneca, *Epistulae morales* 122, in which Seneca moralizes against certain nocturnal Romans and, in the process, articulates some normative day-based conceptions of a Roman *communis vita* (collective life).

We may begin by considering a page from the Senecan dialogue *On Tranquility of Mind* (c. 48–55 CE), in which the author advises his friend Serenus on how to calm the fluctuations that leave his moral life unsettled and to arrive at the *stabilem animi sedem* ("stable state of mind") that the Greeks call εὐθυμία and Seneca calls *tranquillitas* (*Tranq.* 2.3).[2] One of Seneca's main pieces of advice to Serenus is to avoid extremes. For example, we should mix solitude with sociality and relieve our mental concentration with occasional joking or play; these, he suggests, have *naturalem quandam voluptatem* (*Tranq.* 17.3–4) ("a certain natural pleasure"). Seneca proceeds to rehearse, approvingly, how Romans of various periods have decreed certain times for rest or relaxation:[3]

> Legum conditores festos instituerunt dies, ut ad hilaritatem homines publice cogerentur, tamquam necessarium laboribus interponentes

1 My special thanks to Kassandra Miller and Sarah Symons for their generosity, patience, and helpful critiques.

2 For an introduction to and translation of the work, see Fantham 2014, 182–215.

3 Translations are mine throughout.

 | DOI:10.1163/9789004416291_008

> temperamentum; et magni, ut dixi, viri quidam sibi menstruas certis diebus ferias dabant ...
>
> *Tranq.* 17.7

> The founders of our laws instituted festival days so that people would be publicly compelled toward light-heartedness, as if inserting among our labors a mandatory moderation. And, as I said, some great men would grant monthly holidays to themselves on certain days ...

In this opening sketch of temporal habits, we see Seneca move from the level of public policy by lawgivers to the private rules that certain individuals have imposed on themselves. In the section that follows, Seneca continues to move from the categorization of days within the larger time-schemes of the festival calendar and the subdivided month to the subdivision of the day itself:

> ... quidam nullum non diem inter otium et curas dividebant. Qualem Pollionem Asinium oratorem magnum meminimus, quem nulla res ultra decumam detinuit; ne epistulas quidem post eam horam legebat, ne quid novae curae nasceretur, sed totius diei lassitudinem duabus illis horis ponebat. Quidam medio die interiunxerunt et in postmeridianas horas aliquid levioris operae detulerunt. Maiores quoque nostri novam relationem post horam decumam in senatu fieri vetabant. Miles vigilias dividit, et nox immunis est ab expeditione redeuntium.
>
> *Tranq.* 17.7

> [S]ome men divided every day without fail between leisure and cares. Of this sort, we remember, was Asinius Pollio, the great orator, whom no matter detained beyond the tenth [hour]. He did not even read letters after that hour, so that no new care could arise. Instead, for those two hours he laid aside his fatigue from the whole day. Some men have eased up[4] in the middle of the day and have postponed some lighter activity for the afternoon hours. Even our ancestors forbade any new motion to be made in the senate after the tenth hour. The soldier divides up watches, and those returning from an expedition have the night off.

4 Better: "have unharnessed (themselves)." I follow *OLD* s.v. *interiungo* 2 in taking this (as well as the similar use at Sen. *Ep.* 83.3) to be a metaphorical use of a verb commonly used (intransitively) to describe a loosening of the yoke or harness on animals.

We can take our lead from several distinct aspects of this passage: (1) here, we see Seneca emphasizing the social and ethical significance of time-ordering, both for society as a whole and for individuals. (2) We see him home in on the day—and the day characterized in distinctively Roman terms—as a category that affords some very specific examples of his point about relaxation. (3) We see him using the composition of the day as a whole, as well as its routinization, to suggest whole ways of life. (4) While the passage is shaped by Seneca's priorities as a philosophical moralist, he refers to a range of patterns from across the social domain, and these belong to an open-ended repertoire of available diurnal patterns, each with its own emphasis. (5) The passage's relatively diverse collection of examples, ranging from public policy to private practice, and from the ancient past up to more recent historical settings, may provoke us to ask how much license the moralist is prepared to allow for day-ordering by private individuals. What would constitute a problem-case in the addition of leisure to one's schedule? For Seneca elsewhere places limits. In one letter, for example, he banishes day-end leisure entirely: *nullus mihi per otium dies exit* ("No day of mine concludes in leisure"), he writes; *partem noctium studiis vindico* (*Ep.* 8.1) ("I reclaim part of the nights for my studies"). That allusion to nocturnal work, or lucubration, is further echoed in letter 122, with its criticism of Romans who are also awake at night, but for entirely opposite ends, transferring to the day the normally nocturnal activity of sleep.

I will proceed by elaborating the five points just mentioned, beginning with the social and ethical dimensions of time-use in general and culminating in a reading of Seneca's counterexample in letter 122. Throughout this discussion I will be concerned with the representation of "diurnal selves" in Roman literature—that is, the portrayal of a person's social and ethical life through reference to the temporal patterning of their days. But analyzing letter 122 will allow us to see how Seneca enforces a norm that is diurnal in a more precise sense, invoking a range of natural and cultural factors to problematize daytime sleep and nocturnal living. In the process, we will also witness a confrontation between competing groups in the Roman elite staged through a contrast between their uses of short time.

1 Time and Ethos

In our opening passage from *On the Tranquility of Mind* (henceforth *Tranquility*), the temporal structures and strictures that Seneca invokes are quite diverse. Although they all belong within the domain of "sociocultural time" as opposed to natural time,[5] they come from diverse areas of social practice and range between public and private, traditional and recent, and social and moral. Yet Seneca is also at pains to package the examples together: they all involve an ordering of time with a broad socioethical benefit. Even if the motivations for rest-time vary from case to case, taken together the examples illustrate both the possibility of organizing time in this way and a generally positive outcome.

In Seneca's sketch, there is some quite pronounced time-ordering terminology, such as the repeated term *dividere* ("divide"), and several distinct aspects of time-ordering are touched upon within just a few lines. These can be charted under the four parameters used by Zerubavel in his influential analysis of "sociotemporal order" and "regularity," in which he articulates the ways in which recurrent temporal structures come to define a social reality.[6] The first of these parameters is "rigid sequential structures," and indeed the daily habits described by Seneca are defined by the movement from work to relaxation, always in this order. The second is "fixed durations," and in Seneca we see work lasting only a limited part of the day, such as when people "ease up" (cf. *interiunxerunt, Tranq.* 17.7) in the middle of the day, senatorial business is not usually allowed to proceed much beyond the tenth hour, and soldiers' vigils are limited by division of labor.[7] Zerubavel's third parameter is "standard temporal locations," and we see this most fully in Seneca's focus on *duabus illis horis* ("those two hours") between the tenth hour and day's end, as well as in his mention of the *certis diebus* ("certain days") that certain great men set aside for holidays each month. The fourth, "uniform rates of recurrence," is a central characteristic of at least some of the habits involved, which Seneca presents as occurring *nullum non diem* ("every day without fail"). This scarcely exhausts the details and nuances of the passage, but it shows us some of the principal ways in which Seneca ascribes significance to time-ordering.

Seneca ties this sociotemporal order, with its qualified inclusion of rest, to the social and ethical dimensions of specific Roman lives. And in fact, various modern critical terms, such as "time orientation," "time sense," and "chronotype," have been coined to capture the way in which specific relations to time are manifested in people's attitudes and habits.[8] One of the most successful

---

5 For the term's implications, see Sorokin 1943, 158–225.

6 Zerubavel 1981, 1–30.

7 The concern with duration in military watches, for example, is made explicit by how these were later defined by the military writer Vegetius (c. 4th century CE): *in quattuor partes ad clepsydram sunt divisae vigiliae, ut non amplius quam tribus horis nocturnis necesse sit vigilare* (*Mil.* 3.8) ("Watches are divided into four by the clepsydra, so that no one is forced to stand watch for more than three nocturnal hours").

8 On time orientation, a fundamental essay is Jones 1988; for early critical discussions of influential work on these topics generally, see Bergmann 1983, 465–473; Adam 1990, 96–98.

applications of such concepts to ancient Rome is Andrew Riggsby's study entitled "Pliny in Space (and Time)." Riggsby argues against the idea of a general culture-wide conception of time or space: in Pliny's letters, for example, two distinct cognitive models of time-use can be traced, depending on whether the emphasis is on Pliny's habits at one of his villas—where he mentions clocks but equivocates on the precise time of activities and is "not controlled by the clock"—or on his daily activities in the city of Rome—where "his commitment to his social and moral obligations is measured by duration," and so he describes activities in terms of what Riggsby calls "stopwatch time."[9]

In Seneca's writings, too, people's ethics are frequently projected through time orientations. In *On the Brevity of Life*, he presents two particularly detailed and contrasting time orientations which center upon people's basic relationships to time's three parts (the future, present, and past). The time-wasters he refers to as *occupati* ("busy-bodies") are so focused on the future that they squander the present and fail to accumulate a past on which they can reflect.

> Abit igitur vita eorum in profundum; et ut nihil prodest, licet quantumlibet ingeras, si non subest quod excipiat ac servet, sic, nihil refert quantum temporis detur, si non est ubi subsidat, per quassos foratosque animos transmittitur.
>
> *Brev.* 10.5

> Therefore, their life vanishes into the depths. Just as nothing is useful, however much of it you pile up, if there is nothing beneath that can receive it and preserve it, so too, it does not matter how much time is given, if there is nowhere it can settle: it passes through their leaky, perforated minds.[10]

By contrast, those who study philosophy are aware of the uncertainty of the future, the sole availability of the present, and the sanctity of the past, and as a result, *soli vivunt; nec enim suam tantum aetatem bene tuentur. Omne aevum suo adiciunt; quidquid annorum ante illos actum est, illis adquisitum est.* (14.1) ("They alone are alive. For they do not only preserve their own lifetime well. They annex all of time to their own. Whatever years have been completed prior to them, they have appropriated.")

In the writings of Pliny, Seneca, and others we see time serving as a privileged domain in which certain lives—in Greek, βίοι; in Latin, *genera vitae*—can

9 Riggsby 2003, 180 and 182.

10 On the likely allusion to the sieves of the Danaids here, see Williams 2003, 182.

be defined and also contrasted with one another. This relationship between lower-level time-behavior and higher-order life-narratives is characterized by Luckmann in terms of two complementary processes, which he terms "hierarchization" and "dramatization":

> Biographical schemes ... perform two kinds of closely related functions. They "convert" short-term into long-term relevance and vice versa. The meaning of a large variety of ordinary, everyday actions and experiences is subordinated to biographical significance, and biographical significance is infused into certain kinds of actions and experiences directly. Both procedures, hierarchization and dramatization, fuse short- and long-term sequences of action into meaningful components of an entire course of life.
>
> LUCKMANN 1991, 162

If we adapt Luckmann's category of "biographical schemes" to refer to the various Roman literary discourses in which a person's form of life is portrayed, we will indeed find that these discourses sometimes "hierarchize" in a way that relegates short time to a lower value, but sometimes "dramatize" in a way that endows short time with a life-defining significance.

2 Dimensions of the Day at Rome

In *Tranquility*, Seneca zooms in on the day, and so it becomes relevant for us to consider diurnal temporality more closely. Our considerations can be guided in part by structural aspects of the day that are universal across cultures. Moore, in his path-breaking book *Man, Time, and Society* (1963), observes that the day is two main things at once. First, the day is "a cycle of sequential activities"—a description that evokes at once the day's iteration, its perceptible process, and human wakefulness and action. Second, the day is "the fixed boundary within which the demands of the body, the mind, and the social system must be satisfied by activities"—a description that makes us think about the day's finitude and relative brevity, as well as its totality.[11] It is also clear that basic structural aspects of the day, particularly its length and the alternation of light and darkness, have conditioned the circadian rhythms of many biological organisms,

11 Moore 1963, 16.

from plants to shellfish to humans.[12] Beyond these basic dimensions of diurnal time, it can be useful to ask ourselves how the day is defined in a given context—for example, how physiotemporal factors (such as astronomical events) are combined with biotemporal factors (such as circadian rhythms) and sociotemporal factors (such as social traffic, scheduling, time-signaling, etc.)—and how the day can serve as a basis for coordinating these different factors.[13] We must also register daylight's social function, as the enabling condition for the vivid human interaction that serves as the basis of social reality.[14] Then there is the sense of the repeated cycle of the day—whether this is to characterize daily time as "timeless" (Giddens), as background "normalcy" (Zerubavel), or as a "coercive" sequence (Berger and Luckmann).[15]

In the ancient city of Rome these basic structural features of the day were worked out in their own way according to historical and cultural particularities. The days described in the *Tranquility* passage above, for example, involve sequences, durations, time-locations, and cycles that are specifically Roman. Seneca alludes to Asinius Pollio's work as an orator in the law courts, to the introduction of business in the course of a day in the senate, and to military routines, whether in the nightly watches or in the day of an expedition followed by a night of rest for the soldiers involved.[16] He refers to numbered daylight hours and the nocturnal military watch as units of duration, as well as to specific hours in the day (the tenth, eleventh, twelfth) and to specific days within the month, and so he locates us within the Roman day with its seasonally variable hours tracked on sundials and, in military watches, tracked with clepsydrae. Distinctively Roman, too, are the cycles implied in the annual, monthly, and everyday alternations of work and leisure to which Seneca alludes—where the rates of recurrence are constrained by the various civic rhythms we find encoded in surviving calendars, in Ovid's *Fasti*, and in numerous evocations of Roman forensic, political, and military rhythms.[17]

Within the history of western "clock time,"[18] the ancient Roman day is distinctive in numerous ways. For example, Varro's patriotic account of Roman

12 For an overview of modern experiments and discoveries, see Aveni 2002, 13–32, and Wittmann 2016, 77–80.

13 These three levels of time are distinguished by Zerubavel (1981, 2); on coordination, see Zerubavel 1981, 159.

14 Berger and Luckmann 1966, 19–44.

15 Berger and Luckmann 1966, 13; Zerubavel 1981, 21; Giddens, 1984, 35.

16 For the testimonia on these various strictures and divisions in the Roman public day, see Marquardt 1886, vol. 1, 250–269, "Das tägliche Leben."

17 See overall Rüpke 2011.

18 On "clock times," with a focus on the historically-specific configurations of how diurnal time is measured and conceived of, see especially Glennie and Thrift 2009, 22–64.

timekeeping in his lost *Antiquitates rerum humanarum* (*Human Antiquities*) appears to have made much of the Romans' only gradual acquisition and employment of sundials, water clocks, and hours in specific spaces such as temples, the *comitium*, and the lawcourts—a narrative of Roman archaism and imperialist exceptionalism all at once.[19] Varro's work also addressed the Romans' distinctive definition of the civil day as proceeding from midnight to midnight—a notion that set them apart from other nations such as the Athenians (sunset to sunset), the Babylonians (sunrise to sunrise), and the Umbrians (midday to midday).[20] In other ways, too, the signifying possibilities of the Roman day were overdetermined, due to such factors as the greater reliance on natural light they shared with other pre-modern cultures, the religious dimensions of daylight connected with Jupiter as sky-deity, and the relative weakness of the week.[21] And while some calendrical rhythms might reinforce the sense of iteration and everydayness, more often, as Beard observed in an influential 1987 essay, the calendar turned individual days and their festive rituals into pageants of Roman myth and history of multiple periods—the day as "a complex of times," to quote her title.[22] Recent work on technologies of timekeeping has added to our sense of Rome's culturally distinctive diurnal time-culture.[23]

This specific time-culture offered numerous opportunities for social and ethical differentiation of people's time orientations, such as was mentioned above in Riggsby's analysis of Pliny the Younger's ways of representing his villa days and his city days.[24] In the present context we cannot explore all these aspects of Roman diurnal time. I limit myself to focusing on the literary phenomenon that is especially relevant to our topic of "diurnal selves," namely, the description of people's whole days as a basis for characterizing their life as a whole—their identity.

19 For this Varronian legacy, see Plin. *HN* 7.212–215, Cens. 23–24.

20 See esp. Gell. *NA* 3.2, and Bilfinger 1888.

21 See, e.g., the constraints on exposure to daylight for the *flamen dialis*, Gell. *NA* 10.15. On the nundinal week, see Ker 2010.

22 Beard 1987, 12.

23 See, e.g., Hannah 2009; Wolkenhauer 2011; Jones 2016; and the contributions of Wolkenhauer and Miller to the present volume.

24 Riggsby 2003. See also Wolkenhauer's contribution to this volume.

3 Diurnal Pattern and Form of Life

The *Tranquility* passage shows Seneca defining a certain idealized time orientation at the level of the day and connecting this with his broadly positive characterization of certain Roman traditions and the habits of specific Roman men. He treats these routines as models both for how Serenus, his addressee, should approach the ordering of his own days and, by implication, for the ordering of his life in its entirety. The basis for such inferences from day to life is laid out especially clearly by Seneca in *On the Brevity of Life*, cited above. There he imagines confronting an old man who persists in public life and pointing out the poverty of his time-control:[25]

> Repete memoria tecum, quando certus consilii fueris, quotus quisque dies ut destinaveras processerit, quando tibi usus tui fuerit, quando in statu suo vultus, quando animus intrepidus, quid tibi in tam longo aevo facti operis sit, quam multi vitam tuam diripuerint te non sentiente quid perderes, quantum vanus dolor, stulta laetitia, avida cupiditas, blanda conversatio abstulerit, quam exiguum tibi de tuo relictum sit; intelleges te immaturum mori.
>
> *Brev.* 3.3

> Recall to yourself from your memory when you were secure in your plans, how few days proceeded as you had intended, when you enjoyed use of yourself, when your face was in its natural position, when your mind was unintimidated, what you have accomplished in such a long lifetime, how many people have torn your life away without you noticing what you were losing, how much you have lost due to empty grief, stupid cheerfulness, greedy desire, flattering conversation—how little of your own [time] has been left to you. You will understand that you are dying before your time!

An injury to the order of one's day is an injury to one's entire moral autonomy and one's life itself. And indeed, the showcasing of life, or the self, within the framework of the day is one of the most prevalent motifs in Seneca's discussions of time as a moral concern.[26]

The *Tranquility* passage also invites more specific comparisons. The example of Asinius Pollio here, for instance, has close parallels in biographic

25 On this passage see Williams 2003, 133–134.

26 See, e.g., Sen. *Ep.* 83.1–2, with Foucault 1997.

sketches of other famous figures. An anecdote from early in the philosophical tradition relates how Plato's student Xenocrates πολλάκις τῆς ἡμέρας ἑαυτῷ ἐμελέτα, καὶ ὥραν μίαν, φασίν, ἀπένεμε σιωπῇ (D.L. 4.11) ("attended to himself many times in the day and, they say, allotted one hour to silence"). In a slightly different version, Xenocrates does this while διαιρῶν ἕκαστον μέρος τῆς ἡμέρας εἰς πρᾶξίν τινα (Stob. 3.33.11) ("differentiating each part of the day for some action"). And in the tradition on Scipio Africanus the Elder, we hear: *Scipionem hunc Africanum solitavisse noctis extremo, priusquam dilucularet, in Capitolium ventitare ac iubere aperiri cellam Iovis atque ibi solum diu demorari quasi consultantem de republica cum Iove* (Gell. *NA* 6.1.6) ("It was the habit of this Scipio Africanus repeatedly to go into the Capitolium at the conclusion of night, before dawn arrived, and to command that the chamber of Jove be opened up, and there to linger alone for a long time, as if consulting about the republic with Jove"). In these examples, the day-frame provides the basis for a more general characterization of the person through his temporal idiosyncrasies and also includes the combination of public and private life that is central to most of the examples in the *Tranquility* passage.

In its most elaborate forms, the description of a person's daily life presents a seemingly exhaustive, sequential account of a person's day from start to finish. One of the best known such passages is Horace's sketch of his daily routine at the end of his *Satire* 1.6 (*Serm.* 1.6.111–112, 128–129), which begins, *quacumque libido est, / incedo solus* ("Wherever it pleases, I walk on my own ..."), and concludes: *haec est / vita solutorum misera ambitione gravique* ("This is the life of those released from wretched and grievous ambition"). Juvenal, in his *Satire* 1, complains about the form of life that Roman clients must endure in the city of Rome, announcing cynically, *ipse dies pulchro distinguitur ordine rerum: / sportula, deinde forum, iurisque peritus Apollo ...* (1.127–128) ("The day itself is differentiated by a splendid ordering: allowance, then forum, and Apollo the lawyer ..."). In the arc of the first through the third *Satires*—with *Satire* 3 as its centerpiece—Juvenal presents, as Braund observes, a "twenty-four hour catalogue of horrors, starting with insomnia and the duties of the morning and finishing with the dangers of the evening and the night."[27] My preferred term for this literary phenomenon, "diurnal *pattern*," seeks to register several things at once: the aesthetic qualities of the structured day as a design or pattern, whether positive or negative; the dynamic of iteration, of pattern-forming; the totality of the day, as the blueprint or pattern containing all the defining

27 Braund 1996, 34.

qualities of a person's life; and the exemplary potential it has to serve as a pattern for others—again, whether with positive or negative associations.

One of the earliest extended diurnal patterns connected with a specific named person is Ischomachus's first-person description of his day in Xenophon's *Oeconomicus*. Socrates asks this "gentleman" farmer a series of questions, to which he expects to receive a series of answers: πῶς ὑγιείας ἐπιμελῇ; πῶς τῆς τοῦ σώματος ῥώμης; πῶς θέμις εἶναί σοι καὶ ἐκ πολέμου καλῶς σῴζεσθαι; τῆς δὲ χρηματίσεως πέρι καὶ μετὰ ταῦτα … ἀρκέσει ἀκούειν (*Oec.* 11.11) ("How do you take care of your health? How do you take care of your bodily strength? How is it granted to you to protect yourself gloriously in war? As for your money-making, it will please me to hear about it also after these things"). But Ischomachus responds by beginning to describe a typical day's routine:

> Ἐγὼ τοίνυν, ἔφη, ὦ Σώκρατες, ὁ Ἰσχόμαχος, ἀνίστασθαι μὲν ἐξ εὐνῆς εἴθισμαι ἡνίκ' <ἂν> ἔτι ἔνδον καταλαμβάνοιμι, εἴ τινα δεόμενος ἰδεῖν τυγχάνοιμι. κἂν μέν τι κατὰ πόλιν δέῃ πράττειν, ταῦτα πραγματευόμενος περιπάτῳ τούτῳ χρῶμαι· ἂν δὲ μηδὲν ἀναγκαῖον ᾖ κατὰ πόλιν, τὸν μὲν ἵππον ὁ παῖς προάγει εἰς ἀγρόν, ἐγὼ δὲ περιπάτῳ χρῶμαι τῇ εἰς ἀγρὸν ὁδῷ ἴσως ἄμεινον, ὦ Σώκρατες, ἢ εἰ ἐν τῷ ξυστῷ περιπατοίην. ἐπειδὰν δὲ ἔλθω εἰς ἀγρόν, ἄν τέ μοι φυτεύοντες τυγχάνωσιν ἄν τε νειοποιοῦντες ἄν τε σπείροντες ἄν τε καρπὸν προσκομίζοντες, ταῦτα ἐπισκεψάμενος ὅπως ἕκαστα γίγνεται, μεταρρυθμίζω, ἐὰν ἔχω τι βέλτιον τοῦ παρόντος. μετὰ δὲ ταῦτα ὡς τὰ πολλὰ ἀναβὰς ἐπὶ τὸν ἵππον ἱππασάμην ἱππασίαν ὡς ἂν ἐγὼ δύνωμαι ὁμοιοτάτην ταῖς ἐν τῷ πολέμῳ ἀναγκαίαις ἱππασίαις, οὔτε πλαγίου οὔτε κατάντους οὔτε τάφρου οὔτε ὀχετοῦ ἀπεχόμενος, ὡς μέντοι δυνατὸν ταῦτα ποιοῦντα ἐπιμέλομαι μὴ ἀποχωλεῦσαι τὸν ἵππον. ἐπειδὰν δὲ ταῦτα γένηται, ὁ παῖς ἐξαλίσας τὸν ἵππον οἴκαδε ἀπάγει, ἅμα φέρων ἀπὸ τοῦ χώρου ἄν τι δεώμεθα εἰς ἄστυ. ἐγὼ δὲ τὰ μὲν βάδην τὰ δὲ ἀποδραμὼν οἴκαδε ἀπεστλεγγισάμην. εἶτα δὲ ἀριστῶ, ὦ Σώκρατες, ὅσα μήτε κενὸς μήτε ἄγαν πλήρης διημερεύειν.
>
> *Oec.* 11.14–18

> "Well, Socrates," said Ischomachus, "it is my habit to rise from bed at a time when I can catch someone still indoors, if I happen to be in need of seeing him. And if I need to do anything in the city, I take care of these things and, in the process, use this as my walk. But if there is nothing necessary in the city, my slave leads my horse to the farm, while I use my journey to the farm as a walk—perhaps better, Socrates, than if I had walked around in the colonnade. But when I arrive on the farm, if my men happen to be planting, clearing, sowing, or harvesting crops, I inspect how

> each of these is proceeding and make adjustments, if I can do better than at present. After this, I usually mount my horse and do equestrian maneuvers as similar as possible to those necessary in war, avoiding neither side-movement nor incline nor ditch nor channel, though as much as possible I take care not to lame my horse as it does these things. When these things are done, my slave lets the horse have a roll and then leads him homeward, at the same time bringing with him anything from that place that needs to be taken into the city. As for me, now walking toward home and now running, I would clean myself off with a strigil. Next, Socrates, I eat lunch, just enough to get through the day neither empty nor too full."

Ischomachus does not get to proceed through the entirety of his day, because Socrates already expresses wonder at how Ischomachus can accomplish all the things he had asked about ἐν τῷ αὐτῷ χρόνῳ (11.19–20) ("at one and the same time").

Ischomachus's answer, in fact, employs several devices to indicate how his diurnal pattern can characterize his broader form of life. One device is his emphasis on multitasking, here facilitated by the agency of the slave. Another is his use of qualifying expressions that allow for variation: for example, εἴ τινα δεόμενος ἰδεῖν τυγχάνοιμι (11.14) ("if I happen to need to see anyone"). A third device is his use of contrastive foils (e.g., when he contrasts his country walk favorably with strolling through a colonnade in the city). A fourth is Ischomachus's overall emphasis on economic productivity: for example, he claims that his lifestyle makes it μᾶλλον εἰκὸς τὸν οἶκον αὔξεσθαι (11.12) ("more likely that [his] household will be increased"). Such devices meant that Ischomachus's diurnal pattern presents his entire ethos to Socrates—his examined life, as it were.

Ischomachus's diurnal pattern was only directly influential on a few Roman authors,[28] but the many other surviving Greek and Roman literary representations of diurnal patterns almost always share the same function as Ischomachus's account, of defining an entire form of life, as well as some of the specific devices just mentioned. This connection is not one that can be taken for granted, and certainly the phenomenon involves a careful sifting out of some aspects of daily time. After all, there are aspects of diurnal or quotidian time that might undermine its capacity to serve as a basis for higher-level narrative about a person. Daily time is ephemeral, trivial, repetitive, forgettable,

28 Notably, Columella, *On Agriculture* 11.1.

background, and so on. But the composer of a diurnal pattern exploits qualities of diurnal time that permit an interface with higher-level narrative, in a virtuoso performance of Luckmann's "dramatization" of higher-order significance in small time.[29] Everyday time, if its pitfalls are avoided, has the potential to drive longer narratives, appealing to the various ways in which quotidian iteration or everyday interaction can serve as the ultimate basis for social reality.[30]

4 A Repertoire of Patterns

Our *Tranquility* passage characterizes some personal schedules in terms that evoke the broader discourse on diurnal patterns representing specific forms of life. This is most obviously the case for Asinius Pollio, *oratorem magnum … quem nulla res ultra decumam detinuit* (*Tranq.* 17.7) ("the great orator, whom no matter detained beyond the tenth [hour]"). Although the sense is not entirely clear, Seneca clearly wants us to understand that Pollio's routine was either instrumental to, or compatible with, his greatness as an orator (something to which Serenus himself evidently aspired; cf. *Tranq.* 1.3: *eloquentiae famam*). Quintilian quotes a well-known, "urbane" saying about Asinius Pollio, that he was a man *omnium horarum* ("of all hours"), since he was *seriis iocisque pariter accommodato* (*Inst.* 6.3.110) ("suited to serious matters and joking equally"), which, together with Seneca's testimony, suggests that hour-management was a motif in the characterization of Asinius's personality. Seneca's mention of *quidam* ("certain men") who *nullum non diem inter otium et curas dividebant* ("divided every day without fail between leisure and cares") or those who *medio die interiunxerunt* (*Tranq.* 17.7) ("have eased up in the middle of the day") seems to suggest that these habits were a secret to these men's success, too. His description of military custom, referring metonymically to the *miles* ("soldier") who *vigilias dividit* ("divides up watches"), envisages the army's vigilance overall in terms of one man's idealized practice of diligent time-management.

But as the Ischomachus example from Xenophon shows (above), there are other, more detailed and extensive diurnal patterns to be found across Greco-Roman literature. These contain more detail, about both times and activities, than the thumbnail day-sketches given by Seneca in *Tranquility*.

---

29 Luckmann 1991, 162.

30 For qualification of the everyday as a modern condition, and on the modern "critique of everyday life" as an emancipatory response, see Highmore 2002, 1–36.

They approximate the detail of Ischomachus's account while covering the day in its entirety.

Cicero, for example, describes his own daily routine at the end of a letter written from his home on the Palatine in Rome in 46 BCE and addressed to his Epicurean friend Paetus in Naples:[31]

> haec igitur est nunc vita nostra: mane salutamus domi et bonos viros multos, sed tristis, et hos laetos victores, qui me quidem perofficiose et peramanter observant. ubi salutatio defluxit, litteris me involvo; aut scribo aut lego. veniunt etiam qui me audiunt quasi doctum hominem quia paulo sum quam ipsi doctior. inde corpori omne tempus datur. patriam eluxi iam et gravius et diutius quam ulla mater unicum filium.
>
> *Fam.* 9.20.3

> So this is my life now. In the morning, at home, I greet many good men, though gloomy, and also these happy victors, who indeed visit me with much dutifulness and affection. Whenever the morning greeting has ebbed away, I immerse myself in letters: I either write or read. Those men also come who are listening to me like an educated person because I am a little more educated than they themselves [are]. From that point onward all my time is given to the body. I have now mourned the fatherland both more seriously and for a longer time than any mother [mourns] her only son.

Here Cicero describes the depleted form of life he is living amid the uncertainties of Roman public life following Julius Caesar's victory in the civil war. Two main paradigms are operative in the letter. One of these is the daily routine of the republican statesman that is now manifested in the form of a *salutatio*—and this dominated by the triumphant Caesareans. The other paradigm is the so-called "Epicurean day," a model of private life and fine banqueting, the aimless structure of which has been seen as inspiring Horace's diurnal pattern (*Serm.* 1.6, above), and which is later mocked by the Stoic Epictetus (Arr. *Epict. diss.* 3.24.38–41).[32] We also see here the same devices we observed above in Xenophon: multitasking, as Cicero accomplishes both public and private goals simultaneously; qualifying expressions, as when the *salutatio* is implied to last

31 On Cicero's letter as an early instance of Roman daily round descriptions, see Dionisotti 1982, 124 n. 84. On the rhetoric of Cicero's Paetus letters, see Leach 1999.

32 Armstrong 1986, 278.

a sometimes shorter, sometimes longer time; contrastive foils, as we are made aware of differences from his former lifestyle; and economic productivity, here strikingly negated through Cicero's seemingly sumptuous banqueting. Certainly, Cicero's diurnal pattern in the present letter displaces the earlier, more public routine of *salutatio* followed by *deductio* ("escort") into the forum (cf. *Att.* 1.18.1).[33] His quasi-Epicurean routine here, however, will be displaced in turn by a later letter describing his more extreme daily solitude following the death of his daughter Tullia: *cumque mane me in silvam abstrusi densam et asperam, non exeo inde ante vesperum* (*Att.* 12.15.1) ("After I have pushed myself into a thick and rough wood in the morning, I do not emerge from there until evening").

The range of whole-day descriptions (or prescriptions) extends from a fragment of Lucilius (1252–1254 Krenkel) to early Christian documents such at the *Rule of St. Benedict*, often involving greater detail than in Cicero. The mode is most often epistolographic, as in Cicero: letters on daily routine arise in Seneca (*Ep.* 83), Pliny the Younger (*Ep.* 3.1), and others. Satires and epigrams are sometimes also built upon the structures of diurnal time, as in Horace (*Serm.* 1.6) and Martial (4.8).[34] We find the day used for characterizing an emperor's form of life (Suet. *Vesp.* 21–22), the regimen which allowed a doctor to live out a healthy old age (Galen *ST* v.4 Kühn), or the ideal routine of an overseer on a Roman farm (Colum. *RR* 11.1.14–17, drawing upon Xenophon). This is only the tip of the iceberg, and needless to say, each instance within this repertoire could add to our sense of what such representations can do. Here, however, I wish to limit my focus to Seneca's letter 122, which brings together a range of Roman ideas concerning the Roman sociocultural day as a normative diurnal pattern shared by most people in the city—but not all.

5 Diurnal Romans

Epistulae morales 122 begins with a basic observation about seasonal change.[35] Evidently the letter was written during the fall: *detrimentum iam dies sensit; resiluit aliquantum* (*Ep.* 122.1) ("The day has now felt a decrease. It has ebbed somewhat"). Seneca then starts to comment on the appropriate human response to

33 On the walking routine in *Att.* 1.18, see O'Sullivan 2014, 6, 55, 84.

34 A representative discussion is Gibson and Morello's (2012, 115–123) on Pliny's routine descriptions (*Epp.* 3.1, 3.5, 9.36, 9.40).

35 On *Ep.* 122, see the commentary of Summers (1910); also text, translation, and philosophical commentary by Inwood (2007).

this change in nature: *ita tamen ut liberale adhuc spatium sit si quis cum ipso, ut ita dicam, die surgat* (1) ("but not without there still being a generous span if someone should rise with the day itself, as it were"). Having pointed to the possibility of coincidence with sunrise, however, he pushes for an earlier waking: *officiosior meliorque si quis illum expectat et lucem primam excipit* (1) ("More dutiful, and better, is he who waits for the day and seizes upon first light"). But then, over the course of the next few sentences, he moves forward through the stages of the day and traces the successive risings of various groups of people:

> turpis qui alto sole semisomnus iacet, cuius vigilia medio die incipit; et adhuc multis hoc antelucanum est. sunt qui officia lucis noctisque perverterint nec ante diducant oculos hesterna graves crapula quam adpetere nox coepit.
>
> *Ep.* 122.1–2

> Disgraceful is he who lies half-asleep when the sun is high, who begins to be awake at midday. And still for many this is before light. There are those who have inverted the functions of daylight and night and do not pry open their eyes, which are heavy from yesterday's drinking, before night has begun to approach.

The latter group is then compared explicitly to antipodeans—that is, the people whom ancient geographers supposed to be standing on the other side of the (spherical) earth, with the soles of their feet facing ours—and, implicitly, also to the notorious night-banqueting Sybarites:[36]

> qualis illorum condicio dicitur quos natura, ut ait Vergilius, pedibus nostris subditos e contrario posuit,
>
> nosque ubi primus equis Oriens adflavit anhelis,
> illis sera rubens accendit lumina Vesper,
>
> talis horum contraria omnibus non regio sed vita est. sunt quidam in eadem urbe antipodes qui, ut M. Cato ait, nec orientem umquam solem viderunt nec occidentem.
>
> 122.2–3; with VIRG. *G.* 1.250–251

36 On ancient conceptions of the antipodes, see Moretti 1994. For the Sybaritic motif, Summers (1910, 351), quotes Athen. 6.105, Cic. *Fin.* 2.23, Sen. *Ep.* 77.17. See also Sen. *Contr.* 3.1, Colum. 1.pr.16, with Ker 2004, 220–221.

> Just as is said of the situation of those whom nature, as Virgil says, put underneath our feet on the opposite side:
>
> and when the Rising [Sun] first breathes upon us with his panting horses,
> for them the reddish Evening Star kindles its late lights,
>
> so these people are opposite us—not in their region but in their life. In one and the same city there are certain antipodeans who, as Marcus Cato says, have never seen the sun rising or setting.

In the course of the letter, Seneca will proceed to characterize these "antipodeans in Rome" through the range of their perverse activities. These activities manifest their direct contradiction of natural order in numerous different spheres of life, but focusing most of all on the temporal: *hos tu existimas scire quemadmodum vivendum sit, qui nesciunt quando?* (122.3) ("Do you think these people know how we should live, when they don't know when?")

Although the "antipodeans" will turn out to be engaging in more than one kind of temporal inversion, it is chiefly the inversion of day and night with which Seneca is concerned. To use Zerubavel's terms,[37] the problem is primarily a matter of temporal "location"—as is emphasized by the antipodes analogy. Indeed, many of the features of their behavior described by Seneca are conspicuously normal, both in the content of their activities and in their temporal sequence, their duration, and their recurrence; not only this, but they seem to be synchronized and standardized among themselves.[38] The displacement of these activities into the contrarian temporal location of night is what makes all the difference—or at least, Seneca exploits the rhetorical impact of events that are ordinary in all but the time of day when they occur. Variations in time-location alone, it turns out, are sufficient to prevent the "antipodeans" from participating in Roman collective life.

In an earlier article, I discussed letter 122 in connection with productive lucubration, arguing that in this letter Seneca satirizes the "antipodeans" as a kind of "evil twin" of the nocturnal writer and his "imagined community" of Romans in tune with republican tradition.[39] Here I revisit the letter for what it can show about the Roman discourse on "diurnal selves." I will examine, first, how Seneca represents the relationship between Roman collective life and the natural day; second, the ambitions Seneca ascribes to the "antipodeans" when

37 Zerubavel 1981, 1.
38 On "synchronization" and "standardization" in the context of daily schedules, see Glennie and Thrift 2009, 45.
39 Ker 2004.

they choose to live by night; and third, the multiple argumentative strategies Seneca employs for exiling the "antipodeans" from Roman collective life. This analysis will provide us with a basis for answering the question I mentioned at the beginning: How much license is the moralist prepared to allow for day-ordering by private individuals?

First, Seneca repeatedly characterizes the relationship between the social and ethical norms of the sociocultural day and the day as a rhythm in nature. The letter's opening conceit is that it might seem difficult for the sociocultural day and the natural day to be synchronized, since the seasonal shortening of daylight challenges a person's ability to complete *officia* and other personal tasks that require a *liberale ... spatium* (122.1) ("generous span") of time. But Seneca's point is that it remains possible to coordinate the two, if not actually to synchronize them: rising before dawn on a fall or winter day is treated like a kind of hyper-naturalized behavior, and lucubration in general provides a way to compensate for the slight deficit in daylight: *nullus agenti dies longus est. extendamus vitam: ... circumscribatur nox et aliquid ex illa in diem transferatur* (122.3) ("No day is long for one who is active. Let us extend life: ... night must be hemmed in, and something from it transferred into day").[40] Certainly Seneca's idealization of a Roman *dies publicus* (9) ("public day") and *vita communis* (18) ("collective life") in the course of the letter, manifested in the alternation of daily waking *actus* (3) ("activity") with nocturnal sleep, is consonant with the *debitum ordinem* (5) ("appropriate order") that he sees in nature. His reference to *actus* likely reflects an understanding of night, and especially *intempesta nox* (the time of night when the time cannot be told), as not permitting normative activity.[41] Seneca also refers to "nature's custom" or *naturae consuetudinem* (9) ("natural convention"), and whether this refers to a set of cultural norms that derive from nature (more likely) or a natural order that has been defined by cultural convention (less likely), he characterizes it in terms that make clear he has in mind the sequencing of standard domestic and social tasks during daylight hours—things as "natural" as attending a *salutatio* in the morning (13). The commitments with which Seneca is concerned here are partly a matter of philosophical moralism: when he refers to normative users of the day as *naturam sequentium* (17) ("those following nature") and asserts, *tenenda nobis via est quam natura praescripsit* (19) ("we need to adhere to the path which nature has prescribed"), we should hear an echo of the Stoic ideal of living in

40 On the day-extending (and day-enhancing) function of lucubration, see further Ker 2004, 217–219.

41 See especially Varr. *Ling.* 6.6, with Ker 2004, 216.

accordance with nature (e.g., *VB* 8.2).[42] But this Stoic nature is either compatible with, or determinative of, certain other manifestations of natural order in Roman collective life to which Seneca alludes in the course of the letter, such as drinking after meals rather than before them, dressing in a way appropriate to one's gender and time of life, or cultivating flowers in summer rather than winter (122.6–7). Overall, this amalgam of natural rhythms and social habit defines the moral and social order that serves as a baseline and contrast throughout the letter.

This brings us to the second dimension of the letter, which involves the goal Seneca ascribes to the "antipodeans" as they depart from the temporal norm. He observes, for example, *instituerunt omnia contra naturae consuetudinem velle* (122.9) ("they resolved to wish all things contrary to nature's custom"), and the term *instituerunt* (they resolved) may make us think of the public policies and personal rules described in our *Tranquility* passage (cf. *instituerunt, Tranq.* 17.7), thus portraying the "antipodeans" as lawgivers, or at least as careful stewards of their time. But their actions turn out to be antithetical to public norms and in service of a problematic set of social and moral agendas. Seneca gives the impression that he understands the motives of the "antipodeans," and he distinguishes a number of them. One concerns an overall desire for exclusivity, which they manifest in the area of timing just as in other spheres of their lives: *quomodo cultu se a ceteris distinguunt, quomodo elegantia cenarum, munditiis vehiculorum, sic se volunt separare etiam temporum dispositione* (122.18) ("Just as they distinguish themselves from others in their dress, in the sophistication of their dinners, in the splendor of their vehicles, so they want to be separated also by the arrangement of their times"). Connected to this is their *vitae communis fastidium* (18) ("distaste for collective life"), which is less about any inherent appeal of the night than about a time of day that is shared by fewer people. It is *non quia aliquid existiment noctem ipsam habere iucundius, sed quia nihil iuvat solitum* (14) ("not because they reckon the night itself holds anything more pleasing, but because nothing pleases them that is customary"). The night, however, does have some things to offer. One is that it seems to provide a veil for concealment of guilty actions: *gravis malae conscientiae lux est* (14) ("Daylight is burdensome to a bad conscience"). Another is the scope that night provides for the ostentatious use of artificial light: *omnia concupiscenti aut contemnenti prout magno aut parvo empta sunt fastidio est lumen gratuitum* (14) ("If someone desires or scorns all things according to whether they were

42 Cf. Inwood (2007, 354), on characterization here of the "easy and unimpeded quality of life according to nature."

bought at a high or low price, he feels distaste toward free light"). Seneca seeks to convey the perverse and exceptionalist state of mind of the "antipodeans" by briefly reconstructing how they think:

> lucet: somni tempus est. quies est: nunc exerceamur, nunc gestemur, nunc prandeamus. iam lux propius accedit: tempus est cenae. non oportet id facere quod populus; res sordida est trita ac vulgari via vivere. dies publicus relinquatur: proprium nobis ac peculiare mane fiat.
>
> *Ep.* 122.9

> It is light. Time for sleep! It is rest [time]. Now let us exercise, go for a drive, eat lunch! Now daylight is approaching. Time for dinner! We should not do what the people do. It is a filthy thing to live by a well-worn and common path. Let us abandon the public day: let us have a morning that is private and belongs to us alone.

Here we see the "antipodeans" performing exclusivity by mapping a relatively ordinary sequence of activities—i.e., one lacking any markedly hedonistic content—onto a systematically inverse series of temporal signals. It is a profile in private temporality such as Seneca in another letter ascribed to L. Piso, who *usque in horam sextam fere dormiebat: hoc eius erat matutinum* (*Ep.* 83.14) ("would usually sleep right up to the sixth hour: this was his morning time").[43]

The greatest benefit the "antipodeans" seem to see in their reversal is the notoriety that can come from such performances. Seneca observes: *luxuriosi vitam suam esse in sermonibus, dum vivunt, volunt; nam si tacetur, perdere se putant operam* (122.14) ("Luxury-lovers want their lives to be talked about while they live. For if they are met with silence, they think their effort was wasted"), and he portrays them as striving *non tantum luxuriosam rem sed notabilem facere* (122.13) ("to do something that is not only luxurious but noticed") or *quod excitet famam* (122.14) ("that will provoke talk"). It is not enough for them to be transgressive: *in tam occupata civitate fabulas vulgaris nequitia non invenit* (122.14) ("In a city so preoccupied, no one bothers to tell stories about ordinary wickedness"). As evidence of their success, it seems, Seneca recounts a story he heard told by Albinovanus Pedo, who was a *fabulator elegantissimus* ("most

43 But contrary to expectation, Piso *officium ... suum, quo tutela urbis continebatur, diligentissime administravit* (83.14) ("attended with great diligence to his official duty, namely maintaining the protection of the city").

exquisite storyteller") and claimed to have lived upstairs from Sextus Papinius, one of the "antipodeans":[44]

> "Audio" inquit "circa horam tertiam noctis flagellorum sonum. quaero quid faciat: dicitur rationes accipere. audio circa horam sextam noctis clamorem concitatum. quaero quid sit: dicitur vocem exercere. quaero circa horam octavam noctis quid sibi ille sonus rotarum velit: gestari dicitur. circa lucem discurritur, pueri vocantur, cellarii, coqui tumultuantur. quaero quid sit: dicitur mulsum et halicam poposcisse, a balneo exisse. 'excedebat' inquit 'huius diem cena.' minime; valde enim frugaliter vivebat; nihil consumebat nisi noctem."
>
> 122.15–16

> "Around the third hour of the night," he says, "I hear the sound of flogging. I ask what he is doing: he is said to be receiving his accounts. Around the sixth hour of the night I hear an excited clamor. I ask what it is: he is said to be exercising his voice. Around daylight there is a scurrying, slaves are being called, and butlers and cooks are raising a ruckus. I ask what it is: he is said to have called for honey-wine and groats, (and) to have emerged from the bath. 'This man's dinner exceeded the day,' someone says. Far from it. For he lived most frugally. He consumed nothing except night."

The story plays out as a recurring encounter between surprise and normality—the surprise of the time at which sounds are heard, at nocturnal rather than diurnal hours, and the normality of the routine domestic activities that are rendered strange and unfamiliar primarily by their temporal displacement. The humorous coda to the story is somewhat opaque: it seems the interjector quoted by Seneca makes a joke by using *diem* to refer to what Papinius treats as day, which is in fact the night; Pedo's response trades further upon the ambiguity of reference.[45] Overall the anecdote maps out the defamiliarization that results from normal schedules unfolding at surprising times.

But while Seneca devotes portions of the letter to portraying the intentions of the "antipodeans" as temporal innovators, the third dimension on which I wish to focus concerns the ways in which Seneca seeks to undermine those intentions, reinforcing the naturalist basis of the Roman public day in terms that give a variety of problematic meanings to the behavior of the

---

44 On the unhappy fate of Papinius, see Tac. *Ann.* 6.49 and Ker 2004, 224.

45 The problems in understanding this and a following quip (involving wordplay on *lychnobium*) are only partly resolved by Summers (1910, 356).

"antipodeans." One basic problem, of course, is indicated by the metaphorical term "antipodeans."[46] For while *natura* may have placed the actual antipodeans on the other side of the world, the presence of nocturnal Romans as "antipodeans in the same city" rests upon their behavioral, rather than geographic, avoidance of Roman society. Their *e contrario stare* (122.5) ("standing opposite") and the *novas declinationes* (122.17) ("new declensions" or "new latitudes") of their vices reflects not the natural order of a spherical earth but a direct contradiction of natural order. Ancient thought treated the antipodeans as an imagined, distant audience so remote from us that they would typically not have heard of our fame (cf. Cic. *Rep.* 6.20), and their world proved so hard to imagine as simultaneous with ours that Virgil in *Georgics* mentions an alternative possibility: *aut intempesta silet nox / semper et obtenta densentur nocte tenebrae* (*G.* 1.247–248) ("or there is an unceasing silent dead of night and the darkness is thick with night's fog"). But Seneca's "antipodeans" are remote *despite proximity*, which gives them an altogether problematic cultural subjectivity bordering on social death.

Indeed, Seneca alludes to several conceptual paradigms in quick succession, in an effort to characterize the social impossibility of the "antipodeans." In the process, he gives us further glimpses into Roman cultural norms that, for him, are part of the natural order. The "antipodeans" show many signs, for example, of being dead rather than alive. By living at night, Seneca suggests, *isti … mihi defunctorum loco sunt; quantulum enim a funere absunt et quidem acerbo qui ad faces et cereos vivunt?* (122.10) ("These ones, as far as I am concerned, … are in the place of the deceased. For how are they not at a funeral, and a bitter one at that, when they live by torchlight and tapers?")[47] This is a conceit which occurs also at the end of *On Brevity of Life* (*Brev.* 20.5), referring to the Roman *funus acerbum* ("bitter funeral") held under cover of darkness in the case of premature deaths.[48] Furthermore, Seneca suggests, *in quam* [*sc. mortem*] *se vivi condiderunt* (3) ("while alive they have buried themselves in death"), seeing as they have shut themselves away in their nocturnal world. And life also seems to ebb from their bodies by virtue of their gluttony: *suspectior illis quam morbo pallentibus color est: languidi et evanidi albent, et in vivis caro morticina est* (122.4) ("Their complexion is more suspect than that of the sick and wan: they are floppy, faded, and blanched, and there is carrion on their living

---

46 On ancient conceptions of the antipodes as they pertain to *Ep.* 122, see further Ker 2004, 220–221, and Moretti 1994, 35.

47 Cf. *Ep.* 122.3.

48 Summers 1910, 354; Williams 2003, 254.

bodies"). Seneca's critique also offers some modulations of the conceit later espoused by Pliny as *vita vigilia est* (*HN* pr.18) ("To live is to be awake"). For on the positive side, Seneca holds, we should extend our life through lucubration because *huius* [*sc. vitae*] *et officium et argumentum actus est* (*Ep.* 122.3) ("Action is a duty of life and a sign of life"). On the negative side however, a reverse lucubration by the "antipodeans" has the opposite effect, in their *pravitas.... aversandi diem et totam vitam in noctem transferendi* (5) ("crooked turning away from the day and their transferal of their entire life into night")—effectively, an annihilation of life.

Seneca maps the perversity of the "antipodeans" across other areas of cultural knowledge, too. Their behavior would seem to fly in the face of logic in quite a literal sense. The utterance Seneca ascribes to one, *lucet: somni tempus est* (9) ("It is light: it is time for sleep"), is suggestive of a miscarriage of logic, echoing as it does one of the standard examples from philosophical discussions of propositions and arguments, in which "It is day," if true when uttered, is understood to entail the presence of daylight and the absence of night.[49] The "antipodean" seems to have succumbed to an alternative authority of knowledge, like Trimalchio under the influence of Fortunata: *mero meridie si dixerit illi tenebras esse, credet* (Petron. *Sat.* 37.5) ("If she tells him at high noon that it is dark, he will believe her"). Then there is the term *turba lucifugarum*, which activates not only an association with nocturnal animals—they are, he says, *tam infausti ominis quam nocturnae aves* (3) ("as ominous as nocturnal birds")—but also is suggestive of both the perverse misinterpretation of the significance of *lucet* and an avoidance of social contact, either through guilt or through misanthropy.[50]

The overwhelming emphasis, however, is on the decidedly *temporal* aspects of the exclusivity of the "antipodeans," as Seneca goes out of his way to emphasize the significance of timing. As mentioned above, several other time-inversions are alluded to in the course of his sketch: just as the "antipodeans" contradict nature in their nocturnalism, other ways of living contrary to nature include drinking on an empty stomach (122.6), dressing inappropriately for

---

49 E.g., *si dies est, lucet* ("If it is day, it is light") is an argument from necessity, whereas *si mater est, diligit* ("If she is a mother, she loves") is not an argument from necessity (Cic. *Inv.* 1.86); see further Long and Sedley 1987, section 34.

50 Summers 1910, 355: "The word [sc. *lucifuga*] occurs elsewhere perhaps only in Apuleius. *Lucifugus* however is used by Lucilius, Cicero and others. In Lucil. (468M) it is evidently used as Sen. uses *lucifuga* here: in Cic. (*Fin.* 1.61) where *lucifugi* are classed between *difficiles* and *maledici* the reference is clearly to unsociableness. This explains the application of it in Rutil. Namat. 1.440 to monks, and of *lucifugax* in Minuc. Fel. 8.5 to the Christians."

one's gender and age *ut pueritia splendeat tempore alieno* (7) ("so that their youth should shine beyond its time"), and the cultivation of off-season flowers through artificial hydroponics (8). All such temporal behaviors push them toward a metaphorical antipodes, as if the night were another country. It is useful to see this as an opposite case of what Fabian described in his *Time and the Other* (1983). Whereas Fabian sees the rhetoric of anthropologists as implicitly relegating different cultures to different temporalities by denying them the "coevalness" that is an essential ground for communication, Seneca uses different temporalities as the basis for relegating his fellow Romans to a metaphorical exile. It is useful also to compare Melbin's classic essay, "The Colonization of Time" (1978), which focuses on the modern urban phenomenon of "a spreading of wakeful activity throughout the twenty-four hours of the day."[51] But for Melbin this is a shared transformation: "Whereas other terrestrial creatures are either diurnal or nocturnal, and though humans emerged from the same biosphere, we have transformed ourselves into an *incessant species.*"[52] In Seneca, the colonization is a private matter, originating in perverse socio-ethical behavior, and its outcome is a bifurcation of Rome's human population into a *vita communis* (18) ("collective life") and a *vita in contrarium circumacta* (13) ("life turned around to the opposite").

It is important to note, however, that Seneca does not withhold from his "antipodeans" the status of a community. Their perversity sets them in opposition to "collective life," yet Seneca is at pains to emphasize how *hanc vitam agere eodem tempore multos meminimus* (10) ("we recall that many lived this life at the same time").[53] He allows that there was a particular historical period during his own lifetime, evidently during the age of Tiberius, when these "antipodeans" lived a lifestyle that was nocturnal and perhaps sometimes hedonistic, but often also entirely normal, regular, and standardized. This helps us to recognize the specific emphasis of his critique: within the rhetoric of this letter, at least, he is less interested in what they did with their time (the how, *quemadmodum*) than with their activities' temporal location (the when, *quando*). All on its own, however, this was enough to exclude them from the normal operation of aristocratic and court society. In the letter's longest anecdote, concerning the nocturnal Acilius Butas, an aristocrat of praetorian rank, we learn about Butas's failure to coordinate with any of the daily routines that would have ensured his social and economic survival, including attendance at

51 Melbin 1978, 100; cf. Melbin 1987.

52 Melbin 1978, 112.

53 Cf. *Brev.* 13.

the imperial morning greeting and a day-long recitation (122.10–12). Instead, during both *salutatio* and *recitatio* he is not only absent, but he serves as the butt of jokes that mock his antipodean absence at the necessary times of day.

At the conclusion of the letter, where he reiterates the failure of the "antipodeans" to follow nature, Seneca turns to a separate paradigm from Virgil's *Georgics* that does not concern the geographical model of the antipodes but rather a periodizing account of the ages of man. The final sentence of the letter reads: *illam* [*sc. naturam*] *sequentibus omnia facilia, expedita sunt, contra illam nitentibus non alia vita est quam contra aquam remigantibus* (19) ("For those following nature all things are effortless and uncomplicated, [but] those who push against nature have a life no different than the life of those who row against water"). Summers notes here an echo of Virgil's lines about the post-Saturnian age in which human beings continually need to engage in agricultural labor:[54]

> sic omnia fatis
> in peius ruere ac retro sublapsa referri,
> non aliter quam qui adverso vix flumine lembum
> remigiis subigit, si bracchia forte remisit,
> atque illum in praeceps prono rapit alveus amni.
>
> *G.* 1.199–203

> Thus all things are fated to
> rush toward the worse and slip and are carried backward,
> no differently than someone who only just manages to propel his boat
> with oars against the river's course and, if he happens to relax his arms,
> the vessel sweeps him away on the running stream.

As Virgil announces earlier in the same book, *has leges aeternaque foedera certis / inposuit natura locis* ... (1.61–62) ("nature has imposed these laws and eternal bonds upon certain places ..."). Seneca, in alluding to this passage at the letter's end, just as he had quoted from the same Virgilian book at the letter's beginning, offers a further way for us to understand the failures of his antipodeans. Not only do they inhabit a perverse temporal location within the diurnal cycle, but they also fail to recognize the labor they need to perform in order to conform to the laws of nature in the era they inhabit.

Seneca's view that the "antipodeans" have departed from *vita communis* (122.18) ("collective life") and *naturae consuetudo* (122.9) ("natural custom")

---

54 Summers 1910, 356.

stands in subtle but significant contrast to a remark of Galen in his commentary on the *Prognostics* of Hippocrates.[55] Galen is concerned with a Hippocratic passage that spells out why sleep-patterns in keeping with natural custom are better. The lemma from Hippocrates reads as follows:

> Περὶ δὲ ὕπνων ὥσπερ καὶ κατὰ φύσιν σύνηθες ἡμῖν ἐστι, τὴν μὲν ἡμέρην ἐγρηγορέναι χρή, τὴν δὲ νύκτα καθεύδειν. εἰ δ' εἴη τοῦτο μεταβεβλημένον, κάκιον γίνεται. ἥκιστα δ' ἂν λυπέοι, εἰ κοιμῷτο τὸ πρωῒ εἰς τὸ τρίτον μέρος τῆς ἡμέρης. οἱ δ' ἀπὸ τουτέου τοῦ χρόνου ὕπνοι πονηρότεροί εἰσιν.
>
> *In Hip. prog.* III.10, 129 K

> Concerning sleeps, just as our shared custom is also according to nature, we should be awake for the day and sleep for the night. But if this should be changed, it becomes worse. But it would be least harmful if someone should sleep in the morning up to a third part of the day.[56] But sleeps beyond this time [*sc.* a third part of the day] are more burdensome.

The evaluation of sleep-quality based on time of day is here expressed in terms that have a moral resonance but also a medical significance, which already throws into relief the more exclusively moral terms in which Seneca, like Hippocrates, ranked the various degrees of oversleeping (cf. *officiosior meliorque; turpis*, *Ep.* 122.1). Galen approves of the medical analysis overall: Hippocrates ὀρθῶς ("correctly") explained good sleep pathologies in terms of their being κατὰ φύσιν ("in accordance with nature"), and bad pathologies for sleep παρὰ φύσιν ("contrary to nature"), and ditto in correlating good and bad pathologies with custom or lack of custom, since τὸ μὲν γὰρ σύνηθες ἀγαθόν, τὸ δὲ ἀσύνηθες μοχθηρόν (*In Hip. prog.* III.10) ("shared custom is good, and what lacks shared custom is wretched").

Then, however, Galen pivots:

> ἀλλ' ἐπὶ τῶν Ἱπποκράτους χρόνων οὐκ ἄλλο μὲν ἦν τὸ κατὰ φύσιν, ἄλλα δὲ τὰ ἔθη, νυνὶ δὲ ἔμπαλιν οἱ πλούσιοι δρῶσιν ἐν ἄλλοις τέ τισι καὶ κατὰ τοὺς ὕπνους, τῆς μὲν ἡμέρας κοιμώμενοι, νύκτωρ δὲ ἐγρηγορότες. ἐπὶ τούτων οὖν ὡς ἂν παρὰ φύσιν εἰθισμένων ζῆν οὐκ ἀληθής ἐστιν ἡ εἰρημένη πρὸς Ἱπποκράτους

55 The importance of this passage for ancient medical thought about sleep has been emphasized to me by both Ralph Rosen and Kim Beerden; my discussion here goes beyond Ker 2004, 220 n. 49.

56 This is more likely to mean "up to one third of the day" than to refer to a specific third part such as a third hour or the third quarter.

διδασκαλία. κυριώτερον γὰρ ἔν γε τοῖς νῦν χρόνοις ἐστὶ τὸ ἔθος τῆς φύσεως, οὐκ ἐπὶ τῶν πλουσίων γυναικῶν μόνον, ἀλλ' ἤδη καὶ ἀνδρῶν οὐκ ὀλίγων.

In Hip. prog. III.10

> In the time of Hippocrates, however, it was not the case that according-to-nature was one thing and customs were another—whereas now the wealthy act contrarily, in other things and especially as regards sleeps. So as regards these ones, as far as they are accustomed to live in whatever way would be contrary to nature, the teaching by Hippocrates is not true. For in our times now, custom has greater authority than nature, and not in the case of wealthy women only, but even indeed quite a few men.

There are several close correspondences here to what Seneca writes about his "antipodeans in Rome"—especially in the focus on a fairly numerous elite group, their intentional search for whatever is contrary to nature, and their application of this idea to sleep among various other domains of exclusivity. But in Seneca this group is presented as a fleeting and freakish subculture that stood in contrast to *naturae consuetudo* (122.9) ("natural custom") and established its own private schedule. Galen, by contrast, strikingly concedes that custom has come adrift from nature and has acquired an independent authority. For Galen, as an observer of his own times, nature is one thing and συνήθης ("shared custom") is quite another. He doesn't necessarily approve of this shared custom, since it is contrary to nature and has detrimental effects on sleep-patterns and health; he also maligns it as primarily the province of wealthy women. But he credits the shared custom with a reality of the kind Seneca continues to withhold from the nocturnal schedules of his "antipodeans."

6 Conclusion

We began by considering Seneca's examples of the various temporal policies and rules by which Romans had made time for leisure within their days. That motivated us to survey some of the ways in which Roman authors represent the moral and social implications of time, and especially of the day, including the phenomenon of the diurnal pattern, through which forms of life—diurnal selves—could be defined and presented. Seneca's letter 122 showed us an extreme problem-case, an instance in which innovative uses of time within the structure of the day could be understood as undermining certain basic norms—norms of nature and equally norms of Roman culture. Better, perhaps, we can say that the letter showed us how Seneca could disrupt the story his rivals may have aspired to tell about themselves and their nightly routines, by satirizing their pattern in a way that fundamentally undermined, or exiled, their form of life.

References

Adam, Barbara. 1990. *Time and Social Theory*. Philadelphia: Temple University Press.

Armstrong, David. 1986. "Horatius Eques et Scriba: Satires 1.6 and 2.7." *TAPA* 116: 255–288.

Aveni, Anthony. 2002. *Empires of Time. Calendars, Clocks, and Cultures*. Boulder: University Press of Colorado.

Beard, Mary. 1987. "A Complex of Times: No More Sheep on Romulus' Birthday." *Cambridge Classical Journal* 33: 1–15.

Berger, Peter, and Thomas Luckmann. 1966. *The Social Construction of Reality: A Treatise in the Sociology of Knowledge*. New York: Penguin Books.

Bergmann, Werner. 1983. "Das Problem der Zeit in der Soziologie. Ein Literaturüberblick zum Stand der 'zeitsoziologischen' Theorie und Forschung." *Kölner Zeitschrift für Soziologie und Sozialpsychologie* 35: 462–504.

Bilfinger, Gustav. 1888. *Der bürgerliche Tag. Untersuchungen über den Beginn des Kalendertags im classischen Altertum und im christlichen Mittelalter*. Stuttgart: W. Kohlhammer.

Braund, Susanna, ed. 1996. *Juvenal: Satires Book I*. Cambridge: Cambridge University Press.

Dionisotti, Anna C. 1982. "From Ausonius' Schooldays? A Schoolbook and Its Relatives." *Journal of Roman Studies* 72: 83–125.

Fantham, Elaine, trans. 2014. "Seneca, On Tranquility of Mind." In *Hardship and Happiness*, edited by Elizabeth Asmis, Shadi Bartsch, and Martha C. Nussbaum: 182–208. Chicago: University of Chicago Press.

Foucault, Michel. 1997. "Self Writing." In *Ethics: Subjectivity and Truth. Essential Works of Foucault 1954–1984*, edited by Paul Rabinow, translated by Robert Hurley, 1: 207–222. New York: New Press.

Gibson, Roy, and Ruth Morello. 2012. *Reading the Letters of Pliny the Younger: An Introduction*. Cambridge: Cambridge University Press.

Giddens, Anthony. 1984. *The Constitution of Society: Outline of the Theory of Structuration*. Cambridge: Polity Press.

Glennie, Paul, and Nigel Thrift. 2009. *Shaping the Day. A History of Timekeeping in England and Wales 1300–1800*. Oxford: Oxford University Press.

Hannah, Robert. 2009. *Time in Antiquity*. New York: Routledge.

Highmore, Ben. 2002. "Introduction: Questioning Everyday Life." In *The Everyday Life Reader*, edited by Ben Highmore, 1–34. New York: Routledge.

Inwood, Brad. 2007. *Seneca: Selected Philosophical Letters.* Oxford: Oxford University Press.

Johnson, William. 2010. *Readers and Reading Culture in the High Roman Empire: A Study of Elite Communities.* Oxford: Oxford University Press.

Jones, Alexander, ed. 2016. *Time and Cosmos in Greco-Roman Antiquity*. Princeton: Princeton University Press.

Jones, James M. 1988. "Cultural Differences in Temporal Perspectives." In *The Social Psychology of Time*, edited by Joseph E. McGrath, 21–38. Newbury Park, CA: Sage Publications.

Ker, James. 2004. "Nocturnal Writers in Imperial Rome: The Culture of Lucubratio." *Classical Philology* 99: 209–242.

Ker, James. 2010. "Nundinae: The Culture of the Roman Week." *Phoenix* 64: 360–85.

Krenkel, Werner. 1970. *Lucilius. Satiren*. 2 vols. Leiden: Brill.

Kühn, Karl G. 2001. *Claudii Galeni Opera Omnia*. 20 vols. Hildesheim: Olms.

Leach, Eleanor. 1999. "Ciceronian 'Bi-Marcus': Correspondence with M. Terentius Varro and L. Papirius Paetus in 46 B.C.E." *TAPA* 129: 139–79.

Long, Anthony A., and David Sedley. 1987. *The Hellenistic Philosophers.* Vol. I. Cambridge: Cambridge University Press.

Luckmann, Thomas. 1991. "The Constitution of Human Life in Time." In *Chronotypes: The Construction of Time*, edited by John B. Bender and David E. Wellbery, 151–166. Stanford: Stanford University Press.

Marquardt, Joachim. 1886. *Das Privatleben der Römer*. Edited by August Mau. Leipzig: S. Hirzel.

Melbin, Murray. 1978. "The Colonization of Time." In *Timing Space and Spacing Time: Human Activity and Time Geography*, edited by Tommy Carlstein, Don Parkes, and Nigel Thrift, 100–113. New York: Wiley.

Melbin, Murray. 1987. *Night as Frontier: Colonizing the World After Dark*. New York: Free Press.

Moore, Wilbert. 1963. *Man, Time, and Society*. New York: Wiley.

Moretti, Gabriella. 1994. *Gli antipodi: Avventure letterarie di un mito scientifico*. Parma: Pratiche.

O'Sullivan, Timothy. 2014. *Walking in Roman Culture*. Cambridge: Cambridge University Press.

Riggsby, Andrew M. 2003. "Pliny in Space (and Time)." *Arethusa* 36: 167–86.

Rüpke, Jörg. 2011. *The Roman Calendar from Numa to Constantine: Time, History, and the Fasti.* Translated by David Richardson. Malden, MA: Wiley-Blackwell.

Sorokin, Pitrim A. 1943. *Sociocultural Causality, Space, Time. A Study of Referential Principles of Sociology and Social Science*. Durham, NC: Duke University Press.

Summers, Walter. 1910. *Select Letters of Seneca*. London: Macmillan.

Williams, Gareth. 2003. *Seneca: De otio, de brevitate vitae*. Cambridge: Cambridge University Press.

Wittmann, Marc. 2016. *Felt Time: The Science of How We Experience Time*. Cambridge, MA: The MIT Press.

Wolkenhauer, Anja. 2011. *Sonne und Mond, Kalender und Uhr: Studien zur Darstellung und poetischen Reflexion der Zeitordnung in der römischen Literatur*. Berlin: de Gruyter.

Zerubavel, Eviatar. 1981. *Hidden Rhythms. Schedules and Calendars in Social Life*. Chicago: University of Chicago Press.

CHAPTER 7

Time, Punctuality, and Chronotopes: Concepts and Attitudes Concerning Short Time in Ancient Rome

Anja Wolkenhauer

When I first became interested in the history of time in ancient Rome 15 years ago, the initial question I had was quite a simple one, which, if I remember correctly, was prompted by a failed lunch date in the Hamburg university canteen. It was this: how did the Romans manage to be punctual for a meeting?

The Latin language, notoriously, does not have a term for "punctuality" (whereas the English language has at least four, and "German punctuality" comes close to having the status of a secondary virtue);[1] but the Romans could, at least, appear *praecox, in tempore, serius, or intempestive,*[2] so we can be sure that they had a linguistic and social concept of "being right on time." But at what point was it that they were irrevocably too late, and what did it mean to be on time in Rome? Which criteria helped the Romans to differentiate between being "punctual" and "tardy"? What made the social clock of the Romans tick? And how can "short time" be understood in this context?

1 Punctuality as a Cultural Concept

Today, the concept of punctuality marks the peak of the social regulation of time in Western culture, as well as a commitment to a common norm of time that includes and emphasizes short time periods.[3] Punctuality thus functions

1 "Timeliness," "earliness," "promptness," and "punctuality," all of which refer to the idea of being more than right on time and are often connected with ideas of courtesy and respect towards other people.

2 Cf. *ThLL* s.v. *praecox* (Breimeier 1985), s.v. *intempestivus* (Nielsen 1963).

3 Cf., in contrast, Shaw's (1994, 81) definition: "Punctual behavior is little more than synchronized behavior bolstered by a moral system." Both categories—"synchronized behavior" and a guiding "moral system"—are clearly of less importance in Classic definitions, if they are acknowledged at all. Nevertheless, at the latest since Caesar's reform of the calendar, there is a distinctly-articulated awareness in Rome that a stable norm of time belongs to the infrastructure of an empire. Cf. Wolkenhauer 2011, esp. 208–258.

 | DOI:10.1163/9789004416291_009

as an indicator of character, status, and cultural affinity. The person who is punctual or who completes a task on time shows respect for the other, engages with their cultural norms, and is understood to be polite and reliable. None of these attributions can be applied simply to the analysis of historical eras; studies that concern themselves with aspects of sociology and the history of mentalities, based on the classical texts of Norbert Elias and Jacques LeGoff, point toward great differences between cultures and epochs especially with regard to social time.[4]

Though the Roman concept of punctuality was quite different from ours, there are, in fact, many ways of speaking about punctuality in Latin. Although the Latin language offers no abstract noun meaning "punctuality" or "lateness,"[5] a range of verbal expressions are available that characterize events as being punctual:[6] being right on time may be phrased as *venire ad tempus/ in tempore*, but also in a more general way as *agere diligenter* (i.e., to act with great attention to detail, to be meticulous in every respect).[7] An action that can—or cannot—be performed in an appointed manner is the reference point for these verbal expressions. According to a social, and not further explained, convention, the right moment for this action has arrived or has already passed. This becomes quite clear in the colloquially familiar question which Cicero puts into the mouth of Laelius in his dialogue *De re publica*: *possumus audire aliquid, an serius venimus?* (Cic. *Rep.* 1.13.20) ("Can we hear something about this subject or have we come too late?")[8]

---

4 Elias ([1974–1975]1988); LeGoff ([1960] 1977).

5 The term *mora* is used widely in Classical Latin, but concentrates on the obstacle that inhibits a concrete action. The abstract noun *seritas* is probably to be found only once in late antiquity (Symm. *Epist.* 3.28: but the text is corrupt and *seritas* only a conjecture). *Maturitas* can be used to describe the moment when something arrives at its *kairos*, the right or early moment. *Exactitudo* got its temporal significance only in the Medieval and Neolatin Period. Ramminger (2009), *exactitudo*, URL: www.neulatein.de/words/3/015211.htm (last retrieved: 9.4.2018); *punctualitas* seems to establish itself even later.

6 In his famous study "Über die Zeit" ("On Time"), Elias (1988, 11–14) discusses the problem of the nominalization of the concept of time in depth and emphasizes verbal expressions to stress the connection of time and event.

7 Also *tempori, tempere, temperi* (adv.); in Medieval and Neolatin texts also *tempestiviter.*

8 Cf. Ter. *Heaut.* 2.3.123: *in tempore ad eam veni, quod rerum omniumst primum*; Cic. *Ad fam.* 13.62.1: *Et in Atili negotio te amavi* (*cum enim sero venissem, tamen honestum equitem R. beneficio tuo conservavi*). The Christian reading, in contrast, emphasizes, on the basis of the Parable of the Laborers in the Vineyard, that "too late" is a human and not a divine category: to God, the early and the late laborers are equal. Cf. Ambr. Med. *Luc.* 10.138: *veni et tu. vel sero, vel nocte, vel quacumque hora veneris, Iesum invenies ad suscipiendum paratum nec*

The concept of punctuality manifests itself in the texts as a social contract or as an appointment of the appropriate moment (*kairos*) with regard to the expected action. However, it is not only the abstract noun "punctuality" itself that is missing from Latin antiquity, but there are also no texts defining the term; comments made by Macrobius and Ambrosius, authors of late antiquity, though, come closest to such a definition.[9] The latter, in his exegesis of Luke 24.1, implies that *sero* could denote a temporal relationship as well a fixed temporal moment:

> Sero venit, id est: post praestitutum tempus advenit. licet mane sequenti die venerit, sero est tamen, ubi agendi tempus emensum est.
>
> AMBR. MED. *Luc.* 10.151

> He was late, i.e., he arrived after the prearranged time. Even though he may come in the early morning on the following day, it is nevertheless [too] late as soon as the time to act has passed.

In addition to the focus on action, here the second vital difference from the modern idea of punctuality becomes apparent: as far as I know, there is not a single piece of Roman evidence that links punctuality or lateness with a reference to an exact time, as would be commonplace today. Our concept of punctuality naturally presupposes a large number of clocks and watches keeping the exact same time on a detailed scale. This cultural prerequisite is the only reason why it is possible in the first place to be "five minutes late." In the Latin texts of antiquity, however, which mention actions that are deemed as (un) punctual, the time norm is based on the (im)possibility of an action, but not expressed through any reference to the clock: one could be too late to hear a famous orator in Rome, but one could never be five minutes late. On this basis, punctuality only gradually seems to have acquired a moral quality in the antique texts; time's role in forming social hierarchies seems not to have been reflected upon.[10]

---

minorem serius venientibus mercedem solventem; nam et qui sexta hora venit, non est mercede fraudatus et qui undecima plenitudinem mercedis accepit.

9 Macrob. *Sat.* 6.8.7–13 offers an in-depth discussion on the determinations of timeliness, esp. on *maturare*; the term punctuality, however, does not feature in this discussion.

10 In order to assess this more accurately, related cultural concepts such as waiting also would have to be researched (see the current project of Puff [forthcoming]). For the moral evaluation of punctuality, see below.

At this point, further research into the conceptual and terminological history of punctuality would be necessary. Research on punctuality and dealing with short periods of time has mostly been done in the areas of sociology, psychology, and medicine, and is thus highly focused on the present.[11] The added difficulties of a historical analysis of human dealings with "time" are the reason why the question of punctuality has been raised but rarely in the research of pre-modern cultures, which focuses on discourse analysis and the history of mentalities. Nevertheless, a range of relevant studies are available that discuss the methods and instruments of timekeeping,[12] chronology (especially the calendar)[13] and the social discourses that are connected to them.[14]

This book's starting point is the term "short time." It raises the question of what "short time" means from a Roman viewpoint and how its length and precise determination are dependent on certain places and social contexts. In the following, the phenomenon of "short time" is analyzed in two relevant ways that differ from each other as to methodology: first, the history of timekeeping in Rome, according to Pliny the Elder, is reinterpreted in a close reading with a special focus on the idea of exactitude in measuring time. Second, the Roman "chronotopes" (i.e., fixed cultural connections of certain places with certain temporal structures) of the *forum*, *urbs*, and *villa* are highlighted.[15] Their exemplary analysis makes clear that relevance, exactitude, and length of "short time" are cultural as well as scientific concepts that vary in changing contexts.

11 Based on Durkheim, Zerubavel (1982) here coins the term "sociotemporal order," which understands the differentiation and standardization of temporal orders (i.e., the fixation of points in time, duration, chronology, and repetition) to be a consequence of a rising cultural complexity. In more recent in-depth studies, Levine (1997 and 2005), for example, has researched the periods of time that are socially acceptable for the completion of certain tasks (i.e., those time periods that precede the "too late") in a culturally comprehensive fashion. Clayman (1989) has analyzed the observance of the "punctual ending" on social occasions of communication.

12 Gibbs 1976 (groundbreaking); Winter 2013; Talbert 2017.

13 Rüpke 1995; Hannah 2005; Feeney 2008; Bonnin 2015.

14 Wolkenhauer 2011.

15 This term originates from a famous study by Michail Bachtin, who used it as an analytic term to specify the intertextual relationship between space and time as an indicator of literary genre. We use it in a slightly different manner as explained above (Bachtin 1975; cited according to the German edition 2014).

2 What Did "Short Time" Mean in Ancient Rome?

To qualify time as "short" implies that it is somehow measurable and can be contrasted with a notion of "long" time. The terms "short" and "long" presuppose an equivalency between space and time, as is familiar to us at the latest since the Aristotelian theory of time. In Latin, as well as in Greek, the terminology of space is often applied to time: a *spatium breve* may be a short distance either in time or in space, and the Greek σταθμός equates time and space in the idea of a day's journey.

If, however, "short time" is basically understood as measurable, then its more precise definition is closely connected to the history of time measurement: we give names to what we think we are measuring (e.g., temporal units), and we develop our ability to fix accurately in time those things we desire to regulate. The histories of timekeeping terminology, technology, and mentalities are closely connected. Thanks to the report that has been handed down to us in Pliny's *Naturalis Historia*, the history of daytime measurement in Rome becomes transparent for us from about 450 BCE onwards (Table 7.1).[16] Initially, there was a crier daily calling out the apparent progress of the Sun according to its passing behind four central buildings at the *forum Romanum*.[17] About 200 years later, a sundial was erected which probably marked 12 fixed points within the diurnal day.[18] A further 100 years passed before an improved central sundial was installed, which continued to offer 12 points, though more precisely fixed.

Some years later, about 160 BCE, a water clock was set up in a central administration building of the city. During this period, the number of marked units of time within one calendar day rose from two to 24, whereas the length

16 Plin. *HN* 7.212–215; see the in-depth study in Wolkenhauer 2011, 70–93.

17 For this kind of measurement, see Nilsson 1920, esp. 17–19.

18 The most important argument for the twelvefold division of the Roman clock is the conventional partition of the day into twelve stages, which was commonplace in the Eastern Mediterranean, the area from which many clocks came to Rome. Linguistic usage, too, offers evidence—for example, a remark made by Censorinus, who (though late and only in hindsight) closely connects the usage of sundials and the twelvefold partition of the day: *in horas XII diem divisum esse noctemque in totidem vulgo notum est; sed hoc credo Romae post reperta solaria observatum* […] *horarum nomen non minus annos trecentos Romae ignoratum credibile est.* (Cens. 23.6 and 8) ("That the day is divided into 12 hours, and the night into as many, is generally known; but this was only observed in Rome, I do believe, after the introduction of the sundial […] It is probable that the term 'hour' remained unknown in Rome for about 300 years.") The 300 years probably refer to the time period between the foundation of the city and The Law of The Twelve Tables; there is no plausible reason for Michel's (1970, 119) approach of referring the time designation to the timespan stretching between Censorinus and Augustus.

TABLE 7.1 History of time measurement in Rome according to Pliny, *Naturalis Historia* 7, 212–215

Time	Place	Number of fixed moments during the day[a]
Around 450 BC	Law of The Twelve Tables, observation of sunrise and sunset, displayed on the *rostra* at the *forum*	2, later 4
Since 338 BCE	Crier, standing beneath the *Curia Hostilia* at the *Comitium*, calling out the apparent progress of the Sun according to its passing behind four central buildings	4
Since 295 BCE	Sundial erected between the *spolia hostium* in front of the temple of *Quirinus* on *Collis Quirinalis*	max. 12
Since 262 BCE	Sundial with *lineae horarum*, erected on a column near the *rostra* at the *forum*; probably assisted by a crier	12
Since 164 BCE	Sundial with *lineae horarum*, erected right next to the earlier one; probably assisted by a crier	12
Since 159 BCE	Water clock set up inside the *Basilica Aemilia et Fulvia*, which was a central administration building at the *forum* and residence of the *praetor urbanus*	probably 24

a Not every sundial displays hour-lines, and if they do, then their number is not always eleven. The manner of marking the hours, to my knowledge, is not discussed in the texts of antiquity and, hence, we must rely on material evidence. Winter (2013, 160–61) offers an exemplary overview.

of the marked units of time was reduced from more or less 12 hours to one. In other words, the shortest time period marked according to the clock and other timekeeping devices is reduced during these first three centuries of instrumental timekeeping in Rome from 12 hours to one. From Pliny's point of view, an endpoint of technical development had been reached two hundred years before his lifetime, and the temporal penetration of day and night had reached a degree that still applied in his day.

In Rome, the 24 hours of the day were fixed upon around 160 BCE. Soon after, the numerical numbering of the hours (*hora prima, secunda, tertia*. etc.) must have been established. It is only this convention that introduces the counting of the hours as we understand it today into Roman culture. In superimposing a uniform grid over the entire day, it enables us to analyze the fine structuring of the day more closely. This step in the development of the order of time, however, is neither mentioned by Pliny the Elder, nor by any other author, but seems to belong to the givens that no one explicitly justifies. The differentiation of the hours is foreshadowed in archaic Roman literature, which distinguishes between winter and summer hours.[19] From about 70 BCE onwards, there is a lot of evidence for numbered hours occurring almost simultaneously in Varro, Cicero, Caesar, and Catullus.

The briefest unit of time that becomes accessible in this way is the seasonal Roman hour with its changing length from ca. 45 to 75 minutes, which is connected to certain moments of the day via the number it is given. From now on, *hora* is used as a short, though variable, measure unit, which is only given an exact length when the date is added. The numbering of the hours is principally used for the day as well as the night (*hora secunda noctis, hora secunda diei*). Its usage, however, varies greatly in different chronotopes (as, e.g., the *forum*, where it gains great importance).[20] Usually, time is not quantified using any unit smaller than the hour, and this remains valid for most Europeans until early modern times.[21] Shorter time determinations do not appear in narrative texts of Roman antiquity,[22] and units of time shorter than the half-hour were not marked on sundials; as a rule, *clepsydrae* alone were able to measure shorter time periods. Thus, the hour (*hora*) appears to be the smallest established and generally used time unit in ancient Rome.

---

19 A Plautine joke thrives on *hora* describing a definite and simultaneously variable size: the Athenian Simo accuses his slave Pseudolus of gulping down large quantities of wine in just one hour. Pseudolus answers: "*hiberna*" *addito*. (Plaut. *Pseud.* 1304) ("Even a short winter hour is enough time for me to do that!"); Wolkenhauer 2011, 141.

20 See p. 225ff. below for the chronotope *forum*. We have just started to conduct new research on the history of the numbered hours, but nevertheless here are some first observations. The term *hora* is missing entirely, as Censorinus already noticed, from The Law of the Twelve Tables. Later, it is successively introduced into annalistic literature, in which the four-hour rhythm of the watch still shows through distinctly. At the latest since Cicero the numbered hours were used for all moments of the diurnal day, though the first half of the day predominates. The hours of the afternoon and, most of all, of the night clearly fade into the background behind it.

21 For a general overview, see Dohrn-van Rossum 1992.

22 They appear only (rarely) in scientific texts like Manil. 3.258–261; Plin. *HN* 6.39.220; cf. Lucil. at Gell. 3.14.

3 A Desire for More Accurate Timekeeping? Rereading Pliny's Sundials

The measuring of short periods of time, the striving for an exact measuring of time, and the establishment of punctuality as a cultural virtue are, irrespective of time and place, closely interrelated. Thus, it makes sense to look at Pliny's history of the clock more closely with this aspect in mind. While speaking about the transition from the first to the second public sundial in the *forum Romanum*, Pliny says:[23]

> M. Varro primum statutum in publico secundum rostra in columna tradit bello Punico primo a M'. Valerio Messala cos. Catina capta in Sicilia, deportatum inde post XXX annos quam de Papiriano horologio traditur, anno urbis CCCCLXXXX. nec congruebant ad horas eius lineae, paruerunt tamen ei annis undecentum, donec Q. Marcius Philippus, qui cum L. Paulo fuit censor, diligentius ordinatum iuxta posuit, idque munus inter censoria opera gratissima acceptum est.
>
> PLIN. *HN* 7.214

> Marcus Varro reports that the first public clock was placed on a pillar next to the speaker's platform, erected during the First Punic War by the consul Manlius Valerius Messala, after it had been brought to Rome from Catania in Sicily because of the capturing of that town. This happened 30 years after the recorded arrival of the Papirian clock in the year 490 of the City of Rome. The hour-lines of this sundial did not correspond to the hours of the city of Rome; nevertheless, they went by it for 99 years, until Q. Marcius Philippus, who was censor together with L. Paullus, placed a more carefully constructed sundial directly next to it. Among all the acts of their time in office, this has been received as the most welcome gift.

Accordingly, two timekeeping devices were set up in the Roman *forum* within a hundred years. Both stood directly next to each other, one of them positioned on top of a pillar, so that, despite the *forum*'s forest of statues, they were still noticeable enough to give a part of this place its name, *ad solarium*.[24] The first sundial was a piece of loot from Catania, Sicily; the second was made a century later specifically for Rome on behalf of the censors in office. The expression Pliny uses, *diligentius ordinatum*, actually has two meanings—namely, that the

23 See Wolkenhauer 2011, 86–91.

24 Auct. *Ad Her.* 4.14.5; Cic. *Quinct.* 18.59.

new sundial has been both commissioned and set up by the Romans in a way that fits better for Rome.[25]

Because the next public timekeeping device is also associated with the name of a censor, it can be assumed that the censors of this period were assigned the task of inner-city timekeeping analogously to their other infrastructural duties. Pliny assumes that the reason for setting up another sundial directly next to (*iuxta*) the first one can be found in the Romans having noticed (how?) the inaccuracy of the older sundial, which was designed for the latitude of Catania. Therefore, they wanted to make clear that they had the necessary technical knowledge to design a sundial that was aligned to the geographical position of Rome.

Two different points of reference manifest themselves here. The first is the long-familiar competition with the Greek model. Pliny connects his historical lecture with this *topos* via *imitatio* and *aemulatio*. The first sundials, like many other cultural goods, came to Rome from the Eastern Mediterranean. Here, they were studied, imitated, and finally surpassed, as Cicero registers in the memorable phrase *accepta fecerunt meliora*.[26] The second point of reference is the articulated need for supraindividually marked accuracy, e.g. by using a time-keeping instrument. This need was prompted by a surprisingly marginal detail, because the maximal difference of the length of day between Catania and Rome is just under half an hour. Even though there is not enough evidence, one can safely assume in any case that the instances in which the shadow did not reach the terminal lines were only noticeable on a few days a year and only under close observation.[27] The difference would have had little to no effect on social life, especially as long as there were no competing sundials in Rome.[28]

---

25 For the tecnical use, see Keudel (1978), especially 939,21–27, and 940, 27–37 (without mentioning Pliny). If Pliny had wanted to highlight just one of these aspects, he could have chosen a more common expression like *creatum* or, on the contrary, *erectum*. Alexander Jones (above, pp. 137–138) seeks to relate *ordinatum* to the execution, rather than to the placement, of the sundial, but I do not find further arguments for this decision.

26 Cic. *Tusc.* 1.1: *meum semper iudicium fuit omnia nostros aut invenisse per se sapientius quam Graecos aut accepta ab illis fecisse meliora, quae quidem digna statuissent, in quibus elaborarent*. For the topos and its meaning in Cicero, see Zimmermann 1999, esp. 243f.

27 On the accuracy of the Catania dial within Rome, see Jones's contribution to this volume.

28 It is important to bear in mind that this refers only to the chronotope of the *forum*. For scientific purposes, the answer might be different. As Alexander Jones clearly shows in this volume (p. 135), sundials could be used to mark smaller units of time (down to a quarter of an hour) by good observation, and Houston (1989) convincingly argues that science without a measurable short time unit is partly impossible and wholly different from modern science. For a long time, however, Rome made no efforts in this direction. Time seemed then to be mostly a social, not a scientific concept.

As the position of the sundial in the *forum* gives rise to the suggestion that it was primarily important in forensic contexts, it is there that we will look for the specific occasions which created the need to fix time more precisely in this place. But this research yields little: ever since the introduction of the Twelve Tables, the schedule of jurisdiction and political action is marked by the first and the last hour of the day, while *clepsydrae* mark time for the respective trials.

However, the time specifications necessary for this process—i.e., sunrise and sunset—are made independently of the clock, through observation of the sky. Further relevant moments of the day are "read" from the shadows made by the buildings around the *forum*, which, naturally, always depict "local time." Hence, the deviations of the old sundial were temporally insignificant and, as far as we know, of no consequence for judicial processes. The usage of the timekeeping device offers no clue as to why a more precise device should have been necessary.

Nevertheless, Pliny insists on this point and emphasizes it especially. I thus suggest we understand his statement—namely, that the new sundial would have been more carefully constructed, *diligentius ordinatum*, and would have been most welcome to the citizens of Rome, *inter opera gratissima*—less as marking some advancement in social life (as research has always done until now),[29] but rather as a strong indication of an important change in the history of mentalities. The setting up of *both* sundials jointly speaks for this: if it was only more precise timekeeping that was of interest, the old sundial would likely have been demolished, and the new one would have taken its place. But as no serious shortcomings were being corrected, the responsible censors apparently found it necessary to erect the new sundial right next to the old one, in order to visually display progress. Therefore, I propose taking the second sundial as a strong indication of the impalpable need for temporal accuracy. It demonstrates the desire for a kind of punctuality that does not exist, but which is imaginable: the juxtaposition of old and new clock, resembling an experimental setup, mostly aims at displaying the difference between the two. The second sundial is a tangible metaphor of progress. This step was not

29 Even the most recent monographs on timekeeping in antiquity adopt Pliny's judgment without challenging it and do not question the *telos* of the argumentation. Bonnin (2015, 63–65) conclusively presents in his account how irrelevant the deviation of the Catanian clock was to everyday life, and he points to the fact that the Romans dealt with much larger differences (up to a month) in their calendar calculations. Talbert (2017, 141–146), in his research on very small portable sundials, asserts that their symbolic value was esteemed more highly than their temporal exactitude.

directed at isolated administrative areas within the Roman state, but at all citizens who visited the *forum* and its central sundial during the course of a day.

By about 160 BCE, the Romans were capable of noticing when a sundial deviates by around half an hour, even if the sources remain silent about how the Romans noticed and communicated this difference. This intervention by the censors sends a public message that this temporal inaccuracy was socially uncomfortable and in need of change. The exactitude of the timekeeping here first becomes recognizable as a culturally-relevant factor in Rome. Only as an aside and, as it were, as a confirmation, it should be noted that 150 years later emperor Augustus repeated this setup on a very similar occasion: in order to highlight his correction of the Julian calendar, the *horologium Augusti* was built towering over the entire Campus Martius. Its gigantic size compensated for the rather small problem it removed, as it celebrated the elimination of a measuring inaccuracy amounting to less than a minute per day, which is not very much if you remember—as Augustus and his contemporaries surely did—that, a generation earlier, Julius Caesar had corrected three months and abolished the leap month.[30]

The erection of the second sundial implies that a new unit is marked on the mental scale: the half-hour, *hora dimidia(ta)* or *semihora*[31]—i.e., precisely the time interval by which the initial sundial was off and to which the Roman time was adjusted. But even though this may be important within the history of mentalities, the half hour is not reflected in literature. Differently put, we cannot confirm the relevance of the "half hour" in daily life.[32] Whether this unit of time really was irrelevant, or whether its chronotope is simply situated outside of the texts available to us, is a question that cannot actually be solved.[33] I will, nevertheless, pursue the concept of "short time" a little bit further and will try to narrow it down via two related terms and the study of different chronotopes.

30 Wolkenhauer 2011, 208–258.

31 Gell. 3.14 mentions, among other things, the half-hour in his differentiation of *dimidius* and *dimidiatus*. He here refers to Lucilius 571–73 Krenkel, who uses *dimidium horae* in the medical context of malaria. On short time in Galenic medicine, see Miller's contribution to this volume.

32 In his speech *Pro Rabirio*, Cicero repeatedly refers to his limited speaking time of just half an hour. This, however, is clearly due to the chronotope of the *forum*; for more, see below. All other (overall, rare) evidence originates from technical texts and didactic poetry. Cic. *Rab. perd.* 2.6 and 3.9.

33 The entirely unmarked usage of the term in a text in Petron 69.4, which is strongly characterized by colloquial speech, could point at the use in daily life (*lucernam de sinu fictilem protulit et amplius semihora tubicines imitatus est*).

4 Chronotopes Related to Roman Literature

The previous statements apply merely to two timekeeping devices, located in one place and described in one text. This is not a very broad foundation for what is to follow. In order to widen the scope, the above-mentioned term chronotope will serve as a methodological guide. Other disciplines have formulated similar categories in their research—for example the culturally specific "social time," which captures larger temporal fields, but does not offer historical depth, or the term "multitude of time" in analogy with "multitude of language."[34] For the study of time in antiquity, which requires methods that are characterized by discourse analysis and take into account both texts and artifacts, experimental procedures like those of Robert Levine are not applicable. Nevertheless, the issues raised by other "modern" disciplines, as well as their terminology, are indeed stimulating and helpful.[35]

This study therefore defines the term chronotope, with reference to recent time research from a broader cultural studies perspective, as a "fixed cultural connection between certain places and certain temporal structures, which is expressed mostly in texts." Objects and cultural practices additionally supplement the texts. We may say that the chronotope represents the whole of handed-down time-referent information connected to a certain place and conveyed via different methods of transmission.

Knowing the shape of the chronotope means understanding the social rules which were valid at a certain moment of place and time. The term helps make the point that for each place, each era, and each context a specific way of regulating time applies, which defines and orients the respective chronotope. With this term, we can capture spaces of very different sizes, which relate to each other as partial quantities or intersections, in each case shaped by what is technically possible, by the place, by its function, and by the social norms connected to this function. To illustrate this more clearly with an example: at the heart of the lunch-date problem, referred to at the beginning, was the question of whether the time set for the appointment was to be understood academically[36]

34 For his groundbreaking global study about the geography of time, Levine (1997) uses a quite similar approach, but a different metaphor in the title of the Preface, noting that "time … is talking with an accent." He here refers, among other things, to the study of Hall (1984), who also labeled the social encoding of time with the linguistic metaphor "silent language."

35 Thus, e.g., the study of Polinskaya (2018), which discusses the possibilities and limitations of a "Kartographie der Zeit" in the literature of antiquity.

36 In most German and some other European universities, lectures do not start at the time set (usually the full hour), but generally a quarter of an hour later.

(i.e., *cum tempore*) or, according to the conventions outside of university, *sine tempore*, and how long one has to wait until one is justified in becoming angry and leaving. It was thus primarily an issue of social conventions. The mere existence and the sufficient exactness of our watches was not at issue.

The chronotopes of antiquity have also been the subject of other studies using alternate terminology.[37] In what follows, two central Roman chronotopes will be sketched by way of an example: the *forum*, which has been focused on just now as the first place of central time measuring in Rome, and the Roman *villa*, which offers an alternative to city life, both as a place and as a literary *topos*.

4.1 *The Forum*

The *forum* was the political center of the city of Rome. The sundials were set up in the *forum*, in the very center of the city, in the exact same spot in which the fixed moments of the day had been proclaimed since the earliest times. Here was the seat of the senate, here judgment was spoken and trade was conducted, but most of all, this place symbolized the heart of the state.[38] That is why the things that happened here in the course of the day had to be fixed precisely in time, and this not merely due to, but precisely *because* the greatest power as well as the highest level of regulation realized itself in this place. As the table above shows, the place in which Roman time was fixed remained stable across all historical changes and ways of measuring time. Only the number of fixed moments of time between sunrise and sunset rose considerably. This is a clear indicator for a heightened structuring of the diurnal day. A similar kind of logic brought forth turret clocks in the Medieval Period, which were not "accurate" in the modern sense, but which controlled the regulation of time in church parishes. Hence, according to my thesis, the second sundial's established accuracy is primarily symbolic in nature: it is a symbol of progress and temporal exactitude.

For a single area of forensic life, temporal determinations have been handed down to us that come in under the aforementioned threshold of half an hour. They are not dependent on the central sundial and are not concerned with keeping continual time, but with recording the short time periods of judicial speeches with the help of timers (i.e., outflow clocks; *ampullae, clepsydrae*).[39]

---

37 See n. 46.

38 Winter (2013 1, 239–249) summarizes the current state of research on the presence of clocks in Rome in an illustrative way.

39 Limitations of speaking time are common in the Roman judicial system and are mentioned repeatedly in literature. The content-related reasons for these limitations are, on the one hand, to reduce the duration of the lawsuit adequate to the occasion, and, on the

The formal procedure of the law court was made up of the speeches of prosecution and defense and the examination of the witnesses, which relate to each other in a precisely defined temporal sequence. Only in this context are units of time shorter than the hour regularly mentioned. The procedure is very well documented in Greek literature, especially in the comedies; in Latin literature, there are noticeably fewer references to it and they are restricted to forensic texts of the early Imperial Period, namely those by Pliny the Younger, Martial, and Tacitus.[40]

The Greek "measure" of speech made use of the little jug χοῦς as its smallest unit, with an outflow time of about three to six minutes.[41] The younger Pliny complains about half-*clepsydrae* (i.e., around 8 minutes) as the smallest time unit in judicial speech.[42] In his groundbreaking study, James Ker has stressed that only one single case is known for Rome in which the speaking time given in hours correlates with the allotted *clepsydrae*. This rare synchronism offers the chance to calculate an absolute value for the *clepsydra*-time (and at the same time leads us to wonder why the synchronism is to be found only once).[43] In one of his letters, Pliny the Younger describes an important judicial speech that he gave before the emperor and the senators, and for which he records the time allotted to him according to both scales:

> Utcumque tamen animum cogitationemque collegi, coepi dicere non minore audientium assensu quam sollicitudine mea. Dixi horis paene quinque; nam duodecim clepsydris, quas spatiosissimas acceperam, sunt additae quattuor. Adeo illa ipsa, quae dura et adversa dicturo videbantur, secunda dicenti fuerunt. Caesar quidem tantum mihi studium, tantam etiam curam—nimium est enim dicere sollicitudinem—praestitit,

---

other, to distribute the speaking time fairly between prosecutor and defense. Examples are given in Cic. *Rab. Perd.* 6 and 9; Tac. *Dial.* 38.1. It also becomes clear in Cicero's forensic speeches that the numbering of the hours expands into the night. They prove that it was conceivable for a private person to operate a water clock in the first century BCE and use it to read the exact time down to the hour, and that this could lead to judicially relevant evidence. For more, see Wolkenhauer 2011, 117–119; the Greek practice has been thoroughly researched by Winter (2013, 147–153).

40 Riggsby (2009) and Ker (2009) have analyzed in detail the conflict of authority between judge and clock.

41 Winter (2013, 148–149): three to four minutes; Hannah (2013): six minutes.

42 Plin. *Ep.* 6.2.5: *Nam, postquam obiit ille, increbruit passim et invaluit consuetudo binas vel singulas clepsydras, interdum etiam dimidias et dandi et petendi. Nam et qui dicunt, egisse malunt quam agere, et qui audiunt, finire quam iudicare*; see Riggsby 2009, 273f.

43 For synchronisms as instruments of historical time research, see Wolkenhauer 2016–2017, 178–181.

> ut libertum meum post me stantem saepius admoneret voci laterique consulerem, cum me vehementius putaret intendi, quam gracilitas mea perpeti posset.
>
> PLIN. *Ep.* 2.11.14–16

> When I had gathered together courage and thoughts, I started to speak—with as much assent among the listeners as private doubts. I talked for nearly five hours, because to the 12 *clepsydrae* I received, which were most capacious,[44] four more were added. Thus, precisely those things that seemed to me adverse before the speech, then, when I spoke, turned out to be in my favor. The emperor showed me so much affection and care—concern would go too far—that he repeatedly cautioned my freedman, who stood behind me, that I should look after my voice and my lungs, as he believed that I exerted myself more than my feeble constitution could endure.

Since the trial, as the letter records, took place in January, and we can assume that Pliny's addressee correlates the date of the letter with the length of the hours in the usual way, we can translate the temporal determination: the length of a Roman hour amounts to a maximum of 50 minutes in January, and thus we arrive at nearly 250 of today's minutes for the length of the entire speech (i.e., an out-flow time of about 15 minutes *pro clepsydra*). This still leaves many questions open: first of all, the volume of the *clepsydra* can vary between different trials, as long as the same unchanging measure is used within a single trial. This can also explain the specification *spatiosissimas*—i.e., in these cases, the *clepsydrae* were bigger than usual and could hold more water. The double time specification of hours and *clepsydrae* could also indicate the fact that the conversion of the *clepsydra* time to "normal hours" did not happen very often and is therefore not as self-evident as we might think, and that Pliny here wants to bridge a gap for his readers in order to make them aware of his speech's importance. We could understand this, then, as a reference to the narrow limits of the chronotope of rhetoric and, simultaneously, as a hint at the fact that, in certain chronotopes, temporal exactitude was given a special value.[45] Allotting

---

44 It means that 16 *clepsydrae*, which usually amount to a smaller figure, eventually amounted to nearly five hours. This must be at least one of the reasons why Pliny here mentions both time indications.

45 Cf. also Cicero, who pointedly characterized the forensic speech as *clepsydram latrare vel dicere,* ("speaking after the clock") (*De or.* 3.138; *Tusc.* 2.67). It remains a mystery to me why there are so few *clepsydrae* mentioned in literature, even though rhetoric represents such a well-recorded segment of it.

and keeping exact speaking times becomes a proof of professional competence and rhetorical skill, and thus must be especially rehearsed.[46] Suetonius's remark that Augustus dated his letters not only according to the day but also to the hour points in a similar direction.[47]

So far, we have seen that regular "short times" of an hour and, in specific chronotopes, of half an hour, were established. The half- and the quarter-hour as small units of time are specific to the chronotope of the court of law and judicial speech. But other social spaces—such as medical practice, religious cults, or administration (especially in the departments of water and postal service)—have developed very specific chronotopes that are worth considering.[48] Each of these chronotopes has its own time and therefore probably its own ideas of short time and of punctuality. Perhaps even smaller units, such as *quadrans horae*, or qualitative determinations of short time (e.g., *punctum*, *scrupulus*, or *momentum temporis*) are significant in these cases.[49] The less a chronotope is oriented towards an event and the more it is influenced by the clock, the more likely it is that punctuality plays a role there, both as a term and as a cultural concept. The expansion of timekeeping in Rome, which can be tracked with Pliny's help during the time until 160 BCE for which few texts survive, can be followed during the subsequent centuries on a broader textual basis, with respect both to the history of the term and to discourse analysis of its qualitative development.

4.2 *Urbs and Villa*

Beyond the *forum*, which we have encountered as the central place for the social regulation of short time, there are other chronotopes connected to

---

46 Quint. 11.1.46–48.

47 Suet. *Aug.* 50.1: *Ad epistulas omnis horarum quoque momenta nec diei modo sed et noctis, quibus datae significarentur, addebat.* Suetonius does not comment further on this exactitude, but the fact that Augustus customarily used the exact date and time in his letters, is characteristic of Augustus. When he later became a model for a "good emperor," this small detail probably became exemplary as well.

48 Michel (1970) discusses the difference between mathematically exact time specifications and behavior in everyday situations. Remijsen (2007, 129) analyzes the conventions of time within the postal service and refers to the focus on the concept of the hour in religious and academic contexts. In the epitaphs, Luciani (2009) has impressively analyzed a chronotope close to everyday life that attaches great importance to exact dating. The term chronotope is implied by all three. On the value attached to temporal precision in astrological texts, see Heilen's contribution to this volume.

49 The data situation is generally bad. *Punctum temporis* is very often given as "the shortest moment" in Lucretius, or "the moment right now," especially in Cicero; *momentum temporis* is given the aspect of "the moment in which something important starts to happen," and is thus characterized by event time rather than clock-time. It is often used by Livy.

other timekeeping devices. Each of them is characterized by a specific mode of regulating time. City and country, *urbs* and *villa*, are two closely interrelated chronotopes that are often referred to in literature. First, the chronotope of the *urbs* shall briefly be described on the basis of Martial's famous "hour epigram":

> Prima salutantes atque altera conterit hora,
> exercet raucos tertia causidicos,
> in quintam varios extendit Roma labores,
> sexta quies lassis, septima finis erit,
> sufficit in nonam nitidis octava palaestris,
> imperat extructos frangere nona toros:
> hora libellorum decuma est, Eupheme, meorum,
> temperat ambrosias cum tua cura dapes
> et bonus aetherio laxatur nectare Caesar
> ingentique tenet pocula parca manu.
> tunc admitte iocos: gressu timet ire licenti
> ad matutinum nostra Thalia Iovem.
>
> MART. 4.8

> The first and the second hour of the day fatigue the greeting clients. The third exercises the booming voices of the lawyers. Until the [end of the] fifth, Rome pursues various activities. The sixth offers rest to the exhausted; the seventh will end this rest. The eighth serves for sport, shining with sweat, until the [end of the] ninth; the ninth commands us to drop onto the set-up couches. The hour of my little books is the tenth, Euphemos, when your benevolence offers divine dishes, and the good emperor relaxes with celestial nectar and holds a modestly measured cup in his powerful hand. Then let the jocular poetry come to him: because my Thalia dreads to approach morning Jupiter with her casual step.

There is a rich corpus of literature on this text, from which we, however, here only choose a small selection, i.e. the literature focusing on structuring temporal order.[50] The epigram's framework is provided by the numerically-counted hours of the day, which were shown as an important and useful method of dividing time according to imperial standards. Specific tasks and places are assigned to each hour: during the first and second hour of the day, the morning

50 The text is based on Shackleton Bailey 1990. I have tried a first interpretation of its time structure, reviewing the literature, in Wolkenhauer 2011, 109–14; for an in-depth analysis of this epigram, see Moreno Soldevila 2006, 138–148.

salutatio takes place in the house of the *patronus*; during the third, the law court proceedings start in the *forum*; the two following hours are general working time, without any specification. The sixth and the seventh hours of the day are intended for midday rest, which is generally spent at home. Hours 8 and 9 are again dedicated to the public sphere and the *palaestra*, 9 and 10 to the banquet in the late afternoon (which, in this special case, takes place in the palace of the emperor).

As precise and tight as this schedule may seem, it still exhibits flexibility when confronted with the question of the smallest unit of time. The hour specifications mostly refer to pairs of hours, or at least to periods of time that reach beyond the hour. The course of the day is thus marked by time specifications as much as by the connection with specific places and actions; put differently, the Roman hour seems to hold a middle position between event time and clock time.

If we should want to reduce the epigram to a general and even normative statement, it would sound somewhat like the following: at this or that time of the day, a Roman can be found in this or that place: in the early morning, he is at home for the *salutatio*; then he is in the *forum*; noon is spent at home; afterwards, he can be found in the *palaestra*; in the late afternoon, he is visiting; and he arrives back home before sundown. It is the socially-fixed connection of time and place that makes it possible to meet this person. This connection is normative; i.e., there is a right and a wrong moment for an action: one does not bathe in the morning, and one does not eat out late into the night.[51] Simultaneously, however, this norm is not connected with punctuality according to the clock, but rather determines a chronological sequence of events.

Many parallels reveal that Martial's epigram does not construct something extraordinary but, indeed, reflects how the Roman upper class regulated time socially. Here, time is always connected to a specific place, while the units of this social clock, in turn, seem to exceed the limit of the unit (i.e. the hour) that the sundial in the *forum* measured. Martial does not offer us novel precision, but he does depict a method of interpersonal coordination that is less familiar to us today: social conventions, linked to specific spaces, can (to a degree) compensate for a lack of temporal accuracy.

For the members of the Roman upper class, who are the authors and points of reference for our texts, life in the *urbs Roma* alternates with life in the *villa suburbana*. Pliny the Younger, who in a series of letters presented his own country life and that of his friends as an example (*exemplum*), highlights how the strict schedule of the city can be broken when in the country. An excerpt

51 Evidence in Stein-Hölkeskamp 2005, 112–116.

from letter 9.36 describes the course of a summer day and points out how the country chronotope can be modified in comparison to the urban one, which is the measure of all things:

> Quaeris, quemadmodum in Tuscis diem aestate disponam. Evigilo cum libuit, plerumque circa horam primam, saepe ante, tardius raro. Clausae fenestrae manent; mire enim silentio et tenebris ab iis quae avocant abductus et liber et mihi relictus, non oculos animo sed animum oculis sequor, qui eadem quae mens vident, quotiens non vident alia. Cogito, si quid in manibus, cogito ad verbum scribenti emendantique similis, nunc pauciora nunc plura, ut vel difficile vel facile componi tenerive potuerunt. Notarium voco et die admisso quae formaveram dicto; abit rursusque revocatur rursusque dimittitur. Ubi hora quarta vel quinta—neque enim certum dimensumque tempus—ut dies suasit, in xystum me vel cryptoporticum confero, reliqua meditor et dicto. Vehiculum ascendo. Ibi quoque idem quod ambulans aut iacens; durat intentio mutatione ipsa refecta. Paulum redormio, dein ambulo, mox orationem Graecam Latinamve clare et intente non tam vocis causa quam stomachi lego; pariter tamen et illa firmatur. Iterum ambulo ungor exerceor lavor. Cenanti mihi, si cum uxore vel paucis, liber legitur; post cenam comoedia aut lyristes; mox cum meis ambulo, quorum in numero sunt eruditi. Ita variis sermonibus vespera extenditur, et quamquam longissimus dies bene conditur. Non numquam ex hoc ordine aliqua mutantur; nam, si diu iacui vel ambulavi, post somnum demum lectionemque non vehiculo sed, quod brevius quia velocius, equo gestor. Interveniunt amici ex proximis oppidis, partemque diei ad se trahunt interdumque lasso mihi opportuna interpellatione subveniunt.
>
> PLIN. *Ep.* 9.36.1–5

> You ask me how I divide the day in my Tuscan country seat. I wake up whenever I want to, mostly about the first hour, often earlier, rarely later. The windows remain closed, because in the wonderful silence and in the dark, hidden from those who would call me away, and free and left to myself, I do not follow the eyes with the mind but the mind with the eyes, which see the same as reason does so long as they do not look at anything else. When I am currently working on something, I think about it word by word, in the same way as someone who writes or edits, now more, now less, depending on the degree of difficulty or ease with which the words can be put together and remembered. I call for the notary and dictate

> what I have contrived, after the light of day has been allowed to enter; he leaves and is called again and sent away again. As soon as the fourth or fifth hour arrives—there is actually no fixed and measured time—I move to the terrace or into the gallery, just as the day suggests; I think over and dictate the rest. I mount a chariot. There I keep doing the same thing that I have been doing, walking and lying down; my attention is maintained, animated by the variety. I sleep a little, then I go for a walk, and soon I read a Greek or Latin speech aloud and with emphasis, not so much because of the voice, but because of the stomach; thus, the voice is strengthened, too. Then I go for another walk, anoint myself, exercise, wash. During dinner, when I am with my wife or a few people, a book is read to me; after dinner, there is a comedy or a lyre player, and then I go for a walk with my companions, among whom are also educated men. Thus, the evening is stretched out among diverse conversations, and even though it is a very long day, it is quickly lived through.[52] Sometimes something changes in this order, because when I have stayed in bed for a long time or have gone for a walk, I do not take the chariot after my nap and reading, but a horse, which is shorter as it is faster. Friends from the closest villages join me and claim a part of the day for themselves, and sometimes, when I am exhausted, they come to my aid with a convenient break.

Pliny describes a course of day that is mostly characterized by event-time. Only the first and the fourth hour of the urban frame of reference are mentioned, as if they should illustrate how unimportant this grid is in the country: during the first hour of the day, he is still lying in bed instead of receiving the *salutatio*, whereas in the fourth, he is already listening to the afternoon *otium*, instead of performing his judicial and political duties. This is also manifest in the many relativizations of the numerical (i.e., urban) hour specifications (*cum libuit, circa, saepe, raro, vel*).

Pliny makes the differences between both chronotopes visible for his readers, not only implicitly but also explicitly. There are two striking statements: *neque enim certum dimensumque tempus* ("there is actually no fixed and measured time in the country") and *non numquam ex hoc ordine aliqua mutantur* ("sometimes something is changed in this order"). This kind of flexibility, which concerns the duration and, more importantly, the order in which the events happen in time, is unusual. It clearly and fundamentally differentiates the chronotope of the *villa* from the other outlined chronotopes.

52 Pliny describes the subjective perception of an eventful day as short and pleasant; Seneca takes up the same observation in his important critique *De brevitate vitae*.

Pliny's description sounds a little bit like "anything goes." The social regulation of time in the *villa* seems to be far more flexible than the one in the *urbs*; it is connected to a great tolerance concerning time. Time specifications falling short of the hour are not even to be expected there, so their absence should not actually be surprising. The astonishing flexibility of the chronotope, however, could help us to better understand a phenomenon of material culture: the famous inaccuracy of the Pompeian and Herculanean timekeeping devices.[53] The majority of them do not accurately depict local time, whether because of a lack of artisanal skill or because they are imported from the Eastern Mediterranean—or because the exact measurement of short times was not a relevant category within this chronotope. If the "rule of the clock" was so much more relaxed there, this implies that event time was plainly preferred to measured time in the *villa* chronotope. This results in a lack of exact hour determinations, complemented by natural time indicators (sunrise and sunset), and the shortest time is the time of a short action—as, e.g., when Pliny uses the horse instead of the chariot to accelerate the chain of events, or when he quickly washes before he goes to dinner.[54]

If not only fractions of hours but even the hour itself did not present a relevant unit in a certain chronotope, then the clocks there did not have to display it. This means that we are less concerned with technological ignorance here, but rather with a society that saw no problem in timekeeping devices that had more to do with representation than with punctuality, and with a social context which was inclined to prefer event time to clock time—at least in the *villa*.

The observation that whenever Latin literature speaks of country life it does not refer to numbered hours speaks for the fact that this is not simply the individual view of one author, but a profound cultural and literary convention. The epic, too, which is usually unrelated to the urban world, only knows natural timekeepers and events that are typical for a certain time, but no numbered hours. To put it in a nutshell: no senate session begins with "rosy-fingered Eos," and no shepherd drives his animals to the pasture *hora secunda*; in Cicero's

53 More than 30 clocks survive from Pompeii and Herculaneum; they thus offer an intriguing density of time measurement devices considering the size of the towns, a density which was only reintroduced almost 2000 years later. For evidence of the clocks, see Gibbs 1976, 414; Winter 2013, 483–501; for the inaccuracy, see Schaldach 2001, 27–28; Wolkenhauer 2011, 89–91 and 100–101; Talbert 2017, 141–146.

54 Cf. Levine 2005, 365: "When event time predominates, scheduling is determined by activities. Events begin and end when, by mutual consensus, participants feel the time is right." Event time is not an exclusive characteristic of pre-modern cultures. In German, for example, it is established in the idiom to talk to someone "eine Zigarettenlänge" ("for the length of a cigarette").

speeches we can almost exclusively find numerically-counted, "urban" hours; in Virgil's epic, on the contrary, Sun and Moon prevail.

5 Conclusion

From the outlined analyses here, the following results can be drawn and can serve as a basis for future research. The history of the use of timekeeping devices in Rome shows that, between 450 and 150 BCE, the inner structure of a day increases successively and arrives at the measured hour as its smallest unit, a span that lasts between 45 and 75 minutes in Rome and lacks a uniform and stable length due to the seasonal changes. From 160 BCE onwards, the practice of numbering the hours (*hora prima, secunda,* etc.) spread in the city. Scientific approaches that make use of this detail in order to research the conceptual history (or *Begriffsgeschichte*) and the pragmatics of numerically counting hours will help to clarify our ideas about how the Romans dealt with the socially-recognized order of time.

It is mistaken to assume that clocks are only designed according to the needs of time measurement. As in the case of the famous two sundials of the *forum*, the technological improvement of a timekeeping device is not always linked to the concept of "better timekeeping," but may also be understood as a metaphor of political order and cosmic harmony. Within historical time, the concept of "punctuality" (i.e., the orientation on a socially dominant sequence of events) becomes a discernible criterion of the Roman way of life. Defined as a virtue (*virtus*), it valorizes adherence to temporal norms and the measurement of short periods of time.

In order to differentiate between various "local times," the term chronotope has proved to be useful. It helps us to keep in mind that every place on earth has its own time-ordering structure. Chronotopes such as the *forum*, the *villa*, and its counterpart, the *urbs*, can be identified and described based on their temporal specifics. We saw that time specifications shorter than the hour are only to be expected in small, specialized chronotopes (e.g., in the context of judicial speech). The chronotope of the *urbs* seems to work with units of time up to two hours and with marked time limits, but also includes place as a second important factor. The chronotope of the *villa* is characterized by temporal flexibility, which proves to be close to concepts of event time. Further chronotopes could be added and are partly discussed in this publication from different perspectives.[55] Each of them poses different questions concerning

55 See especially the contributions of Heilen (astrology) and Miller (medicine).

the measurement and the assessment of short time units. In most cases, one will find that though the Latin language *does* include terms that could be used to identify units of time shorter than the hour (like *dimidium*, *quadrans* or *dodrans horae*), these were used quite rarely. Simultaneously, the fact that each chronotope follows its own rules also implies that areas of dense and low temporal specificity alternate with each other within Roman culture, and that some of them aim at a kind of timekeeping that is exact and captures short periods of time, whereas others are neither able to do this, nor do they desire to or even think it useful. It turns out that questions concerning the shortest relevant time, punctuality, and tardiness are only to be asked (and answered in a fruitful way) with regard to the technical, spatial, social, and functional conditions that the respective chronotope implies.

References

Bachtin, Michail M. 2014. *Chronotopos. Aus dem Russischen von M. Dewey, mit einem Nachwort v. M.C. Frank und K. Mahlke*. 3rd ed. Frankfurt am Main: Suhrkamp.

Bonnin, Jérôme. 2015. *La mesure du temps dans l'antiquité*. Paris: Les Belles Lettres.

Breimeier, Heinrich. 1985. "Praecox." *Thesaurus Linguae Latinae*. Leipzig: Teubner.

Clayman, Steven E. 1989. "The Production of Punctuality: Social Interaction, Temporal Organization, and Social Structure." *American Journal of Sociology* 95 (3): 659–91.

Dohrn-van Rossum, Gerhard. 1992. *Die Geschichte der Stunde. Uhren und moderne Zeitordnungen*. Munich: C. Hanser.

Elias, Norbert. (1974–1975) 1988. *Über die Zeit. Arbeiten zur Wissenssoziologie 2*. Edited by Michael Schröter. Frankfurt am Main: Suhrkamp.

Feeney, Denis. 2008. *Caesar's Calendar. Ancient Time and the Beginnings of History*. Berkeley: University of California Press.

Gibbs, Sharon L. 1976. *Greek and Roman Sundials*. New Haven, CT: Yale University Press.

Hall, Edward T. 1984. *The Dance of Life: The Other Dimension of Time*. New York: Anchor Press.

Hannah, Robert. 2005. *Greek and Roman Calendars. Constructions of Time in the Classical World*. London: Bristol Classical Press.

Hannah, Robert. 2013. "Clocks." *The Encyclopedia of Ancient History*. Edited by Roger S. Bagnall. Malden, MA: Wiley-Blackwell.

Houston, George W. 1989. "The State of the Art: Current Work in the Technology of Ancient Rome." *The Classical Journal* 85: 63–80.

Ker, James. 2009. "Drinking from the Water-Clock: Time and Speech in Imperial Rome." *Arethusa* 42: 279–302.

Keudel, Ursula. 1978. "Ordino." *Thesaurus Linguae Latinae*. Leipzig: Teubner.

Krenkel, Werner. 1970. *Lucilius. Satiren*. 2 vols. Leiden: Brill.

LeGoff, Jacques. (1960) 1977. "Zeit der Kirche und Zeit des Händlers." In *Schrift und Materie der Geschichte: Vorschläge zur systematischen Aneignung historischer Prozesse*, edited by Claudia Honegger, 393–414. Frankfurt am Main: Suhrkamp.

Levine, Robert. 1997. *A Geography of Time*. New York: Basic Books.

Levine, Robert. 2005. "A Geography of Busyness." *Social Research* 72 (2): 355–70.

Luciani, Franco. 2009. "Ultimi minuti di vita: Le suddivisioni dell'hora nelle epigrafi funerarie latine." In *Temporalia. Itinerari nel tempo e sul tempo*, edited by Franco Luciani, Chiara Maratini, and Annapaola Zaccaria Ruggiu, 121–44. Padua: Sargon.

Michel, Henri. 1970. "La notion de l'heure dans l'antiquité." *Janus* 57: 115–24.

Moreno Soldevila, Rosario. 2006. *Martial, Book IV: A Commentary*. Leiden: Brill.

Nielsen, Palle W. 1963. "Intempestivus." *Thesaurus Linguae Latinae*. Leipzig: Teubner.

Nilsson, Martin P. 1920. *Primitive Time-Reckoning: A Study in the Origins and First Development of the Art of Counting Time among the Primitive and Early Culture Peoples*. Lund: C.W.K. Gleerup.

Polinskaya, Irene. 2018. "The Distant Present and the Near Past: The Mounting of Time in the Herodotean 'Aiginetan Logos' (5.54–97)." In *Zeitmontagen. Formen und Funktionen Gezielter Anachronismen*, edited by A. Junghanß, B. Kaiser, and D. Pausch, 209–35. Stuttgart: Franz Steiner Verlag.

Puff, Helmut. Forthcoming. *Toward a History of Waiting.*

Ramminger, Johann. 2009. "Exactitudo." *Neulateinische Wortliste. Ein Wörterbuch des Lateinischen von Petrarca bis 1700.* [Online dictionary] accessed June 5, 2019, www.neulatein.de/words/3/015211.htm.

Remijsen, Sofia. 2007. "The Postal Service and the Hour as a Unit of Time in Antiquity." *Historia* 56 (2): 127–40.

Riggsby, Andrew M. 2009. "For Whom the Clock Drips." *Arethusa* 42 (3): 271–278.

Rüpke, Jörg. 1995. *Kalender und Öffentlichkeit. Die Geschichte der Repräsentation und religiösen Qualifikation von Zeit in Rom*. Berlin: de Gruyter.

Schaldach, Karlheinz. 2001. *Römische Sonnenuhren. Eine Einführung in die antike Gnomonik*. 3rd ed. Frankfurt am Main: Harri Deutsch.

Shackleton Bailey, David R. 1990. *M. Valerii Martialis Epigrammata, Post W. Heraeum*. Stuttgart: Teubner.

Shaw, Jenny. 1994. "Punctuality and the Everyday Ethics of Time. Some Evidence from the Mass Observation Archive." *Time & Society* 3 (1): 79–97.

Stein-Hölkeskamp, Elke. 2005. *Das römische Gastmahl. Eine Kulturgeschichte*. Munich: C.H. Beck.

Talbert, Robert J.A. 2017. *Roman Portable Sundials: The Empire in Your Hand*. Oxford: Oxford University Press.

Winter, Eva. 2013. *Zeitzeichen: zur Entwicklung und Verwendung antiker Zeitmesser.* 2 vols. Berlin: de Gruyter.

Wolkenhauer, Anja. 2011. *Sonne und Mond, Kalender und Uhr: Studien zur Darstellung und poetischen Reflexion der Zeitordnung in der römischen Literatur.* Berlin: de Gruyter.

Wolkenhauer, Anja. 2016–2017. "Vorüberlegungen für eine Zeitgeschichte von Nacht und Schlaf in Rom." *Römische Quartalsschrift für christliche Altertumskunde und Kirchengeschichte* 111: 172–86.

Zerubavel, Eviatar. 1982. "The Standardization of Time: A Sociohistorical Perspective." *American Journal of Sociology* 88 (1): 1–23.

Zimmermann, Bernhard. 1999. "Cicero und die Griechen." In *Rezeption und Identität. Die kulturelle Auseinandersetzung Roms mit Griechenland als europäisches Paradigma*, edited by Gregor Vogt-Spira and Bettina Rommel, 240–48. Stuttgart: Franz Steiner Verlag.

CHAPTER 8

Short Time in Greco-Roman Astrology

Stephan Heilen

It is the merit of the Stoics to have distinguished the vast variety of ancient forms of divination in two handy categories, technical (τεχνικός) and non-technical (ἄτεχνος).[1] This dichotomy was adopted and Latinized by Cicero who speaks, in his treatise *On divination*, of *divinatio naturalis* and *divinatio artificiosa*, respectively.[2] Astrology clearly belongs to the technical category because it is the most sophisticated form of ancient divination,[3] the only one that requires a substantial amount of exact science in the modern sense, namely astronomy and mathematics.[4]

There are two reasons why astrology deserves close attention in the context of the present volume. Firstly, its close relationship to accurate time-keeping: astrological interpretations require accurate determination of the underlying astronomical parameters, namely planetary longitudes and horizon phenomena such as risings, culminations, and settings of the luminaries (Sun and Moon), planets, and specific parts of the zodiac. These astronomical data were not determined (and, in many respects, could not be determined) through direct observation of the sky, but rather through computations based on planetary tables, which require a chronological input. Therefore the astrologer's astronomical accuracy requires accurate time-measurement. Secondly, astrology had a massive cultural impact: compared to our modern Western societies with their enlightened rejection of all forms of "superstition," astrology was omnipresent in ancient daily life. It pervaded all geographical areas of the Roman Empire at all social levels and extended to many more fields of life than modern readers might expect.

1 For details, see Wardle 2006, 126–27.

2 Cf. Cic. *Div.* I.11 (brother Quintus speaking): *Duo sunt enim divinandi genera, quorum alterum artis est, alterum naturae*; cf. ibid. I.34, I.72, I.109–110, I.116, I.130, II.26–27, II.100. Cicero assigns oracles based on prophetic frenzy as well as dreams to natural divination, and all other forms—such as extispicy, brontology, auspices, lots, etc.—to technical divination. Since Latin *ars* equals Greek τέχνη, the conceptual equivalence of "technical" and "artificial" divination is obvious.

3 It is so assigned by Cic. *Div.* I.12 (even if Cicero's knowledge of astrology reflects an earlier stage of its development, due to Cicero's dependence on Panaetius).

4 Cf. Neugebauer 1957, 171: "Compared with the background of religion, magic and mysticism, the fundamental doctrines of astrology are pure science."

 | DOI:10.1163/9789004416291_010

In the interest of clarity, the following analysis will employ three semantically distinct yet interrelated technical terms: "chronological" (i.e., regarding time, generally speaking), and—more specifically—"chronometrical" (i.e., regarding the measurement of time) as opposed to "chronographical" (i.e., regarding the notation of time). Chronographical precision is, of course, easier to achieve than chronometrical accuracy.

For the following analysis, it seems useful to introduce a distinction in three levels of chronometrical precision to be investigated in Greco-Roman astrological sources: standard (what is common practice?), high (what does Claudius Ptolemy, as the leading astronomer among ancient astrologers, deem desirable and technically feasible?), and extreme precision (what, if anything, goes beyond Ptolemy's demands?). The two higher levels of this structure require a further differentiation between theoretical claims and actual practice.

It will be argued (a) that chronographical precision down to one seasonal hour is omnipresent in Greco-Roman astrology; (b) that there is also solid, yet more limited, evidence of high precision in astrological practice; (c) that one even finds rare cases of extreme precision, in theory as well as in practice, motivated by special purposes; and (d) that the latter two categories seem to be the product of a long-lasting ancient controversy about the validity of astrology in which accurate time-keeping was of central importance.

1 Standard Precision

The indispensable astronomical parameters of an ancient horoscope are eight ecliptic longitudes: namely, those of the Sun, Moon, the five known planets, and the ascendant (that is, the part of the zodiac that intersects with the eastern horizon). Elaborate horoscopes determine more astronomical parameters, but these are the essential ones. Two of them are far more sensitive to chronometrical accuracy than the remaining ones: namely, the longitudes of the Moon and, even more so, the longitude of the ascendant.[5] The concept of the ascendant did not exist in Babylonian proto-astrology.[6] It is a Hellenistic invention inspired by a typical feature of Egyptian culture: the consecutive ris-

5 The fact that the planetary longitudes are far less sensitive to chronometrical accuracy is an important reason why quite a number of so-called "sign-entry almanacs" are extant which specify only the calendrical date of a planet's entry into a zodiacal sign. These almanacs, which require far less computational effort and writing space than tables that give day-by-day positions (so-called "ephemerides"), were presumably used by the large number of modest, unambitious astrologers. On both kinds of tables, cf. Jones 1999, I 40–44.

6 At least, no attestation has hitherto been found in the cuneiform record.

ings of 36 relatively small constellations called “decans” which allowed rough timekeeping at night. However, the decans were, for the Egyptians, not just clusters of fixed stars, but also deities. By the Hellenistic period,[7] this gave rise to the belief that the decan-god that happened to be rising at a child's birth watched the nativity and presided over it. Since each Egyptian decan equals, on average, ten degrees of the Greek (ultimately Babylonian) abstract zodiac ($360° \div 36 = 10°$), three of them equal 30 degrees, which is the extension of one zodiacal sign. Thus, the decans and the zodiacal signs were, in Hellenistic Egypt, associated with each other, and the idea that an astrological influence was exercised by the rising decan was extended to the rising zodiacal sign, and eventually to the single rising degree. The Greek term for this rising part, either decan or sign or degree, was ὡροσκόπος which, literally translated, means “hour-watcher” or, perhaps more appropriately, “time-span-watcher,” because the Greek noun ὥρα does not initially have the technical meaning “hour” (either equinoctial or seasonal) but “any period of time fixed by natural laws and revolutions,” whether of the year, month, day, or shorter periods, down to just a few seconds.[8] It soon came to be valued as the astrologically most important and influential single parameter in each set of birth data, and for this reason, its accurate determination was agreed to be of paramount importance.[9]

On average, each decan needs 40 minutes to rise, each zodiacal sign two hours, and each degree four minutes.[10] The vast majority of extant Greek horoscopes are content with specifying the ascendant to the sign. Some specify it to the degree, but not a single one specifies it to the decan, presumably because the Hellenistic concept of the ascendant was soon emancipated from its obvious origin in the Egyptian decan yet preserved its Greek name ὡροσκόπος.[11]

7 I.e., the reign of the Ptolemies over Egypt (332–30 BCE).

8 *LSJ* s.v. ὥρα, with reference (A.II.1.b) to an example of a very short ὥρα: Sextus Empiricus reminds his reader (while criticizing the astrologers' claim that they are able to grasp the exact time of birth) that when woodcutters fell trees on a mountain-side, μετὰ γὰρ ἱκανὴν ὥραν τοῦ κατενεχθῆναι τὸν πέλεκυν ἐξακούεται ἡ τῆς πληγῆς φωνή (Sext. Emp. *Math.* v.69) (“the sound of the blow is heard a considerable time after the fall of the axe,” trans. Bury 1949, 353). In Latin, the semantic spectrum of *hora* is similarly broad (cf. *ThLL* s.v. *hora* I.B: *quodvis spatium temporis*). For details on the technical term ὡροσκόπος, see Heilen 2015, 637–644; for more details on the semantic spectrum of ὥρα, cf. Sattler's contribution to this volume.

9 It even came to denote, by way of *pars pro toto*, the entire nativity; hence the modern term “horoscope.” This modern meaning is not attested in antiquity.

10 The true rising times vary in the Mediterranean area between about one and three equal hours per sign, depending on the geographical latitude and the distance of the sign in question from 0° Aries. The rising times of decans and single degrees vary accordingly.

11 This emancipation presumably occurred during the gap of at least one century between the origin of the Hellenistic concept of the ascendant, which historians of astrology date

This information is based on 345 extant Greek horoscopes,[12] of which 168 come from documentary sources (i.e., from papyri),[13] while the remaining 177 are so-called literary horoscopes, which were mostly included in ancient astrological manuals to illustrate various techniques. Most of the documentary horoscopes (and some of the literary ones) specify the time of birth, typically with precision to the seasonal hour of day or night, and sometimes with specifications that the horoscope had been cast for the beginning (ἀρχή)[14] or—far less frequently—the end[15] of a certain seasonal hour. In a similar vein, the birth times are given with precision to one seasonal hour in 51 extant Demotic horoscopes that were written on *ostraka* (pot-sherds). There is also a small number of extant Latin horoscopes, each coming from literary sources; some of them specify the hour, some do not.[16]

This typical precision to one seasonal hour corresponds to a level of accuracy that may be assumed to have been, from the beginning of the Common Era, accessible to large parts of the population of the Roman Empire, through sundials during the day and water clocks during the night.[17] It goes without saying that this level of precision is not sufficient if a sign has just begun to rise or has almost entirely risen. Therefore, ambitious astrologers recommended higher chronometrical precision, as we shall see in section 2 below.

---

no later than around the middle of the second century BCE, and the three earliest extant Greek papyrus horoscopes, which were cast for dates within the last decade of the first century BCE. Cf. the respective catalogue entries in Heilen 2015, 204–333, esp. 216, for 'Hor. gr. -9.VIII.14', 'Hor. gr. -6.III.29', and 'Hor. gr. -3.X.2'.

12 See the chronologically differentiated diagram in Heilen 2015, 524.

13 Very rarely from other written sources.

14 See the catalogue entries in Heilen 2015, 204–333, for 'Hor. gr. 46.V.13', 'Hor. gr. 137.XII.4', 'Hor. gr. 145.IV.4', 'Hor. gr. 190.I.25', 'Hor. gr. 227.XII.4', 'Hor. gr. 242.X.4', 'Hor. gr. 243.I.11', 'Hor. gr. 246.X.19', 'Hor. gr. 260.IX.29' (unusually precise, cf. below, table 2), 'Hor. gr. 282.IX.13', 'Hor. gr. 419.VII.2', 'Hor. gr. 428.IX.8', 'Hor. gr. 484.VII.18'. See also the contribution by Jones to the present volume.

15 The only case that I am aware of is Heilen 2015, 279, 'Hor. gr. 219.V.18' (*P. Oxy.* 3298 *recto*, col. I.4): ὥρᾳ ζ' ἐσχάτῃ ('at the end of the 7th hour').

16 For all of these Greco-Roman and Egyptian horoscopes, as well as for 30 extant cuneiform nativity omens (proto-horoscopes), see the same catalogue (Heilen 2015, 204–333) and the English survey of Heilen (2019). For the specification of the hour in Demotic horoscopes, see Ross 2006, 150.

17 Cf. the contributions by Symons, von Lieven and Schomberg, and Jones. It is important to keep in mind that the astrologer would normally not be present at a child's birth but have to rely on chronological information provided by the native's parents or the native himself/herself. Therefore, even if the astrologer were one of the presumably few who had above-average tools for time-keeping at their disposal, these instruments would be useless during most consultations with clients, which would take place some (often much) time after the actual birth in question.

Apart from determining the ascendants of horoscopes, there was another field of ancient astrology in which chronometrical accuracy to the hour was important, namely time-rulership. This doctrine is the basis of so-called elections (i.e., choosing the best time for an action) and also the origin of the modern seven-day-week. I shall first explain the chronological structure of this doctrine and then its astrological function.

Each hour of the day is assigned a specific planet according to the sequence of the planetary spheres in the geocentric universe, starting from the outermost sphere, that of Saturn, and moving down to the sphere of the Moon. This descending order (Saturn, Jupiter, Mars, Sun, Venus, Mercury, Moon) was the rule in ancient enumerations of the planets. Thus, the first hour of the first day is ruled by Saturn, the second by Jupiter, the third by Mars, and so on until the seventh hour is ruled by the Moon. Then the series starts over, with Saturn ruling the eighth hour, and so on. Since the 24th hour will be ruled by Mars, the next planet, the Sun, will rule the first hour of the next day. Whoever rules the first hour is automatically also ruler of the entire day. This is why the first day of the week is Saturday, the next day Sunday, and so on. It takes seven days until the series restarts, because the last hour of the day of Venus is ruled by the Moon, which has the lowest sphere in the geocentric universe; therefore, the first hour of the eighth day will again be ruled by Saturn, thus initiating a new week.[18]

This astrological doctrine claims that the success or failure of any human undertaking will depend on the qualities of its presiding two rulers, the day-ruler and its deputy, the hour-ruler. Jupiter and Venus are considered to be benefic deities, Saturn and Mars malefic, the remaining three ambivalent. If one wishes to choose, for example, a propitious date for a marriage, the best possible hour of the entire week is on Friday afternoon, more precisely: the eighth seasonal hour after sunrise (i.e., about 1–2 p.m.), when Venus is both day-ruler and hour-ruler, while hours ruled by Saturn and/or Mars would be deemed disastrous.

18 While the first three days of the week preserve, in English, the names of the presiding planetary deities (Saturn, Sun, and Moon), the following four are named after the Germanic equivalents of Mars, Mercury, Jupiter, and Venus in Norse myth. For the ancient astrological week, cf. Boll 1912 (esp. 2556.25–2561.6 and 2570.49–2578.67) and further literature mentioned by Hübner (1982, 53, n. 1). Boll (1912, 2577.30–34) rightly emphasizes that the first day of the week must originally have been Saturday (cf. ibid. 2557.38), but relatively soon Sunday became the first day, so as to comply with the Jewish week and with Greco-Roman solar theology.

The doctrine of day-rulership is attested in a variety of texts from the late first century BCE onwards.[19] The historian Cassius Dio reports that at his time (around 200 CE) the astrological week had spread to all parts of the Roman Empire.[20] This testifies to the enormous success of the astrological concept of time-rulership, especially if we consider that other, pre-existing concepts of time, such as the traditional eight-day week of the Romans, were suppressed by the new planetary week. Moreover, the Christian Church fathers fought in vain against it, even after Christianity had become the State Church of the Roman Empire.

What matters for our purpose is the close succession of good and bad hours in this system. In particular, every hour of benefic Jupiter is flanked by two hours of the malefic planets Saturn and Mars. This obviously required accurate time-keeping for those who believed in time-rulership.

While there are quite a number of ancient inscriptions commemorating the succession of days of the astrological week, only one extant inscription details the rulers of the single hours. This serendipitous archaeological find is *CIL* IX.5808, the fragment of an inscription from about 100 CE (maybe earlier) which was found in the ruins of Potentia, a small city on the Adriatic coast.[21]

---

19 The earliest attestation is in the elegies of Tibullus (Tib. 1.3.18), who mentions only Saturday (since Saturn is a malefic astrological deity, Tibullus cannot travel on this day). The first attestation of the entire astrological week is, with a high degree of probability, in the *Fasti Sabini* (*CIL* IX.4769, ed. Mommsen 1883, 453), a calendar from the late 1st c. BCE. It does not mention the planets, but its underlying concept of a seven-day week (the days are marked again and again from A to G) can only refer to the astrological week (cf. Boll 1912, 2573.52, and Rüpke 1995, 95–97, esp. 97). The first known literary treatise on the astrological week is lost, namely Plutarch's Διὰ τί τὰς ὁμωνύμους τοῖς πλάνησιν ἡμέρας οὐ κατὰ τὴν ἐκείνων τάξιν ἀλλ' ἐνηλλαγμένως ἀριθμοῦσιν (Plut. *Quaest. symp.* 4.7) (*Why the weekdays, although they are named after the planets, do not follow their cosmic order*). On these and many other sources, including archaeological evidence, cf. Boll 1912, 2558–2560 and 2573–2576. See also Bezza 1995, 475–517, and (esp. for the hour-rulers) various texts in the appendices of the *Catalogus Codicum Astrologorum Graecorum* (*CCAG*): some of this material is of early origin, such as that attributed to Serapion (*CCAG* I.99–102, V.1.179–180, V.3.87), while some is later, by Christian authors (*CCAG* IV.99, VII.88–95, XI.2.125–133).

20 Cass. Dio XXXVII.18.1.

21 For the date (c. 100 CE), cf. Alfieri 1948, 114, n. 1. This date hinges on the assumption that the inscription was erected together with a list of Roman consuls (*fasti consulares*) which was found on the same site. This list exhibits very similar epigraphic characters and records consuls who held office in the decades around 100 CE. Cf. Paci (1995, 9–11), who advises caution because of the difficulty of dating the astrological inscription palaeographically (he thinks it may date back to the Augustan Period), but agrees with Alfieri that this inscription must have been erected, perhaps in the market-place of Potentia, together with a now-lost calendar (like the *Fasti Sabini*, cf. n. 19 above) that accompanied the extant consular *fasti*. The astrological inscription (*CIL* IX 5808) was overlooked by Boll (1912).

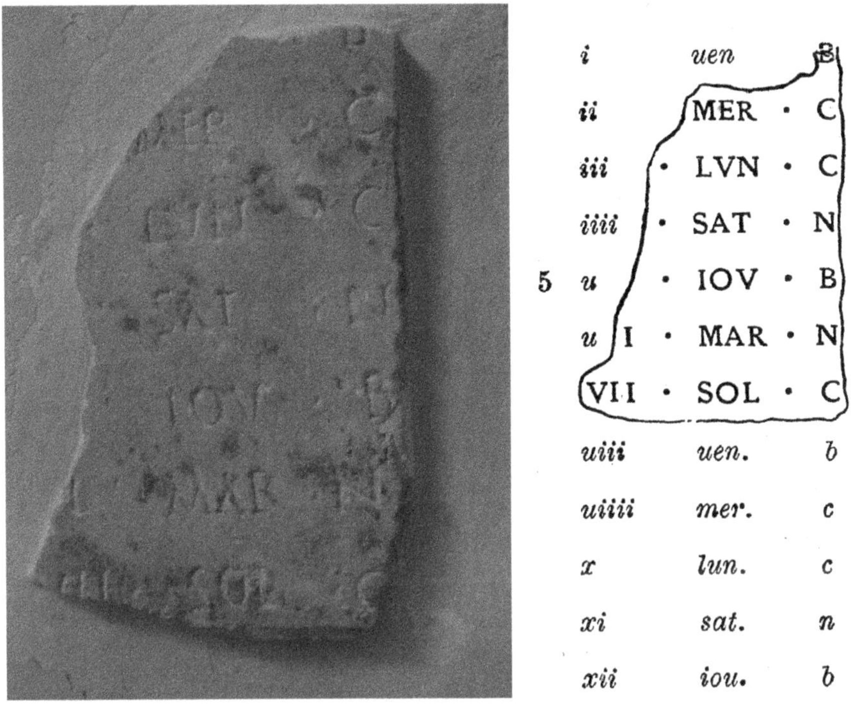

FIGURE 8.1 a) CIL IX.5808 © *Famiglia Leopardi Recanati.* b) Graphic reconstruction (*Mommsen* 1883, 556)

Fig. 8.1a above features the first photograph ever published of this relatively small fragment,[22] whose original is not accessible to the public.[23] Figure 8.1b is a graphic reconstruction of the three fragmentarily preserved columns.

22 The dimensions are 12.4 cm (max. height) × 8.4 cm (max. width). Since the extant text comprises a bit more than six rows out of twenty-four (for a full 24-hour day) and a bit less than the full width of one day's data, one can estimate that the complete marble slab once measured about 50 cm (height) × 70 cm (width). If day-hours and night-hours were listed in two contiguous columns (as assumed by Mommsen [1883, 453], in his graphic reconstruction, of which only the left part is reproduced in fig. 8.1b), that would lead to total dimensions of about 25 × 140 cm, which seems unlikely. It is impossible to tell whether the almost perfect vertical right edge is the original edge or the product of a break.

23 After its discovery in 1820, the fragment was taken from the archaeological site to the noble mansion of Count Monaldo Leopardi in Recanati, where it was immured in one of the inner walls. While parts of the Palazzo Leopardi are now a museum devoted to Monaldo's son, the famous poet Giacomo Leopardi (1798–1837), the inscription is not accessible because it is in one of the private rooms where the noble descendants still live. I thank Dr. Sara Eugeni of the Casa Leopardi for her repeated kind help.

Each hour is marked as either B(ONA) or C(OMMUNIS) or N(OXIA), meaning, respectively, "good," "ambivalent," or "noxious" (hour). The fragment belongs either to Friday morning (the 2nd to the 7th hour from sunrise) or to the night between Monday and Tuesday (the 2nd to the 7th hour after sunset). Since an incomplete list would be practically useless, it is safe to assume that the inscription originally comprised all seven days of the week. This inscription is important both because of its early date and because of its epigraphic nature, which implies a bigger technical and financial effort than copying the data by hand on papyrus. It is probable, yet beyond proof, that this inscription was on public display for the benefit of the citizens of Potentia.[24] Some individuals—in the Roman capital, at least—would not even need public lists of planetary hours because they privately owned astrological books for instant consultation whenever the need occurred. We know this from Juvenal, who warns his reader away from women like the one he describes (his satire would be pointless if it had no counterpart in the social reality of his time):

> Illius occursus etiam vitare memento,
> in cuius manibus ceu pinguia sucina tritas
> cernis ephemeridas, quae nullum consulit et iam
> consulitur, quae castra viro patriamve petente
> non ibit pariter numeris revocata Thrasylli.
> ad primum lapidem vectari cum placet, hora
> sumitur ex libro; si prurit frictus ocelli
> angulus, inspecta genesi collyria poscit;
> aegra licet iaceat, capiendo nulla videtur
> aptior hora cibo nisi quam dederit Petosiris.
>
> IUV. *Sat.* VI.572–583

> But beware of ever encountering one whom you see clutching a well-worn calendar in her hands as if it were a ball of clammy amber; one who inquires of none, but is now herself inquired of; one who, if her husband is going forth to camp, or returning home from abroad, will not bear him company if the numbers of Thrasyllus[25] call her back. If she wants to drive as far as the first mile-stone, she finds the right hour from her book;

24 Cf. n. 21 above. It cannot be excluded, however, that the inscription was erected in a private context. Cf. Rüpke 2004, 362 (on archaeological finds of Roman Fasti): "More striking is their frequent appearance as a large mural decoration for private apartments." Rüpke gives several examples of such wall paintings from the 1st c. BCE to the 2nd c. CE, all from Latium (for details, cf. Rüpke 1995, 42, 52, 83, 86).

25 The famous court astrologer of Tiberius.

> if there is a sore place in the corner of her eye, she will not call for a salve until she has consulted her horoscope: and if she be ill in bed, deems no hour so suitable for taking food as that prescribed to her by Petosiris.[26]
>
> RAMSAY 1928, 129–31

Apart from the election of propitious hours for whatever kind of action, the doctrine of time-rulership was also, but less frequently, used to make predictions regarding the lives of newborn children, depending on their day and hour of birth.[27] One late antique text on this topic is the so-called Calendar of the year 354 CE, which contained illustrations for the single weekdays (five out of seven survive in Early Modern copies).[28] Each of them features the respective deity set in the upper register of a frame of sacred monumental architecture. Each deity is standing upright, adorned with its canonical iconographic attributes (e.g., Saturn with priestly veil and sickle) and flanked by two vertical slabs inscribed with three columns of text containing the respective data of the twelve nocturnal (left) and diurnal (right) hours. Each hour is marked with the usual qualifying letters B, C, and N (see above). The lower register contains another large slab inscribed with the astrological effects of the respective day ruler. In Saturday's case, it says:

> Saturni dies horaque eius cum erit nocturna sive diurna, omnia obscura laboriosaque fiunt: qui nascentur periculosi erunt; qui recesserit non invenietur; qui decubuerit periclitabitur; furtum factum non invenietur.[29]
>
> DIVJAK AND WISCHMEYER 2014, 125

26 This sage and King Nechepsos are the two legendary Egyptian founders of astrology. Cf. Boll (1912, 2572–2573), who shows, in his discussion of Juvenal's passage, that Petosiris may well have been credited in antiquity with being the inventor of the planetary hours.

27 The importance of both kinds of applications is emphasized by various astrologers, e.g., Paul of Alexandria (4th c. CE): Tὸν πολεύοντα καὶ διέποντα ἐφ' ἑκάστης γεννήσεως καὶ ἐπὶ πάσης ἡμέρας σκοπεῖν δεήσει (Paul. Alex. 21 41.18–19 Boer). ("It will be necessary to look at the day-ruler and the hour-ruler in each nativity and for every day," trans. Holden 2012, 36).

28 See Divjak and Wischmeyer 2014, 111–15. The illustrations regarding Jupiter's and Venus's days (Thursday and Friday) are lost.

29 The German paraphrase, ibid. 135, contains a misunderstanding of *periculosi*: "In den Tagen und Stunden des Saturn Geborene sind *periculosi*, bilden also eine Gefahr für ihre Mitwelt und sich selbst." This clearly contradicts the astrological qualities and effects of Saturn. For the correct interpretation see Müller, *ThLL* X.1 col. 1452.34–1456.67 s.v. *periculosus* (impr. 1997), esp. 1455.67–1456.9 (part II.A) *de eis, qui periculo expositi sunt*, with numerous attestations from Cato the Elder onwards, including (ibid. 1456.5–6) the passage of the Calendar of 354 that is here in question, and with a reference to the opposite health prediction of the same calendar for those born during a day or hour of Venus: *vitales*.

When it is Saturn's day and hour, either nocturnal or diurnal, everything becomes dark and toilsome; those who are born [i.e., in such an hour] will be endangered; a slave who ran away then will not be found; who has fallen ill then will be endangered; an object that was then stolen will not be found.

In comparison, the entry for Friday says the following:

Veneris dies horaque eius cum erit nocturna sive diurna, sponsalia facere, pueros puellas in disciplina mittere utile est. qui nascentur vitales erunt; qui recesserit invenietur; qui decubuerit convalescit; furtum factum invenietur.[30]

DIVJAK AND WISCHMEYER 2014, 130

When it is Venus's day and hour, either nocturnal or diurnal, it is useful to marry or to send children to school. Those who are born will be vital; a runaway slave will be found; who has fallen ill will recover; a stolen object will be found.

One finds similar tenets in Greek theoretical texts, too—for instance, in chapter V.56 of Rhetorius (sixth century CE), which is part of an epitome of the lost *Treasures* of Antiochus of Athens (second century CE).[31] This text predicts, for example, that when Mars rules both the day and the hour of nativities, the natives "meet their end from iron or fire or the wrath of a noble or the bursting forth of blood" (Holden 2009, 42).[32] In practical applications, however, the day and hour rulers are only rarely mentioned. Among the 345 extant Greek horoscopes mentioned above, only one nativity mentions both of them. This is *P. Berlin* 9825, cast for Thursday, 18 November 319 CE (day-ruler Jupiter, hour-ruler Saturn).[33] Two other nativities—namely, a Greek horoscope graffito from

30 Cf. ibid. 136 (German paraphrase and notes).

31 Rhet. V.56 (Cumont 1921, 125–26) = Antioch. *Epit.* IIIb.21. For details, cf. Heilen 2015, 342–43.

32 Rhet. V.56.3 (my paragraph count refers to the late David Pingree's forthcoming edition of Rhetorius, which I am preparing for publication): Ἄρεως πολεύοντος ὥρᾳ α′ ε′ ϛ′ η′ ἀπὸ σιδήρου ἢ πυρὸς τελευτῶσι ἢ χόλῳ μεγιστάνου ἢ αἵματος ἀναφορᾷ. Note that Cumont (1921, 125, lin. 26) erroneously expanded the manuscripts' abbreviation for ὥρᾳ as ὡροσκόπου. The initial words Ἄρεως πολεύοντος ὥρᾳ α′ ε′ ϛ′ η′, which I have rephrased in the interest of clarity, read thus in Holden's (2009) translation: "When Mars is day-ruler and the chart [is set for the] 1st, 5th, 6th, or 8th [hour]." Mars is both day-ruler and hour-ruler in the first and eighth hours of Tuesday (= Mars-day).

33 Edited by Greenbaum and Jones (2017). Cf. Heilen 2015, 290 ('Hor. gr. 319.XI.18–19').

Dura Europus (Syria) cast for Saturday, 9 January 219 CE,[34] and *P. Oxy. astron.* 4274, cast for Saturday, 14 June 503 CE[35]—each mention only the day-ruler, as does a group of five interrogation horoscopes for dates ranging from 475 CE to 487 CE, all five of which are likely to be by the same author, the so-called court astrologer to emperor Zeno.[36] In such cases the hour-ruler is implied in the combined data of day-ruler and hour of birth. Another interrogation horoscope by the same author mentions only the hour-ruler.[37]

Interestingly, the data of the Berlin papyrus (*P. Berlin* 9825) imply—as recomputation of the astronomical data reveals—that Thursday (Jupiter's day) was by the author conceived of as beginning with sunset on the previous day (Wednesday). The Calendar of 354 also has the 24-hour day begin with sunset. This suggests that the underlying rationale for assigning day-rulers, which is based on the first diurnal hour,[38] had been forgotten or neglected by many (though not all) practitioners in this late period.[39]

2 High Precision

For both internal and external reasons, there must have been a discourse on chronometrical accuracy among astrologers from relatively early times onwards. Internal reasons are technical problems in the determination of certain data that the astrologers themselves deemed indispensable for casting horoscopes; external reasons are arguments against the validity of astrology that were advanced by its critics. One example for each category will suffice.

First, an example of an internal problem. All ancient astrologers agree on the importance of certain planetary dignities in the interpretation of horoscopes.[40] One category of dignities whose examination was obligatory is that of the

34 Cf. Neugebauer and Van Hoesen 1959, 54–55 (version b) and 167.

35 Edited by Jones (1999, I 279–81 and II 416–17). Cf. Heilen 2015, 311 ('Hor. gr. 503.VI.14').

36 See Heilen 2015, 101–02 (table 1), entries 1, 2, 4, 5, 8; cf. ibid. 98, n. 413, and the respective entries in the catalogue, ibid. 204–333.

37 No. 7 of the table referred to in the previous note. Cf. Neugebauer and Van Hoesen 1959, 143, n. 3.

38 This is confirmed by (e.g.) Vett. Val. 1.9 and Paul. Alex. 21.

39 The "day" is measured from sunrise in no. 7 of the court-astrologer (see n. 37 above). His other horoscopes cannot be compared because they specify only the day-rulers.

40 On the most important dignities (houses, exaltations, triangles, terms) see Neugebauer and Van Hoesen 1959, 2–13, and (updated and more detailed) Heilen 2015, 713–40. The importance of the terms, as attested by ancient astrological manuals and papyrus horoscopes, is corroborated by further archaeological evidence which is reviewed ibid. 579 with n. 996.

TABLE 8.1 The "Egyptian terms" of the planets. Read thus: the first six degrees of Aries are assigned to Jupiter (♃), the next six to Venus (♀), the next eight to Mercury (☿), the next five to Mars (♂), the last five to Saturn (♄), etc.; ARI = Aries, TAU = Taurus, GEM = Gemini, CNC = Cancer, LEO = Leo, VIR = Virgo, SCO = Scorpius, SGR = Sagittarius, CAP = Capricornus, AQR = Aquarius, PSC = Pisces.[a]

ARI	♃ 6	♀ 6	☿ 8	♂ 5	♄ 5
TAU	♀ 8	☿ 6	♃ 8	♄ 5	♂ 3
GEM	☿ 6	♃ 6	♀ 5	♂ 7	♄ 6
CNC	♂ 7	♀ 6	☿ 6	♃ 7	♄ 4
LEO	♃ 6	♀ 5	♄ 7	☿ 6	♂ 6
VIR	☿ 7	♀ 10	♃ 4	♂ 7	♄ 2
LIB	♄ 6	☿ 8	♃ 7	♀ 7	♂ 2
SCO	♂ 7	♀ 4	☿ 8	♃ 5	♄ 6
SGR	♃ 12	♀ 5	☿ 4	♄ 5	♂ 4
CAP	☿ 7	♃ 7	♀ 8	♄ 4	♂ 4
AQR	☿ 7	♀ 6	♃ 7	♂ 5	♄ 5
PSC	♀ 12	♃ 4	☿ 3	♂ 9	♄ 2

a I use these capitalized three-letter codes for the abstract zodiacal signs in analogy with the non-capitalized codes (Ari, Tau, Gem, etc.) established for the homonymous (not identical!) visible constellations in 1922 by the International Astronomical Union.

so-called "Egyptian terms." They were considered to be a product of divine revelation to revered Egyptian authorities of the past.[41] In this system, which has no astronomical foundation and lacks any discernible rationale, each zodiacal sign is subdivided into five sections that are assigned to the five planets.[42] The terms are irregular with respect both to their succession within each zodiacal sign and to their respective extensions, as Table 8.1 shows.[43]

One of the central tasks in casting a horoscope was to determine the terms into which the longitudes of the heavenly bodies and the ascendant fell at the time of birth. The problem is that the terms are small—some no more than two degrees of arc—while the longitude of the most important single parameter, the ascendant, changes quickly, on average one degree every four minutes. Therefore, a two-degree term equals an eight-minute time window for the ascendant, while a three-degree term equals a twelve-minute time window, and so on. It is obvious that chronometrical accuracy to about one hour, as

41 The truth is, however, that they are of Babylonian origin, as a recently discovered cuneiform text shows. See Jones and Steele 2011.

42 The luminaries Sun and Moon are not assigned any terms.

43 For the ancient sources of the data contained in this table, see Heilen 2015, 720.

was the standard (to judge from the chronographical data), was insufficient because it entailed a high risk of getting the term of the ascendant wrong, with detrimental consequences for the validity and reliability of the prediction.[44]

As to external pressure, it will suffice to mention the "twin argument," which was advanced by critics of astrology from early on:[45] how can twins have different fates if they are born more or less simultaneously, in the same place, and from the same mother? The astrological counter-argument was voiced by Nigidius Figulus, who compared the quickly-revolving cosmos to a potter's wheel: even if one tried to touch its outer rim with one's finger twice, as quickly as possible, the two points touched would not be identical. In the same fashion, Nigidius argued, "the heavens move at such speed that even a tiny interval of time (such as that separating the birth of twins) makes a vast difference to the planetary relationships in the zodiac." (Long 1982, 173, n. 28; I closely follow his wording).[46] The obvious rejoinder was to ask how such a fleeting moment could be grasped by the astrologer. This objection was indeed raised as one of many arguments against astrology by the sceptic philosopher Favorinus of Arelate (mid-second century CE).[47] Another sceptic, Sextus Empiricus (late second century CE), goes an important step further: although he also mentions

44 For instance, the determination of the planet which "rules the nativity" and of this planet's deputy can depend on the term of the ascendant. See Heilen 2015, 1057–75, and Greenbaum 2016, 255–66.

45 Cf. esp. Cic. *Div.* II.90 (he explicitly draws on the second-century BCE Stoic Diogenes of Babylon who is, according to Timpanaro [1998, 383] likely to have adopted the twin argument from his student and later head of the Platonic academy Carneades) and Aug. *De civ.* V.2–6. Cf. Long 1982, 173.

46 Cf. the original Latin of Aug. *De civ.* V.3 (= Nig. Fig. test. 17, Swoboda 1889, 137): *frustra itaque adfertur nobile illud commentum de figuli rota, quod respondisse ferunt Nigidium hac quaestione turbatum, unde et figulus appellatus est. dum enim rotam figuli ui quanta potuit intorsisset, currente illa bis numero de atramento tamquam uno eius loco summa celeritate percussit; deinde inuenta sunt signa, quae fixerat, desistente motu, non paruo intervallo in rotae illius extremitate distantia. 'sic, inquit, in tanta rapacitate caeli, etiamsi alter post alterum tanta celeritate nascatur, quanta rotam bis ipse percussi, in caeli spatio plurimum est: hinc sunt, inquit, quaecumque dissimillima perhibentur in moribus casibusque geminorum.'*

47 His anti-astrological arguments are extant thanks to a summary by Gell. *NA* XIV.1. For the present argument, see ibid. XIV.1.25–26: *'atque id velim etiam,' inquit, 'ut respondeant* [scil. *astrologi*]*: si tam parvum atque rapidum est momentum temporis, in quo homo nascens fatum accipit, ut in eodem illo puncto sub eodem circulo caeli plures simul ad eandem competentiam nasci non queant, et si idcirco gemini quoque non eadem vitae sorte sunt, quoniam non eodem temporis puncto editi sunt, peto' inquit 'respondeant, cursum illum temporis transvolantis, qui vix cogitatione animi conprehendi potest, quonam pacto aut consulto adsequi queant aut ipsi perspicere et reprehendere, cum in tam praecipiti dierum noctiumque vertigine minima momenta ingentes facere dicant mutationes?'*

manifold arguments against astrology, he insists that it is enough to demonstrate the technical impossibility of determining the rising sign of the zodiac in order to refute the validity of astrology altogether. This is because, he says, so many other astrologically relevant data depend on the ascendant that the predictive system collapses if it is deprived of this one datum.[48] It is because of this strategy that he devotes the major part of his attack on astrology to demonstrating the technical impossibility (in his view) of determining any single one of three interrelated data: (a) the exact time of conception,[49] (b) the exact time of birth,[50] and (c) the rising sign.[51] Since knowledge of either (a) or (b) is a necessary prerequisite of (c), and reliable knowledge of (c) is in itself impossible, it is obvious, in Sextus's view, that the astrologers' claims to competently determine (c) are totally undermined. This whole passage, as well as the remainder of the treatise, abounds with forms of ἀκρίβεια ("accuracy"), ἀκριβής ("accurate") and ἀκριβῶς ("accurately").[52]

In view of such criticism, it is not surprising that we find both theoretical and practical evidence of a strong pursuit of accuracy, at least among the most serious and technically competent astrologers.[53] Claudius Ptolemy has an entire chapter titled Περὶ μοίρας ὡροσκοπούσης ("On the ascending degree") in his *Tetrabiblos* (III.3).[54] In this chapter he claims that the ascendant must be determined to the degree of arc, even if this is a challenging task:

48 Cf. Sext. Emp. *Math.* V.50–54 (with the comments of Long 1982, 186), esp. V.54: οὔτε δὲ τὴν γένεσιν τῶν ὑπὸ τὴν ἐπίσκεψιν πιπτόντων λαμβάνειν δυνατόν ἐστιν, ὡς παραστήσομεν, οὔτε τὸ ὡροσκόπιον ἀπλανὲς καθέστηκεν, οὔτε τὸ ἀνίσχον ζῴδιον πρὸς ἀκρίβειαν καταλαμβάνεται. τοίνυν ἀσύστατός ἐστιν ἡ τῶν Χαλδαίων μέθοδος.

49 The moment of conception was in principle (not in practice) privileged over that of birth; cf. Ptol. *Tetr.* III.2.1–3 and the monographic analysis of ancient conception astrology by Frommhold (2004, esp. 10–31).

50 While Sextus mostly speaks of the *time* of birth, in one instance he more specifically denies the astrologers' ability to determine the *hour* (or *time span*, cf. n. 8 above) of birth: οἱ Χαλδαῖοι ἐκπεσοῦνται τοῦ δεόντως τὴν τῆς ἀποτέξεως ὥραν ὁρίζειν (Sext. Emp. *Math.* V.67).

51 Cf. Sext. Emp. *Math.* V.55–64 (a), V.65–72 (b), and V.73–85 (c). The previous chapters are mostly a summary of astrological doctrines (V.4–42).

52 While the present contribution focuses on chronometrical, chronographical, and astronomical exactness denoted by ἀκρίβεια, it may be noted in passing that the semantic range of this word and its derivatives is much broader, including philosophical, rhetorical, sociological, historiographical, medical, and other notions (cf. Kurz 1970).

53 As opposed to the notorious quacks who used to offer their services at the circus and around the temples (and who were ridiculed by Horace, Juvenal, and other ancient authors).

54 Note that the chapter count differs between the authoritative edition of the Greek text (Hübner 1998) and the English translation by Robbins (1940, without paragraph count), where this is ch. III.2. My Greek quotations follow Hübner 1998.

> Ἐπειδὴ περὶ τοῦ πρώτου καὶ κυριωτάτου (τουτέστι τοῦ μορίου τῆς κατὰ τὴν ἐκτροπὴν ὥρας) ἀπορία γίγνεται πολλάκις—μόνης μὲν ὡς ἐπίπαν τῆς δι' ἀστρολάβων ὡροσκοπείων κατ' αὐτὴν τὴν ἔκτεξιν διοπτεύσεως τοῖς ἐπιστημονικῶς παρατηροῦσι τὸ λεπτὸν τῆς ὥρας ὑποβάλλειν δυναμένης, τῶν δὲ ἄλλων σχεδὸν ἁπάντων ὡροσκοπείων, οἷς οἱ πλεῖστοι τῶν ἐπιμελεστέρων προσέχουσι, πολλαχῇ διαψεύδεσθαι τῆς ἀληθείας δυναμένων (τῶν μὲν ἡλιακῶν παρὰ τὰς τῶν θέσεων καὶ τῶν γνωμόνων ἐπισυμπιπτούσας διαστροφάς, τῶν δὲ ὑδρολογίων παρὰ τὰς τῆς ῥύσεως τοῦ ὕδατος ὑπὸ διαφόρων αἰτιῶν καὶ διὰ τὸ τυχὸν ἐποχάς τε καὶ ἀνωμαλίας)—ἀναγκαῖον ἂν εἴη προπαραδοθῆναι, τίνα ἄν τις τρόπον εὑρίσκοι τὴν ὀφείλουσαν ἀνατέλλειν μοῖραν τοῦ ζῳδιακοῦ κατὰ τὸν φυσικὸν καὶ ἀκόλουθον λόγον [...].
>
> PTOL. *Tetr.* III.3.1–2

> Difficulty often arises with regard to the first and most important fact, that is, the fraction of the hour of the birth; for in general only observation by means of horoscopic astrolabes at the time of birth can for scientific observers give the minute of the hour, while practically all other horoscopic instruments on which the majority of the more careful practitioners rely are frequently capable of error, the solar instruments by the occasional shifting of their positions or of their gnomons, and the water clocks by stoppages and irregularities in the flow of the water from different causes and by mere chance. It would therefore be necessary that an account first be given how one might, by natural and consistent reasoning, discover the degree of the zodiac which should be rising [...].
>
> ROBBINS 1940, 229–31

In view of this difficulty, Ptolemy proposes a speculative method for determining the rising degree, which he devised for the majority of practitioners as an alternative to scientific measurement by means of an astrolabe.[55] Nevertheless, it is clear that he rejects the objection raised by sceptics like Sextus Empiricus—namely, that accurate measurement is technically impossible. In Ptolemy's view it is difficult, yet possible, for experts who have access to and know how to use an astrolabe. In similar yet more general terms, Ptolemy had insisted earlier on both the paramount importance and technical possibility of highly accurate measurements in the interest of true astrological predictions:

55 This speculative method is based on the preceding syzygy, i.e., the new or full Moon before the time of birth. The longitude(s) of Sun and Moon during this syzygy must be determined accurately (ἀκριβῶς), as he emphasizes twice (ibid. III.3.3–4).

ὅτι δὲ εὐπροφασίστως μέν, οὐ προσηκόντως δὲ τὴν πρὸς τὸ ἀδύνατον ἔσχε διαβολήν, οὕτως ἂν κατανοήσαιμεν. πρῶτον μὲν γὰρ τὰ πταίσματα τῶν μὴ ἀκριβούντων τὸ ἔργον πολλὰ ὄντα ὡς ἐν μεγάλῃ καὶ πολυμερεῖ θεωρίᾳ καὶ τοῖς ἀληθευομένοις τὴν τοῦ ἐκ τύχης παρέσχε δόξαν, οὐκ ὀρθῶς· τὸ γὰρ τοιοῦτον οὐ τῆς ἐπιστήμης ἀλλὰ τῶν μεταχειριζομένων ἐστὶν ἀδυναμία.

PTOL. *Tetr.* I.2.12

The following considerations might lead us to observe that criticism of the science on the score of impossibility has been specious but undeserved. In the first place, the mistakes of those who are not accurately instructed in its practice, and they are many, as one would expect in an important and many-sided art, have brought about the belief that even its true predictions depend upon chance, which is incorrect. For a thing like this is an impotence, not of the science, but of those who practise it.[56]

ROBBINS 1940, 13

Dozens of references to accuracy (ἀκρίβεια) in the work of Ptolemy's contemporary Vettius Valens of Antioch,[57] as well as significant features of some extant Greek horoscopes by various authors, suggest that Ptolemy's voice is not alone in its pursuit of accuracy but is part of a discourse on this ideal within the astrological community,[58] and this discourse permeated deep into the countryside of Roman Egypt. My criteria for selecting the following horoscopes are explicit claims to accuracy or unusually precise chronographical data regarding the time of birth.[59] These cases may seem few, but we ought to

56 Cf. Robbins 1940 I.2.9 on the frequent mistakes in meteorological predictions by sailors and others: παρὰ μέντοι τὸ μήτε αὐτῶν τούτων τοὺς χρόνους καὶ τοὺς τόπους ὑπὸ ἀπειρίας ἀκριβῶς δύνασθαι κατανοεῖν μήτε τὰς τῶν πλανωμένων ἀστέρων περιόδους πλεῖστον καὶ αὐτὰς συμβαλλομένας τὸ πολλάκις αὐτοῖς σφάλλεσθαι συμβαίνει. ("Yet because they cannot in their ignorance accurately know the times and places of these phenomena, nor the periodic movements of the planets, which contribute importantly to the effect, it happens that they often err," Robbins 1940, 11).

57 Cf. Pingree 1986, 466 (index). To give just one example: Vett. Val. V.3.4: τῇ οὖν γενεθλιακῇ ἡμέρᾳ κατὰ τὸ ἔτος τὸ καταγόμενον ψηφίσαντες ἀκριβῶς τοὺς ἀστέρας τὸν ὡροσκόπον οὕτως εὑρήσομεν. ("So we shall on the birthday, computing the stars accurately with regard to the current year, find the ascendant thus.")

58 Cf. the title of the monograph of Kurz (1970).

59 For detailed information on these horoscopes, see the catalogue entries in Heilen 2015, 204–333, for Hor. gr. 81.III.31, Hor. gr. 150–250a, Hor. gr. 150–250b, Hor. gr. 260.IX.29, and Hor. gr. 497.X.28. The horoscopes from Oxyrhynchus were excavated from the ancient rubbish mounds of that city, located hundreds of miles south of Alexandria. *P. Lond.* 130 is from Hermopolis (see below, section 3.2). The case of Eutocius of Ascalon, a renowned mathematician active in Alexandria around 500 CE, is different.

keep in mind that the extant evidence is a tiny part (probably far less than one thousandth) of the horoscopes that were actually cast in Greco-Roman antiquity. Hence, the absolute number of cases is less significant than the fact that the first four cases constitute about 2.5% of the extant 168 Greek documentary horoscopes.[60] Moreover, several dozens of extant horoscopes do not fulfill the criteria established above, yet specify planetary longitudes to the minute of arc;[61] they are not included in the Table 8.2 because their astronomical precision may have other reasons than the pursuit of accuracy.[62]

In these cases, the expression for accurate computation is invariably ψηφίζειν ἀκριβῶς, and the claims are remarkably bold; in the first case (*P. Lond.* 130) partly because of tiny fractions in the planetary longitudes, which will be discussed below (section 3.2), and in the other three cases, because they specify the ascendant to the minute of arc, which corresponds to a time unit of about four seconds. With regard to *P. Oxy.* XII 1476, Neugebauer comments thus:

> The accuracy in the indication of time, 10 hours and 2 degrees (≈ 8 minutes), is most unusual and obviously senseless. Computation with Theon's tables leads for the latitude of Alexandria to 9;53,36 equinoctial hours after sunrise which is equivalent to 10;6,24 ... seasonal hours or 10 seasonal hours plus 1;34 ≈ 2 time degrees, in excellent agreement with the text.
>
> NEUGEBAUER 1959, 61

In sum, the above quotations show that technical accuracy (ἀκρίβεια) has, by the second century CE, become a key-word, both in philosophical criticisms of astrology and in the apologetic statements and practical applications of its supporters.

60 See "Standard Precision" section above. The extant documentary horoscopes come from a wide variety of authors and are therefore more representative than the extant literary horoscopes (to which the fifth case belongs), which come from just a few authors (121 out of 177 from Vettius Valens alone!).

61 To name only those cast for dates that precede the earliest known application of Ptolemy's astronomy to a horoscope (which is Hor. gr. 327.VIII.17, cf. Jones 2010, 32–33), see the catalogue entries in Heilen 2015, 204–333, for Hor. gr. 46.V.13, Hor. gr. 50–150a, Hor. gr. 75.VII.19, Hor. gr. 95.IV.13, Hor. gr. 100–200a, Hor. gr. 110.III.15, Hor. gr. 114.V.13, Hor. gr. 127.XI.23, Hor. gr. 130.I.14, Hor. gr. 137.XII.4, Hor. gr. 200–300c, Hor. gr. 218.XI.27, Hor. gr. 219.V.18, Hor. gr. 250–300a, Hor. gr. 250–350a, Hor. gr. 250–350b, Hor. gr. 260.IX.29, Hor. gr. 319.XI.18–19.

62 I.e., they may simply be reporting the results of their computations as produced by their astronomical tables, without thinking much about the usefulness of that degree of precision.

TABLE 8.2 Greek horoscopes containing claims to accuracy. Legend: in the date-column, "astron." means: established by astronomical recomputation of the data; "calend.": the birthdate is calendrically expressed in the text; "paleogr.": the text is paleographically dated because no calendrical date is given and the astronomical data do not suffice for recomputation.

Source	Claim to "accurate calculation"	Date	Time (h / m = seasonal hours / minutes)	Ascendant	Precision of planetary longitudes
P. Lond. 130	lines 16–18 ἕκαστον [sc. τῶν ἑπτὰ θεῶν] ψηφίσας ἀκριβῶς[a]	31 March 81 CE (calend. and astron.)	3rd hour of night	SCO 18°	varies between full degrees and fractions of third order
P. Oxy. astron. 4276	col. II.5–6 (on the ascendant) ψηφισθεὶς ἀκρειβῶς	c. 150–250 CE (paleogr.)	1st hour of day	ARI 28° 35′	(lost)
P. Oxy. astron. 4277	frg. 1, col. 1.19 (on the ascendant) ἀκρειβῶς ψηφισθεὶς	c. 150–250 CE (paleogr.)	(lost in lacuna)	LEO 22° 11′	degree and minute of arc
P. Oxy. XII 1476	–	29 Sept. 260 CE	10h 2° (!) of day	AQR 28°	degree and minute of arc
Eutocius of Ascalon (ap. Rhet. VI.52.3–38)[b]	VI.52.5 (on the planets and the centers) ἐψηφίσαμεν ... ἀκριβῶς[c]	28 Oct. 497 CE (calend. and astron.)	7h 12m of day	AQR 2° 12′	degree and minute of arc

a "I have accurately calculated [...] each one (of the seven gods)" (Neugebauer and Van Hoesen 1959, 23).

b The full Greek text of this horoscope will become available in the late David Pingree's edition of Rhetorius (see n. 32 above). A partial English translation is available in Neugebauer and Van Hoesen 1959, 152–53.

c The full sentence reads: κατὰ τὸν χρόνον οὖν τοῦτον καὶ τόν τε τόπον τὸν εἰρημένον (τουτέστι τὴν Ἀλεξάνδρειαν) ἐψηφίσαμεν τάς τε ἐποχὰς τῶν ἀστέρων καὶ τὰ κέντρα ἀκριβῶς. ("So I computed the positions of the planets as well as the centers accurately for this time and for the said place, namely Alexandria.")

3 Extreme Precision

3.1 *The Libri Myriogeneseos*

We shall once more begin with a theoretical source, the *Libri Myriogeneseos* of Ps.-Asclepius.[63] This lost Greek work is known only through Firmicus Maternus, who refers to it several times in his *Mathesis*.[64] The *Myriogenesis* pretended to be a revelation by Hermes (i.e., Mercury),[65] the tutelary deity of astrology, and made birth-horoscope predictions for each single rising minute of arc, without taking the planets into account.[66] In other words, it differentiated 21,600 possible ascendants (12 × 30 × 60), i.e., 900 per hour, 15 per minute, or one every four seconds. No practical application is extant,[67] and although Firmicus twice refers to a Latin version of the *Myriogenesis* from his own pen in 12 books, he apparently never wrote it.[68] All that we can reliably assert about the date of the *Myriogenesis* is that it must antedate Firmicus, who wrote in the 330s CE. The extreme differentiation of ascendants makes it unlikely that the work antedated the Common Era.[69] We cannot exclude the possibility that Firmicus simply invented the *Myriogenesis*, but there is no evidence that he fabricated any of the sources that he mentions, and even if this one were a literary fiction, we would still have, through the fiction itself, an ancient testimony of the theoretical idea of differentiating 21,600 ascendants. The nearest parallel attested in the ancient astrological manuals is Firmicus's own differentiation of predictions for each single rising *degree* of arc,[70] i.e., 360 predictions for time

63 For the formation of this title, compare the title of an apocryphal text of the Old Testament, λεπτογένεσις. This title is attested only in Latin transliteration. Cf. W. Hübner, *ThLL* VII.2, 1179.11–16 s.v. *leptogenesis* (impr. 1974).

64 Firm. Mat. *Math.* III.1.2, V.1.36–38, VI.2.8, VIII.1.10, VIII.18.1. While he speaks of a single book (*liber*) in *Math.* III.1.2, he uses the plural (*libri*, without a numeral) in *Math.* VI.2.8. For more details, see Heilen 2015, 1262–65 (with refutation, ibid. n. 3181, of untenable opinions that had been advanced by some earlier scholars).

65 Firm. Mat. *Math.* V.1.38 *quam* (scil. *Myriogenesim*) *sibi venerabilem Mercurii stellam intimasse professus est* (scil. *Aesculapius*).

66 Cf. Firm. Mat. *Math.* V.1.36 *nam si Myriogenesim Aesculapii legeris* [...], *invenies ex singulis minutis sine aliquo stellarum additamento omnium geniturarum ordines explicatos.*

67 Unless one is inclined to assume that this is the tacit purpose of some of the very precise specifications of ascendants in the horoscopes of table 2 above.

68 Firm. Mat. *Math.* V.1.38; cf. ibid. VIII.1.10 (without specification of the number of books). If we are to believe Firmicus, each book would have included 1,800 predictions. This seems feasible if the predictions are very short and each book long, as some of Firmicus's extant books actually are: the third comprises more than 100 printed pages.

69 Cf. Cumont 1937, 209, n. 1.

70 Firm. Mat. *Math.* VIII.19–30. Cf. ibid. VIII.18.1 *quicquid enim Myriogenesis de singulis minutis pronuntiat, hoc nos de singulis faciemus partibus.*

intervals of four minutes each. Such a μονομοιρία[71] must have existed as early as the second century BCE, perhaps as a treatise attributed to Hermes.[72]

3.2 *P. Lond. 130*

A practical claim to extreme accuracy is contained in the first horoscope of table 2 above. It is a papyrus horoscope for an individual born on 31 March 81 CE.[73] The author of this text is an otherwise unknown Titus Pitenius, who wrote the text in Hermopolis, Lower Egypt, for a certain Hermon.[74] In his introduction, the author emphasizes that he has accurately calculated the astrologically relevant data for each one of the "seven gods" (i.e., for the Sun, the Moon, and the five planets).[75] He further states that he based his computations on so-called "perpetual tables" established by "the Egyptian men of old,"[76] and that he will give these data to the degree and minute of arc.[77] The following exposition of the astronomically and astrologically relevant data is, compared to other extant horoscopes, unusually detailed and elaborate. Among the many details, we find the typical core parameters of horoscopes—that is, the canonical set of eight ecliptic longitudes that are indispensable for astrological predictions. Titus presents these data with various levels of precision, as the Table 8.3 shows.

Our author complies with his introductory claim (i.e., that he will give the data to the degree and minute) only in four cases (Sun, Saturn, Mars, Venus). He falls short of his claim in two cases (Mercury and the ascendant) and goes beyond it in the two remaining cases (Moon and Jupiter). These last two are of special interest to us, not only because precision to sexagesimal fractions of the third order is unique among the 350 extant Greco-Roman horoscopes, but especially because such tiny fractions of a degree are equivalent to very short

71 Cf. Le Bœuffle 1987, 184–85 (no. 799) s.v. *monomoiriai*.

72 Cf. Cumont 1937, 207–16, esp. 208–09.

73 See Neugebauer and Van Hoesen 1959, 21–28 (Greek text, English translation, and commentary).

74 Lines 33–34, 185–87, and 200–02 of the text. We do not know if the native is Hermon himself, or a child born to Hermon, or still another individual. The horoscope must have been cast at least six months after the birth in question, because the text alludes (lines 37–38) to the "deified Titus" (emperor Titus died on 13 September 81 CE). Note that in our main text here, "Titus" always refers to the homonymous author of the papyrus horoscope. Paleographically speaking, the text dates from the late first or early second century CE.

75 Lines 16–18: ἕκαστον ψηφίσας ἀκριβῶς.

76 Lines 1–13: τῶν Αἰγυπτίων οἱ τὸ παλαιὸν ἄνδρες ... διὰ κανόνων αἰωνίων.

77 Lines 19–20: πρός τε μοῖραν καὶ λεπτόν.

TABLE 8.3 Astronomical longitudes in P. Lond. 130

Object	Zodiacal sign	Longitude therein		
		Author's literal expression[a]	Mathematical notation	
			With sexagesimal fractions (in parentheses, if omitted by author)	With non-sexagesimal fractions
Sun	ARI	fourteen degrees and six minutes (that is, the tenth part of a degree)[b]	$14° 6'$	$14° + \frac{1}{10}°$
Moon	TAU	thirteen degrees, and a thousandth part of a degree[c]	$13°$ $(0' 3'' 36''')$	$13° + \frac{1}{1000}°$
Saturn	PSC	six degrees, lacking a sixtieth of a degree[d]	$(5° 59')$	$6° - \frac{1}{60}°$
Jupiter	CNC	six degrees and ten sixtieths of the third order, which are one twenty-one-thousand-six-hundredth part of a degree[e]	$6° 0' 0'' 10'''$	$6° + \frac{1}{21600}°$
Mars	AQR	sixteen degrees and a twentieth[f]	$16°$ $(3')$	$16° + \frac{1}{20}°$
Venus	PSC	sixteen degrees and four minutes, which is the fifteenth part of a degree[g]	$16° 4'$	$16° + \frac{1}{15}°$
Mercury	ARI	ten full degrees[h]	$10°$	
Ascendant	SCO	eighteen degrees[i]	$18°$	

a These expressions are quoted from the translation of Neugebauer and Van Hoesen 1959, 23–24.

b Lines 53–58: Κριοῦ μοίρας … δεκατέσσαρας καὶ προσέτι λεπτὰ ἕξ, τουτέστιν μέρος δέκατον μοίρης.

c Lines 76–80: Ταύρου μοίρας … δεκατρεῖς καὶ ἔτι μέρος χιλιοστὸν μοίρης.

d Lines 97–100: Ἰχθύων μοίρας … ἓξ λειπούσας μέρος ἑξηκοστὸν μοίρης.

e Lines 110–15: Καρκίνου μοίρας ἓξ καὶ ἔτι λεπτὰ τρίτα δέκα, ἃ δὴ μέρος μοίρης ἐστὶν δισμυροχιλεξακοσιοστόν. The last word is morphologically incorrect. It ought to be δισ-μυ-ρι-ο-στο-οχιλιοστοεξακοσιοστόν.

f Lines 127–30: Ὑδροχόου μοίρας δέκα ἓξ εἰκοστόν.

g Lines. 140–45: Ἰχθύων μοίρας … δέκα ἓξ καὶ λεπτὰ τέσσαρα, ἃ δὴ πεντεκαιδέκατον μοίρης μέρος ἐστίν.

h Lines 158–60: Κριοῦ … μοίρας στερεὰς δέκα.

i Lines 166–68: Σκορπίου … μοίρας δέκα ὀκτώ.

time spans: on the date and hour in question, the Moon advanced $\frac{1}{1000}°$ in about 6½ seconds, while Jupiter advanced $\frac{1}{21600}°$ in about 40 seconds.[78]

How did these data originate? Are they to be taken seriously? And if not, what is their purpose? As to the data's origin, the perpetual tables that Titus claims to have employed are no longer extant, but their existence in antiquity is confirmed by Vettius Valens (second century CE), who asserts that, for accurate predictions regarding the periods of life, one must measure the time of every nativity accurately and then derive the planetary longitudes with precision to the degree from "perpetual tables."[79] This shows that such tables allowed computations to the degree (at least), and it seems to imply that they were well known among and available to practicing astrologers.[80] As to our horoscope, Neugebauer comments:

> From our present text it seems [...] to follow that these tables were computed to at least three sexagesimal places. This is not surprising in itself because the fundamental period relations can only be guaranteed if the mean motions are computed accurately enough. Three places are given, e.g., for the Moon in *P. Lund.* 35a and four places in *P. Ryl.* 27. It is no contradiction that often no fractions at all appear in the final results where a higher accuracy is of no practical value.
>
> NEUGEBAUER AND VAN HOESEN 1959, 24

Let us, then, temporarily concede that the "perpetual tables" may have led to data of such precision as in our horoscope. Did Titus Pitenius, then, not omit those tiny fractions because he thought they were of practical value? This is doubtful for various reasons:

a) The text itself does not contain any clue to what might be the practical value of the data's partly extreme precision.[81]

---

78 Computed with *Galiastro* 4.1 for 31 March 81 CE, 9 PM (Alexandria), according to modern time-reckoning in equal hours.

79 Vett. Val. VI.2.2: ἐπὶ πάσης γενέσεως ἀκριβῶς ἀστερίσαντας τὸν χρόνον μοιρικῶς ἐκ τῆς τῶν Αἰωνίων κανόνων πραγματείας χρὴ λαμβάνειν ἀπὸ ἑνὸς ἑκάστου ἀστέρος μοίρας ἐφ' ὃν βούλεταί τις κτλ.

80 Another source is Ptolemy, who criticizes the shortcomings of "perpetual tables" in the *Almagest* (IX.2) in a survey of earlier attempts to develop a consistent theory of planetary motion. His description of such tables is, however, difficult to reconcile with the texts of Titus Pitenius and Vettius Valens (cf. Neugebauer 1975, 789, and Evans 1998, 345).

81 The text complies with the typical format of papyrus horoscopes in reporting data without making predictions, which were usually not put in writing but delivered orally during astrological consultations.

b) The thousands of pages of extant Greco-Roman astrological manuals never differentiate predictions beyond the degree or arc, and the same is true of the rare cases of papyrus horoscopes that contain predictions.

c) Titus Pitenius is content with rather vague data in two respects which, being of central astrological importance, would make high precision desirable: firstly, he specifies the time of birth as "the third hour of night."[82] It would require a much more precise chronometrical input to justify computational accuracy to fractions of the third order in the resulting planetary longitudes. Secondly, Titus calls the midheaven quite vaguely "the meridian at right angles to this" (i.e., to the ascendant) and states that it fell in LEO.[83] For the given ascendant 18° SCO, this would lead to 18° LEO, a gross inaccuracy because the true midheaven would, on the geographical latitude of Lower Egypt,[84] be 23° 40′ LEO.[85]

If, then, Titus's motivation for specifying some planetary longitudes to fractions of third order was, in all likelihood, not a genuine pursuit of astronomical accuracy, what was it? I think the only plausible answer is: to impress the reader with what has the superficial appearance of utmost accuracy. This explanation ties in with observations made by Roger Beck who has analyzed Titus's text from a narratological perspective. Beck emphasizes Titus's "highly elaborate descriptions of planetary journeyings [that] have replaced the usual matter-of-fact listing of celestial longitudes."[86] He pertinently speaks of "poetic astronomy."[87] Not only does Titus present the planets as divine beings, but he also stimulates the reader's fantasy with references to their various colors

82 Lines 40–42: ἐπὶ τρίτης τῆς νυκτὸς ὥρας.

83 Neugebauer and Van Hoesen 1959, 24 (lines 173–176).

84 Neugebauer and Van Hoesen 1959, 24 (lines 201–202 [Hermopolis]).

85 See the commentary of Neugebauer and Van Hoesen 1959, 26. By the way, Titus's inaccuracy is of the same magnitude (5° 40′) as that produced by the old-fashioned numerical procedure for calculating midheaven that is given by Paul of Alexandria: see Paul. Alex. 30 pp. 81–82 Boer, with the astronomical commentary of O. Neugebauer ibid. 140–45, and the table of rising times at Neugebauer 1975, 732 (*clima* of Alexandria, Babylonian system A). The MC is then obtained as the sum of 180° (the rising times from 0° ARI to 30° VIR) + 38° 20′ + ($^{18}/_{30}$ × 35°) = 239° 20′, to be measured from 0° CAP. This yields 29° 20′ LEO, which is 5° 40′ away from the correct MC (23° 40′ LEO).

86 Beck 2013, 397 (abstract).

87 Beck 2013 (keywords). While I altogether appreciate Beck's observations, I find his repeated emphasis on the "story" that is allegedly being told unconvincing. A story needs a plot, and that is absent here. Titus Pitenius rather offers a prose ecphrasis, and his intention seems to be similar to that which Giovanni Pontano, more than a millennium later, has theorized in his poetics of *admiratio* and practiced in his astrological didactic poem *Urania* (cf. Weh 2017, esp. 43–53).

and motions. One of the textual features meant to fascinate the reader is the extreme precision of some data.

One may object, however, that only two out of eight positions are extremely precise, and that some oddities remain: even Neugebauer, the authoritative commentator, could only express his perplexity regarding Titus's references to "a thousandth part of a degree" in the Moon's case because "we have no tables which give decimal fractions of degrees," and to Saturn's position "lacking a sixtieth of a degree" because that is "very unusual terminology for 59 minutes."[88] Hence, we need a more comprehensive explanation than that advanced in the last paragraph. The key to finding it lies in one important aspect of Titus's data which is quite obvious, yet seems to have been overlooked by all modern readers of his text: all six fractions mentioned by Titus are unit fractions. It would be absurd to interpret this fact as a chance product. We must rather conclude that Titus Pitenius has systematically replaced the sexagesimal fractions that resulted from his astronomical computations with unit fractions (the mathematical quality of this conversion remains to be seen). In three out of the six relevant cases he mentions what seem to have been the intermediate results of his computations, i.e., sexagesimal fractions, before converting them (Sun, Jupiter, Venus); in the remaining three cases he passes in silence over the sexagesimal fractions and mentions only the final unit fractions (Moon, Saturn, Mars). The reason for these omissions may be just occasional carelessness, but a more likely explanation will be given shortly.

Let us first glance at an interesting parallel in *P. Paris* 19bis, a fragmentary papyrus horoscope from Thebes which was cast for a birth in 137 CE. Only one planetary longitude is preserved; it reads: "The divine Sun [...] was found in the sign Sagittarius, 16 degrees 35 parts of sixty, which is the half (and) twelfth (part of a degree)."[89] The sexagesimal value 35′ is here converted into a sum of unit fractions, $\frac{1}{2}^\circ + \frac{1}{12}^\circ$. There is good reason to assume that the other longitudes were computed with the same degree of precision,[90] and it is likely

88 Neugebauer and Van Hoesen 1959, 25. Cf. Bouché-Leclercq (1899, 258, n. 3): "L'auteur du papyrus CXXX prétend estimer des fractions invraisemblables de degré" (with reference to line 114) and ibid. 291, n. 1.

89 *P. Paris* 19bis (= Hor. gr. 137.XII.4 in Heilen 2015, 265), col. I.12–14; trans. Neugebauer and Van Hoesen (1959, 42).

90 We have a variant of the same horoscope that is extant in two copies (*P. Paris* 19 and *P. Lond.* 110; see Neugebauer and Van Hoesen 1959, 39–42). This variant likewise specifies all longitudes to the degree and minute (except for the ascendant), but it is otherwise far less detailed. Its longitude of the Sun (13° 23′ SGR, the only figure that can be compared) is different from that in *P. Paris* 19bis (16° 35′ SGR). It is therefore likely that the variant horoscope was cast first, and that its owner went later to another astrologer who cast a second, independent horoscope (ibid. 39), i.e., the one that interests us (*P. Paris* 19bis).

that some or all of these sexagesimal fractions were in a second step converted into (sums of) unit fractions. Interestingly, the author of our horoscope was, in all likelihood, not a native speaker of Greek, as various errors of spelling and grammar indicate.[91] Moreover, he introduced his text with references to revered astrological authorities: Chaldaeans, as well as Egyptians, with special emphasis on the Egyptian gods Amenhotep, son of Hapu, and Imhotep, son of Ptah.[92] In other words, we have here a poorly Hellenized Egyptian astrologer working in Thebes, one of the traditional religious centers of Egypt, who converts sexagesimal fractions into unit fractions, which are typical of Egyptian mathematics.

Why, then, would someone with a Roman name (Titus Pitenius) be so keen on using unit fractions instead of sexagesimal fractions, the typical results of computations based on astronomical tables? The only plausible answer is: to imitate typically Egyptian habits of a remote past when the legendary founders of astrology, especially King Nechepsos and the sage Petosiris, were supposed to have lived.[93] Neugebauer states:

> The primitive, strictly additive, Egyptian way of computing with unit fractions had a detrimental effect throughout, even on Greek astronomy. For example in the *Almagest* one can find many cases where final results are given in unit fractions, abolishing for no good reason a more accurate value determined by consistent sexagesimal computation.
>
> NEUGEBAUER 1975, 559[94]

---

91 The same is true of the author of the earlier variant text (see previous note).

92 See Ryholt 2011, 71, and Heilen 2015, 548 and 553.

93 The figure of Nechepsos is inspired by the historical King Necho II (610–595 BCE); cf. Heilen 2015, 551.

94 Cf. Neugebauer 1975, 65: "This is one of the many cases which one can find in the *Almagest* where final results expressed with unit fractions deviate badly (and unnecessarily much) from the accurately computed results." For an example, cf. ibid. 938, n. 3: "Ptolemy replaced the Babylonian value M = 14;24h by 14;25h because he used ⅟₁₂ as the smallest unit fraction (*Geogr.* VIII 27)." Cf. similar remarks on inaccuracies caused by Ptolemy's use of unit fractions ibid. 81, n. 5; 853, n. 18; 1001, n. 15; and, more generally on Greek astronomy, ibid. 591: "In part it is no doubt due to the influence of the Egyptian arithmetical habits when results of sexagesimal computations are finally given in the form of unit fractions." This happens as early as ca. 300 BCE (ibid. 706 with reference to the astronomical Greek papyrus *P. Hibeh* 27). Although Neugebauer refers to our papyrus in other passages of the same work (ibid. 592; 657, n. 3; 789; 806, n. 7), he never addresses Titus Pitenius's use of unit fractions. On unit fractions in Egyptian mathematics, see Gillings 1972, 74–119; Imhausen 2007; Imhausen 2016, 89–99 and 189–92. For the traditional Egyptian use of two non-unit-fractions, ⅔ and ⅚, see Hughes and Jasnow 1997, 85. In Demotic texts from the 3rd c. BCE to the Roman period one finds complex expressions that seem to be common

Unlike Ptolemy, however, Titus Pitenius did not just comply with a widespread (in Roman Egypt) computational habit. Rather, he manipulated the sexagesimal data that resulted from his tables with the artificial goal of obtaining just a *single* unit fraction of one *degree* of arc in each planet's case. My reasons for this contention are two:

Firstly, it is theoretically possible, yet extremely unlikely, that a set of six sexagesimal results should allow conversion of the respective fractions into six single unit fractions of full degrees of arc. The reason is easy to grasp: if we limit the level of precision to minutes of arc, we have sixty possible results in sexagesimal notation for each astronomical longitude, but only eleven of these possible results (i.e., ca. 18%) either are or can be reduced to single unit fractions.[95] If we raise the level of precision to seconds of arc, which are implied in Titus's longitude of the Moon, we will have 3,600 possible results, but the number of those that can be expressed as single unit fractions of full degrees grows to no more than 44 (i.e., ca. 1% of 3,600).[96] With every higher order of degree-fractions, the percentage of convertible candidates decreases exponentially: at the third order, which is implied in Titus's longitude of Jupiter, we will have 216,000 possible results, of which only 111 (i.e., ca. 0.05%) can be expressed as single unit fractions of full degrees.[97] Moreover, even though the absolute number of reducible fractions will keep growing, they will increasingly cluster near zero.

Secondly, statistically speaking, every second computational result will contain a fraction that is larger than ½ and thus defies expression as a single unit fraction. Titus seems to have obtained one such result in the case of Saturn (5° 59′ PSC). It would have been normal to express the 59 minutes of arc as a *sum*

fractions (cf. Parker 1972, 8–10), but Fowler (1999, 262) more appropriately calls them "incomplete divisions." He is followed by Imhausen (2016, 191) who states: "The evidence from the demotic mathematical papyri clearly proves that the concept of fractions did not change." Caution is advised regarding Knorr (1982) who thought that some late antique Greek (and Egyptian) mathematicians used common fractions.

95 These are the fractions whose denominators are either 1 or integer factors of 60: namely, 2, 3, 4, 5, 6, 10, 12, 15, 20, and 30.

96 These are the fractions whose denominators are those specified in the previous note, as well as 8, 9, 16, 18, 24, 25, 36, 40, 45, 48, 50, 60, 72, 75, 80, 90, 100, 120, 144, 150, 180, 200, 225, 240, 300, 360, 400, 450, 600, 720, 900, 1200, 1800.

97 These are the fractions whose denominators are those specified in the previous two notes, as well as 27, 32, 54, 64, 96, 108, 125, 135, 160, 192, 216, 250, 270, 288, 320, 375, 432, 480, 500, 540, 576, 675, 750, 800, 864, 960, 1000, 1080, 1125, 1350, 1440, 1500, 1600, 1728, 2000, 2160, 2250, 2400, 2700, 2880, 3000, 3375, 3600, 4000, 4320, 4500, 4800, 5400, 6000, 6750, 7200, 8000, 8640, 9000, 10800, 12000, 13500, 14400, 18000, 21600, 24000, 27000, 36000, 43200, 54000, 72000, and 108000.

of unit fractions, as another Greek horoscope actually does: $\frac{1}{2} + \frac{1}{3} + \frac{1}{12} + \frac{1}{15}$ (= $\frac{59}{60}$).[98] To switch from this usual additive syntax to a subtraction from the next higher integer (6° – $\frac{1}{60}$°) is very unusual (if not unique), but it makes sense if your purpose is to employ only single unit fractions.

It is therefore likely that Titus has silently rounded some or all of his computational results to values that allowed for conversion into single unit fractions,[99] and we cannot even exclude the possibility that he invented the Moon's and Jupiter's unit fractions altogether, the latter possibly with a view to those 21,600 predictions in the *Libri Myriogeneseos* of Ps.-Asclepius,[100] whose authority could make this very number appear as a symbol of utmost astrological precision. But even if he only rounded some or all results, this still means that his data's appearance of highest astronomical and, by implication, chronometrical accuracy is bought at the price of hidden inaccuracies.

A comparison with a roughly contemporary horoscope, *P. Oxy.* XXXI 2555 (cast for 13 May 46 CE),[101] illustrates the artificiality of Titus's astronomical data. Those of the Oxyrhynchus horoscope are: Moon: 19° $\frac{1}{2}$ $\frac{1}{3}$ $\frac{1}{12}$ (= 19° $\frac{11}{12}$ = 19° 55′) CAP; Sun: 29° $\frac{1}{2}$ $\frac{1}{3}$ $\frac{1}{12}$ (= 29° $\frac{11}{12}$ = 29° 55′) TAU; Saturn: 2° $\frac{1}{10}$ (= 2° 6′) CAP; Jupiter: 23° $\frac{1}{2}$ $\frac{1}{4}$ (= 23° $\frac{3}{4}$ = 23° 45′) CNC; Mars: 27° $\frac{1}{2}$ (= 27° 30′) GEM. The data regarding Venus, Mercury, and the ascendant are lost. Three of the five extant longitudes are expressed in sums of unit fractions (i.e., only the remaining two are single unit fractions), and none of the ten total unit fractions is smaller than $\frac{1}{12}$, the same unit-fraction beyond which Ptolemy does not go.[102]

In sum, the evidence strongly suggests (a) that Titus Pitenius wished to present himself, despite his foreign name, as a genuine disciple of those "Egyptian men of old" whom he praises in his introduction and whose tables he professes to have used, (b) that he pursued this goal by replacing Greco-Babylonian sexagesimal fractions with typically Egyptian unit fractions, (c) that he did so in a stylized manner (i.e., just a single unit fraction for each planetary deity, instead of more or less complex sums of unit fractions), and (d) that he expected his exposition of the data to look all the more Egyptian, all the more accurate, and all the more impressive, the tinier some of his unit fractions were. The artificial manner in which Titus treats the astronomical data proves that he is not making the typical amateurish mistake of reporting tiny fractions out of

98 The source is Rhetorius, *Epitome* IV.15. See Neugebauer and Van Hoesen 1959, 133 n. 22, and Heilen 2015, 299 (Hor. gr. 400–401[?]).

99 Compare, outside the field of astronomy, complex fractional computations (such as root computation) in which rounding off is often involved (Knorr 1982, 156).

100 See above, n. 63.

101 Cf. Heilen 2015, 223 ('Hor. gr. 46.V.13').

102 Cf. above, n. 95.

ignorance of their practical irrelevance and uselessness. Whether he did or did not understand the technical absurdity of specifying the Moon's and Jupiter's position with that extreme precision, we cannot tell, and the answer does not really matter because his motivation for presenting those data thus is his literary purpose evinced above. This insight allows us to eliminate his horoscope from the astrological evidence for the practical relevance of time-units smaller than one hour.

4 Conclusion

One could adduce other examples from hitherto unexplored areas of ancient astrology, such as Galen's creation of an artificial "medical week," whose exact end point, which was important for the medical theory of "critical days," he defined with precision to six minutes of time;[103] or the earlier unique endeavor of the Roman astrologer Tarutius, an acquaintance of Cicero and Varro, who computed, at the latter's request, the exact times of the conception and birth of Romulus, as well as that of the foundation of Rome, with the following results: Romulus had been conceived in his mother's womb during the third hour of 24 June 772 BCE; he was born on 24 March 771 BCE at sunrise; and he founded Rome on 4 October 754 (*sic*) at the end of the second hour of day.[104] But these are special cases, and it is time to draw conclusions from the more representative examples that have been discussed in the previous sections: chronometrical accuracy and chronographical precision down to one seasonal hour is omnipresent in ancient Greco-Roman (as well as Egyptian) astrology, not only

103 Galen, *De dieb. decr.* III.9. 932–933 Kühn. Thanks to Galen's paramount importance in the history of medicine, this artificial "medical week" had a significant historical influence. For details, see Heilen (2018). Galen's chronographical precision on this occasion is relevant to our purpose because of its implicit claim to extreme medical empiricism: the end-points of medical weeks depend on the time when an illness grows so serious that the patient must take to his or her bed (the so-called "decumbiture"). Since, however, this time cannot be grasped with precision to six minutes by ancient patients or their family members, it makes little sense to determine the end points of the last days of the following medical weeks (i.e., the end-points of the critical days) so precisely. Cf. for example the early modern criticism by Giovanni Manardi (1462–1536), who remarks that this would require determination of the time of the *decubitus* with an astrolabe (Manardi, epistle 15.5 in *Epistolae medicinales*, ed. Basil. 1549, 406–21, here: 414). On Galen, see also Miller's chapter below.

104 For details regarding Tarutius's astronomically astonishing computation and his motivation for departing from the traditional date of the foundation of Rome (21 April 753 BCE), see Heilen 2007, esp. 46–47 and 52–53, and Heilen 2018, 425–26 (section 3.5).

in the casting of horoscopes, but also in the observance of time-rulership for all sorts of activities. Moreover, we find theoretical demands for, and practical attempts at, measuring time more accurately in order to cast horoscopes more accurately. Due to technical difficulties, however, these attempts did, on the whole, not go beyond specifications that the horoscope had been cast for the beginning or the end of a certain seasonal hour.[105] The measure of the half hour, which is attested on some extant sundials, and once in the context of the ancient postal service, is never mentioned in the extant horoscopes, probably because this measure was very rare in normal social contexts.[106] It is only in very rare and late extant cases that we find horoscopes computed to the hour and minute (or hour and degree) of time: that is, with the same degree of chronometrical accuracy and chronographical precision as modern horoscopes.[107] Astrology is the only ancient science whose validity came, in the ancient discourse, to depend more and more on the question of whether its practitioners were able to measure short time accurately. One may therefore conjecture that it played a role in the diffusion of more accurate timekeeping in the ancient world.

References

Alfieri, Nereo. 1948. "I fasti consulares di potentia (Regio V)." *Athenaeum n.s.* 26: 110–134.

Beck, Roger. 2013. "Imagery and Narrative in an Ancient Horoscope: P.Lond. 130 (Greek Horoscopes No. 81)." *Journal for the Study of Religion, Nature and Culture* 7: 397–406.

Bezza, Giuseppe. 1995. *Arcana mundi. Antologia del pensiero astrologico antico*. Milan: Rizzoli.

Boer, Emilie, ed. 1958. *Pauli Alexandrini elementa apotelesmatica*. Leipzig: Teubner.

Boll, Franz. 1912. "Hebdomas." *Realencyclopädie der classischen Altertumswissenschaft*, 2547–78. Stuttgart: Alfred Druckenmüller.

Bouché-Leclercq, Auguste. 1899. *L'astrologie grecque*. Paris: Culture et civilisation.

Bury, Robert G., ed. 1949. *Sextus Empiricus with an English Translation, Vol. IV: Against the Professors*. Cambridge, MA: Harvard University Press.

Catalogus codicum astrologorum Graecorum. 1898–1953. 12 vols. Brussels: Henri Lamertin.

Cumont, Franz, ed. 1921. *Catalogus codicum astrologorum Graecorum. Codicum Parisinorum pars quarta*. Brussels: Henri Lamertin.

105 See above, n. 14–15.

106 Tolsa 2017, 214, n. 27, with reference (for the sundials) to Hannah (2009, 125) and (for the couriers) to Remijsen (2007, 138).

107 See the last two examples of Table 8.2 above.

Cumont, Franz. 1937. *L'Égypte des astrologues*. Brussels: Fondation égyptologique reine Élisa.

Divjak, Johannes, and Wolfgang Wischmeyer, eds. 2014. *Das Kalenderhandbuch von 354. Der Chronograph des Filocalus*. 2 vols. Vienna: Holzhausen.

Evans, James. 1998. *The History and Practice of Ancient Astronomy*. Oxford: Oxford University Press.

Fowler, David H. 1999. *The Mathematics of Plato's Academy. A New Reconstruction*. 2nd ed. Oxford: Oxford University Press.

Frommhold, Katrin. 2004. *Bedeutung und Berechnung der Empfängnis in der Astrologie der Antike*. Münster: Aschendorff.

Gillings, Richard J. 1972. *Mathematics in the Time of the Pharaohs*. Cambridge, MA: MIT Press.

Greenbaum, Dorian. 2016. *The Daimon in Hellenistic Astrology. Origins and Influence*. Leiden: Brill.

Greenbaum, Dorian, and Alexander Jones. 2017. "P.Berl. 9825: An Elaborate Horoscope for 319 CE and its Significance for Greek Astronomical and Astrological Practice." *ISAW Papers* 12. Accessed June 5, 2019, http://doi.org/2333.1/brv15m2n.

Hannah, Robert. 2009. *Time in Antiquity*. New York: Routledge.

Heilen, Stephan. 2007. "Ancient Scholars on the Horoscope of Rome." *Culture and Cosmos* 11 (1–2): 43–68.

Heilen, Stephan. 2015. *Hadriani genitura. Die astrologischen Fragmente des Antigonos von Nikaia. Edition, Übersetzung und Kommentar*. Berlin: de Gruyter.

Heilen, Stephan. 2018. "Galen's Computation of Medical Weeks: Textual Emendations, Interpretation History, Rhetorical and Mathematical Examinations." *SCIAMVS* 19: 201–279.

Heilen, Stephan. 2019. "Hellenistic Horoscopes in Greek and Latin: Contexts and Uses." In *Hellenistic Astronomy. The Science in Its Contexts*, edited by Alan C. Bowman and Francesca Rochberg, 490–508. Leiden: Brill.

Holden, James H. 2009. *Rhetorius the Egyptian. Astrological Compendium Containing His Explanation and Narration of the Whole Art of Astrology, Translated from the Greek*. Tempe, AZ: American Federation of Astrologers.

Holden, James H. 2012. *Paul of Alexandria. Introduction to Astrology. Translated from the Greek*. 3rd ed. Tempe, AZ: American Federation of Astrologers.

Hübner, Wolfgang. 1982. "Zwei griechische Texte über die Tages- und Stundenherrscher." *Zeitschrift für Papyrologie und Epigraphik* 49: 53–66.

Hübner, Wolfgang, ed. 1998. *Claudii Ptolemaei Ἀποτελεσματικά*. Leipzig: Teubner.

Hughes, George R., and Richard Jasnow. 1997. *Oriental Institute Hawara Papyri. Demotic and Greek Texts from an Egyptian Family Archive in the Fayum (Fourth to Third Century B.C.)*. Chicago: Oriental Institute.

Imhausen, Annette. 2007. "Egyptian Mathematics." In *The Mathematics of Egypt, Mesopotamia, China, India, and Islam*, edited by Victor Katz, 7–56. Princeton: Princeton University Press.

Imhausen, Annette. 2016. *Mathematics in Ancient Egypt. A Contextual History*. Princeton: Princeton University Press.

Jones, Alexander, ed. 1999. *Astronomical Papyri from Oxyrhynchus* (*P. Oxy. 4133–4300a*). *Edited with Translations and Commentaries*. 2 vols. Philadelphia: American Philosophical Society.

Jones, Alexander. 2010. "Ancient Rejection and Adoption of Ptolemy's Frame of Reference for Longitudes." In *Ptolemy in Perspective. Use and Criticism of His Work from Antiquity to the Nineteenth Century*, edited by Alexander Jones, 11–44. New York: Springer.

Jones, Alexander, and John Steele. 2011. "A New Discovery of a Component of Greek Astrology in Babylonian Tablets: The 'Terms.'" *ISAW Papers* 1. Accessed June 5, 2019, http://doi.org/2333.1/k98sf96r.

Knorr, Wilbur. 1982. "Techniques of Fractions in Ancient Egypt and Greece." *Historia Mathematica* 9: 133–171.

Kühn, Karl G., ed. 1825. *Claudii Galeni Opera Omnia*. Vol. IX. Leipzig: Knobloch. [Repr. Hildesheim: Olms, 2001, and Cambridge: Cambridge University Press, 2012.]

Kurz, Dietrich. 1970. *AKPIBEIA. Das Ideal der Exaktheit bei den Griechen bis Aristoteles*. Göppingen: Kümmerle.

Le Bœuffle, André. 1987. *Astronomie astrologie. Lexique Latin*. Paris: Picard.

Long, Anthony A. 1982. "Astrology: Arguments Pro and Contra." In *Science and Speculation. Studies in Hellenistic Theory and Practice*, edited by Jonathan Barnes, Jacques Brunschwig, Myles Burnyeat, and Malcolm Schofield, 165–192. Cambridge: Cambridge University Press.

Mommsen, Theodor, ed. 1883. *Corpus inscriptionum latinarum consilio et auctoritate academiae litterarum regiae Borussicae editum. Volum nonum: Inscriptiones Calabriae Apuliae Samnii Sabinorum Piceni latinae*. Berlin: Georg Reimer.

Neugebauer, Otto. 1957. *The Exact Sciences in Antiquity*. 2nd ed. Providence, RI: Brown University Press.

Neugebauer, Otto. 1975. *A History of Ancient Mathematical Astronomy*. New York: Springer.

Neugebauer, Otto, and Henry B. Van Hoesen. 1959. *Greek Horoscopes*. Philadelphia: American Philosophical Society.

Paci, Gianfranco. 1995. "Le iscrizioni romane di Potentia." In *Atti Del XXIX Convegno di Studi Maceratesi, Porto Recanati, 13–14 Novembre 1993*, 1–25. Macerata: Centro di Studi Maceratesi.

Parker, Richard A. 1972. *Demotic Mathematical Papyri*. Providence, RI: Brown University Press.

Pingree, David, ed. 1986. *Vettii Valentis Antiocheni anthologiarum libri novem*. Leipzig: Teubner.

Pingree, David, ed. (forthcoming). *Rhetorii Aegyptii compendium astrologicum: Libri V et VI. Imprimendum curavit S. Heilen*. Berlin: De Gruyter.

Ramsay, George G., ed. 1928. *Juvenal and Persius with an English Translation*. London: William Heinemann.

Remijsen, Sofia. 2007. "The Postal Service and the Hour as a Unit of Time in Antiquity." *Historia* 56 (2): 127–40.

Robbins, Frank E., ed. 1940. *Ptolemy, Tetrabiblos, Edited and Translated*. Cambridge, MA: Harvard University Press.

Ross, Micah Tecumseh. 2006. "An Introduction to the Horoscopic Ostraca of Medînet Mâdi." *Egitto e Vicino Oriente* 29: 147–80.

Rüpke, Jörg. 1995. *Kalender und Öffentlichkeit. Die Geschichte der Repräsentation und religiösen Qualifikation von Zeit in Rom*. Berlin: de Gruyter.

Rüpke, Jörg. 2004. "Fasti." In *Brill's New Pauly. Encyclopaedia of the Ancient World*, vol. 5, 361–66. Leiden: Brill.

Ryholt, Kim. 2011. "New Light on the Legendary King Nechepsos of Egypt." *Journal of Egyptian Archaeology* 97: 61–72.

Swoboda, Anton, ed. 1889. *P. Nigidii Figuli Operum Reliquiae*. Amsterdam: Hakkert.

Timpanaro, Sebastiano. 1998. *Marco Tullio Cicerone, Della divinazione. Introduzione, traduzione e note*. Milan: Garzanti.

Tolsa, Cristian. 2017. "Time of Birth and Ascendant in the Papyrus Horoscopes." *Zeitschrift für Papyrologie und Epigraphik* 204: 209–220.

Wardle, David. 2006. *Cicero on Divination. De divinatione. Book 1. Translated with Introduction and Historical Commentary*. Oxford: Oxford University Press.

Weh, Dennis, ed. 2017. *Giovanni Pontanos Urania Buch 1. Einleitung, Edition, Übersetzung und Kommentar*. Wiesbaden: Harrassowitz.

CHAPTER 9

Hourly Timekeeping and the Problem of Irregular Fevers

Kassandra Jackson Miller

Hourly timekeeping played an important role in certain medical debates of the Roman Imperial period, particularly over how to classify types of intermittent fevers, whose intensities ebb and flow according to recognizable temporal schemes. Such periodicity is now recognized as the signature symptom of malaria, three strains of which are endemic to the Mediterranean: *Plasmodium falciparum*, *P. vivax*, and *P. malariae*.[1] From the late Classical period onward, physicians competed to identify the temporal patterns of these diseases, in order to better predict the outcomes of individual cases. It was important that a doctor learn to prognosticate as accurately as possible, for he could endanger his medical reputation by taking on incurable patients.[2]

---

1 Malarial species belong to the genus *Plasmodium* and are transmitted by mosquito vectors. *P. falciparum* produces a malignant tertian fever (whose intensities peak every 48 hours, or third day inclusive), *P. vivax* a benign tertian, and *P. malariae* a quartan (whose intensities peak every fourth day inclusive). These periodicities result from the fact that, at the end of synchronized cycles of spore-generation, malarial protozoa rupture host blood cells, inducing a paroxysm in the patient. See Scheidel 2001, 75, and Sallares 2002, 9–11.

2 The Hippocratic author of *Prognostic* summarizes this state of affairs thus (all translations are my own): Τὸν ἰητρὸν δοκέει μοι ἄριστον εἶναι πρόνοιαν ἐπιτηδεύειν· προγιγνώσκων γὰρ καὶ προλέγων παρὰ τοῖσι νοσέουσι τά τε παρεόντα καὶ τὰ προγεγονότα καὶ τὰ μέλλοντα ἔσεσθαι, ὁκόσα τε παραλείπουσιν οἱ ἀσθενέοντες ἐκδιηγεύμενος, πιστεύοιτ' ἂν μᾶλλον γιγνώσκειν τὰ τῶν νοσεόντων πρήγματα, ὥστε τολμᾷν ἐπιτρέπειν τοὺς ἀνθρώπους σφέας ἑωυτοὺς τῷ ἰητρῷ.... καὶ τοὺς ἀποθανουμένους τε καὶ σωθησομένους προγιγνώσκων καὶ προαγορεύων ἀναίτιος ἂν εἴη (II.110–112 Littré = 1.1–3.11 Jouanna) ("It seems to me to be an excellent thing that the physician practice prognosis. For if he should prognosticate and foretell, in the presence of patients, things present, things past, and what will be in the future, and if he explains however many things the patients have neglected, he would be more trusted to know the circumstances of those who are sick, with the result that men will dare to entrust themselves to this physician.... [A]nd by knowing and declaring in advance who will die and who will be saved, he can be blameless"). Galen, likewise, declares in *On Periods*: Ἐπεὶ τοίνυν οὐδὲν οὕτω χρήσιμόν ἐστι, καὶ μάλιστα ἐν τοῖς πυρετοῖς, ὡς τὸ προγινώσκειν τὰς ἀρχὰς τῶν ἐσομένων παροξυσμῶν, ἐὰν μέν τις εὐθέως ἀπ' ἀρχῆς οἷός τε ᾖ στοχασμῷ τεχνικῷ τοῦτ' ἐξεργάζεσθαι, μέγιστα τὸν κάμνοντα δι' ὅλου τοῦ νοσήματος ὠφελήσει· (VII.479–480 Kühn) ("Nothing is so useful—and especially in the case of fevers—as prognosticating the beginnings of the paroxysms yet to come. If someone is able

 | DOI:10.1163/9789004416291_011

The Hippocratic writers classified periodic fevers according to the number of whole days that passed between each paroxysm of fever and chills.[3] A τριταῖος ("tertian") fever, for example, could be recognized by the fact that it recurred every two days (or every third day, counting inclusively). Troublingly, however, some intermittent fevers recurred at irregular intervals, which could not be neatly measured in whole numbers of days.[4] These fevers defied reliable prognosis and could make fools of physicians at the bedside.

This was a problem, and one that stimulated debate into the Roman Imperial period. By that time, however, more sophisticated tools had become available to physicians interested in tracking febrile paroxysms. Sundials and water clocks, which could mark time down to the hour, had become increasingly widespread, and were available in formats ranging from the monumental to the portable.[5] In addition, the practice of scientific table-making had made its way from Babylonia to the Greco-Roman world.[6] While, at first, these technologies belonged almost exclusively to the purview of astronomers,

---

straightaway, from the beginning, to accomplish this by means of a technical estimate, he will benefit the patient very greatly throughout his illness").

3 See, e.g., *Nat. hom.* XV.1–29 Littré. The Hippocratic authors of the *Epidemics*, for example, organized their patient case histories according to the ordinal day (counting from the onset of the illness) on which various symptoms occurred. These physicians were particularly interested in developing systems to predict the day on which a disease might undergo a "crisis," and whether that crisis would be good or bad. A crisis was a decisive moment in the course of a disease at which the illness would either be cured or begin a transition into a new phase. In the *Prognostic*, the author offers a more unified theory of critical days. For discussion of such theories in the Hippocratic writers, see Langholf 1990, 78–127; Lloyd 1979, 154–68; Cooper 2011, 127–28.

4 The Hippocratic author of the *Prognostic*, for example, describes a number of acute fevers, none of whose periods can be calculated exactly using whole days: Οὐ δύναται δὲ ὅλῃσιν ἡμέρῃσιν οὐδὲν τουτέων ἀριθμέεσθαι ἀτρεκέως (II.169 Littré = 58.8–9 Jouanna).

5 See, e.g., Jones's contribution to this volume.

6 Babylonian priests were eager to predict the occurrences of celestial omens, such as lunar and solar eclipses. To accomplish this, they maintained meticulous records of astral, planetary, and lunar positions with which they could calculate impending syzygies (such as those between the Sun and Moon). We do not know precisely when or how this Babylonian data was transmitted to the Greek world, but we do know that Hipparchus, who was active in the 2nd century BCE, had access to Babylonian lunar eclipse reports that ranged from the mid-8th to at least the early 4th century BCE. See Jones 1999, 443. Toomer (1985, 193–206) suggests that Hipparchus himself was instrumental in copying sections from the Babylonian archives and conveying them to Alexandria. Additionally, a variety of tables begin to appear in the papyrological record that record the hour of a celestial event's occurrence alongside numerous other factors. Ἐφημερίδες, for example, which list the daily positions of certain celestial bodies, often include the hour of zodiac sign-entry as well as the longitude and time of particular syzygies. The most extensive collection and commentary to date can be found in Jones 1999. Sign-almanacs from the 3rd century CE onward also tend to include a column for the time of sign-entry in seasonal hours. See, e.g., Jones nos. 4192 and 4194–4196a.

they were soon adopted by authors writing on a variety of subjects, including mathematics, astrology, and medicine.

One such author was the renowned physician and philosopher, Galen of Pergamum, who served two emperors in Rome during the latter half of the second century CE. In what follows, I will explore how Galen integrated hours into his own system for classifying irregular fevers, and how he claims, in doing so, to differentiate himself from rival physicians. I focus on a rarely-studied Galenic treatise, *Adversus eos qui de typis scripserunt vel de circuitibus* (*Against Those Who Have Written on Types or on Periods*), henceforth referred to as *On Periods*. This treatise—of which no modern translation has yet been published—is a polemic directed against physicians whose fever-classification systems Galen found unwieldy and impractical.[7] I argue that, in this text, Galen appeals to hourly timekeeping to help him present himself, in contrast to such doctors, as an astute observer and a logical, systematic thinker, who approaches the problem of irregular fevers with a sound scientific methodology.

1 Galen's Fever Typology: an Overview

In order to appreciate the nuances of Galen's strategies, it is important first to become familiar with the principles of Galen's fever classification system, which he develops most extensively in the treatises *De crisibus* (*On Crises*), *De typis* (*On Types*), and *De differentiis febrium* (*On the Differences Among Fevers*). For Galen, fevers ultimately arise because one's internal, life-giving fire has been excessively stoked.[8] This stoking could be caused by a variety of factors, one of which was that certain bodily χυμοί ("humors": specifically, yellow bile, black bile, and phlegm) could become clogged in the arteries and start to ferment.[9] Fevers brought about in this way Galen calls σηπόμενοι ("septic").[10]

7 The sole Greek edition is that of Kühn (1824 [repr. 2001], 475–512).

8 As Wittern notes (1989, 4), Galen is inconsistent in his details. In *On the Differences Among Fevers*, he claims that fevers should be categorized among "external" or "unnatural" fires (e.g., ἀλλοτρίαν τὴν ἔξωθεν, οὐκ ἔμφυτον at VII.374 K), but in his commentary on the Hippocratic *Aphorisms*, he defines fevers as transformations of the inherent fire within living creatures (e.g., at XVIIB 414 and 426 K).

9 Galen also recognized a fourth humor, blood, but did not believe it could ferment to produce an intermittent fever. Instead, he considered fevers that originated in the blood to be "ephemeral" (*De diff. febr.* VII.374–377 K).

10 As Lonie (1981, 29) observes, "The basic feature of [Galen's] explanation was that the temporal discontinuity of the fever is correlated with a spatial discontinuity of the humors in the body. Both the fits of shivering and the period of the onsets were accounted for by *quanta* of putrid humors moving from one part of the body to another. These *quanta* accumulated, putrefied, and were expelled in periods of time which differed according

Different putrid humors yielded septic fevers with different characteristics. Galen famously associated each humor with two of four ποιότητες ("qualities"): hot, cold, moist, and dry. Yellow bile was hot and dry, black bile cold and dry, phlegm cold and moist.[11] Thus, in Galen's view, a fever caused by an excess of yellow bile would overheat and desiccate a patient. To alleviate this condition, Galen recommends forbidding the patient from having contact with other "hot" or "dry" substances and, instead, treating the patient allopathically (in this case, with substances considered "cold" and "moist").[12] For Galen, it was important that a doctor know which humor was to blame for a given fever, so that he could not only predict the course of the illness, but also design a treatment plan for the patient tailored to that humor's qualities.

Another important aspect of septic fevers was their temporality. Galen divided such fevers into two broad categories: the συνεχεῖς ("continuous") and the διαλείποντες ("intermittent"). Intermittent fevers went through cycles of paroxysm and remission, and the lengths of those periods depended upon the type of humor that had been corrupted. Phlegm, for instance, if fermented, produced a fever that underwent a full cycle of paroxysm and remission every twenty-four hours, earning it the name ἀφημερινός ("quotidian"). Fermented yellow bile caused a fever with a forty-eight-hour cycle, which was called τριταῖος ("tertian") because it arrived every third day, inclusive of the day from which the counting began. Septic black bile produced a fever with a seventy-two-hour cycle, called τεταρταῖος ("quartan").[13]

"Quotidian," "tertian," and "quartan" were, for Galen, the three τύποι ἁπλοῖ ("simple fever types"). To account for so-called πλανῆται ("wandering" or "irregular") fevers, which deviated from these types, Galen also incorporated into his system mechanisms for combining and adjusting the three primary types to produce a wider range of periodicities.[14] He made a distinction between τύποι ἑστῶτες ("stationary fever types"), whose paroxysms arrived at precisely the same hour every day, and τύποι κινούμενοι ("moving fever types"), whose paroxysms tended to arrive two to three hours early or late.[15] This allowed Galen to include under the heading of a "moving quotidian," for instance, fevers with a wide range of paroxysmal cycles, extending from roughly twenty-

to the quantity and quality of the particular humor." Lonie goes on to note that sixteenth-century writers found this component of Galen's fever theory to be unpersuasive.

11 *De nat. fac.* II.129–131 Kühn.

12 *De meth. med.* X.590 Kühn.

13 These terms—and the recognition of 1-day, 2-day, and 3-day fever periods—did not originate with Galen. We see them already in the Hippocratic works of the late Classical period.

14 *De cris.* IX. 680 Kühn.

15 *De typ.* VII. 464 Kühn.

one to twenty-seven hours. Galen also dealt with irregular fevers by suggesting that the three simple fever types could synthesize to produce τύποι σύνθετοι ("compound fever types"). Such compounds could either be ὁμογενεῖς ("homogeneous"), i.e., composed of multiple fevers of the same simple type (such as three tertians or two quartans), or ἑτερογενεῖς ("heterogeneous"), i.e., formed by a mixture of different simple types (such as one tertian and one quartan).[16]

Galen's classification system proved so attractive to later physicians that it became the foundation of fever theory first in the Medieval Arabic world and then in the European West, up through the end of the 15th century.[17] His reputation began to crumble during the Enlightenment, but vestiges of Galen's fever classification system remain embedded in modern medical terminology.[18] To this day, fevers are still categorized as "continuous" or "intermittent," and while the terms "quotidian," "tertian," and "quartan" have recently gone out of fashion in the medical world, they are still used on occasion to describe strains of malaria.[19]

2 Hours in Action: a Case Study

An extended passage from *On Crises* offers a glimpse of Galen's classification system in action. Galen's ultimate aim in this passage is to demystify an irregular fever by interpreting it as a homogeneous compound of three

16 *De cris.* VII. 464.18–465.8 Kühn. Galen also speaks often of the "semitertian" fever, though he does not classify it among the simple types. For Galen, a semitertian was the deadly product of a continuous quotidian fever and a tertian, and he distinguishes between multiple kinds (*De diff. febr.* VII. 358–359 K). Another Imperial-period physician, Celsus, describes the semitertian as a tertian that never undergoes complete remission and whose paroxysms last roughly 36 hours (III.3.2).

17 During that time, Galen's writings were widely considered to be one of the two main pillars of ancient medical wisdom, the other being the *Canon of Medicine* (*al-Qānūn fī l-ṭibb*) of the Arabic philosopher Ibn-Sīnā or Avicenna (980–1037 CE).

18 Galen's reputation was shaken in the 16th and 17th centuries, initially under the pressure of attacks launched by the mystic Paracelsus and his followers, but ultimately also due to scientific discoveries, such as those made by Vesalius and Harvey, which disproved central elements of Galen's medical theories. Vesalius's *Fabrica* (1543), for example, demonstrated that Galen's was not the last word on human anatomy, while Harvey's (1628) work on the circulatory system revealed that Galen had been wrong in his interpretations of veins and arteries. On the reception of Galenic medicine from the Renaissance onward, see Temkin 1973; Smith 1979, 1–44; Bates 1981, 45–70; Lonie 1981; Yeo 2005, 435; Nutton 2008, 355–90.

19 On the modern medical definitions and treatments of malaria, see, e.g., Pampana 1963 and the Center for Disease Control's entry for "malaria." On the role that malaria may have played in the ancient world, see Jones 1907 and 1909; de Zulueta 1973, 1–15; Grmek 1989, 245–83; Burke 1996, 2252–81; Scheidel 2001, 76–89; Sallares 2002; Van der Eijk 2014, 112–17.

tertians. Here, Galen's references to hourly timekeeping help him to illustrate the temporal patterns that underlie the illness in question, and to promote the image of himself as an exemplary scientist. The passage is quoted in full to better demonstrate the shape of Galen's argument and the temporal landscape he depicts:

> καίτοι πολλάκις γε σαφεστάτην ἔχουσι τὴν ἀναλογίαν, ὥσπερ ὁ νεανίσκος ὁ περὶ τὰ τέλη τοῦ φθινοπώρου τῇ πρώτῃ τῶν ἡμερῶν ὥρας που πέμπτης ἀρξάμενος πυρέττειν ἅμα ῥίγει βραχεῖ. … ἡμῖν μὲν οὖν ἅπασαν τοῦ πυρετοῦ τὴν ἰδέαν ἐπισκεψαμένοις ἐπιπλοκὴ τριταίων ἐφαίνετο τριῶν. ὅ τε γὰρ ἐν τῇ πρώτῃ τῶν ἡμερῶν ἀρξάμενος ὥρας πέμπτης ὅ τ' ἐν τῇ μετὰ ταύτην νυκτὶ πάλιν ἑβδόμης ὅ τ' ἐπὶ τῆς δευτέρας ἡμέρας ὥρας δεκάτης ἅπαντά τε τὰ τῶν σφυγμῶν ἴδια καὶ τὰ τῆς θερμασίας ἀπέσωζον τριταίων. ἤδη δὲ καὶ τῶν σφυγμῶν ἴδια καὶ τὰ τῆς θερμασίας ἀπέσωζον τριταίων. ἤδη δὲ καὶ τὸ μετὰ ῥίγους καὶ τὸ σὺν ἱδρῶτι τριταίων ἴδιον εἶχον. οὐδεὶς δ' αὐτῶν εἰς ἀπυρεξίαν ἐπαύετο τῷ φθάνειν ἄρχεσθαι τοῦ μετ' αὐτὸν τριταίου τὸν παροξυσμόν, ὡς εἴ γε μηδεὶς ἕτερος ἐπελάμβανεν ὡρῶν που τριῶν ἢ τεττάρων ἄλλων, ἀπύρετος ὁ ἄνθρωπος ἀκριβῶς ἂν ἐγένετο. τριῶν μὲν δὴ τριταίων ἐπιπλοκὴ κατὰ τὴν δευτέραν εὐθὺς ἡμῖν ἡμέραν ἐφαίνετο σαφῶς, τὸ δ' ὅτι καὶ προληπτικῶς παροξυνόντων ἐπὶ τῆς τρίτης πρῶτον ἐγνώσθη. δευτέρας μὲν γὰρ ὥρας ὁ παροξυσμὸς εἰσέβαλλεν, οὐ πέμπτης ὥσπερ ἐν τῇ πρώτῃ, τὰ δὲ τῆς τριταίας περιόδου γνωρίσματα πάντ' ἐναργέστερα τῶν ἐπὶ τῆς πρώτης ἡμέρας ἐκόμισεν. ἐντεῦθεν δὲ παραφυλάττοντες ἅπαντας εὕρομεν τοὺς παροξυσμοὺς ἢ δυοῖν ἢ τρισὶν ὥραις προλαμβάνοντας ὥστε καὶ τὸν ἀνάλογον τῷ πρώτῳ κατὰ τὴν περίοδον ἀπαντῶντα τῇ μὲν πέμπτῃ τῶν ἡμερῶν ἀνίσχοντος ἡλίου γενέσθαι, τῇ δ' ἑβδόμῃ πολὺ θᾶττον ὡς περὶ ὥραν που τῆς νυκτὸς ἐνάτην, εἶθ' ὁ μετὰ τοῦτον ἀνάλογον ὥρας νυκτὸς ἑβδόμης, ὁ δ' ἐφεξῆς τετάρτης ἐγένετο, τοὐντεῦθεν δὲ πάλιν ὁ μὲν ἐφεξῆς τετάρτης, ὁ δ' ἐπὶ τούτῳ πέμπτης, εἶθ' ἕκτης αὖθις, εἶτ' ὀγδόης ὁ δέκατος ἀπὸ τῆς ἀρχῆς ἐγένετο παροξυσμός. ἀνάλογον δὲ τούτῳ καὶ οἱ ἄλλοι δύο μετὰ τὴν ἑβδόμην περίοδον οὐ μόνον οὐ προελάμβανον ἀλλὰ καὶ ὑστέριζον, ὅτε δὴ καὶ σαφέστατ πᾶσιν ἐφάνη καὶ τοῖς ἡμιτριταῖον εἶναι νομίζουσιν αὐτόν, ὡς ἐσφάλλοντο μέγιστα.
>
> IX.680–683 KÜHN

And yet frequently [fevers] have a clear mathematical proportion, just like in the case of the young man who began to have a fever in autumn, on the first day and around the fifth hour together with short shivering spells.... To me, who examined the whole form of the fever, it seemed to be a combination of three tertians.... [This fact] appeared clear to me immediately upon the second day. The first [tertian] was recognized to have also had an early paroxysm on the third [day], for the paroxysm came on in the second hour—not the fifth as on the first [day]—and

> conveyed all the signs of the tertian period more clearly than on the first day. From there, keeping an eye on everything, we found that the paroxysms were coming on two or three hours early, with the result that the mathematical proportion corresponding with the first with regard to its period came into being on the fifth day during sunrise, then much more rapidly on the seventh day, as around the ninth hour of the night. Then the next one according to this mathematical proportion [arose] in the seventh hour of the night. The next one in succession arose in the fourth hour, and thence again the next one at the fourth hour, and the one after this during the fifth hour, and then again during the sixth. Then the tenth paroxysm from the beginning came about during the eighth hour. Analogously to this one, the other two after the seventh period not only did not come early but they also arrived late, which thing is very clear to all, even to those who think that [this fever] is a semitertian, although they are very much mistaken.

In the opening of this passage, Galen makes it clear that he is participating in an ongoing terminological debate over how to define atypical fevers. Other doctors, he claims, find themselves entirely nonplussed by fever periods that do not fit their standard simple types. Galen, on the other hand, asserts that he knows better and can even support his diagnosis by offering physical and temporal evidence for his interpretation. By indicating, in the subsequent case history, the exact days *and hours* at which the patient experienced paroxysms, Galen attempts to demonstrate that, in spite of the apparent irregularity of these paroxysms, an underlying pattern is still discernible. He claims that this pattern can actually be explained as the result of three tertian fevers with different start-dates, whose paroxysmal periods overlap (see Table 9.1). Galen explicates the unusual periods of these tertians by calling upon his distinction between fixed and moving fever-types (see Table 9.2). These tertians, he argues, are moving, and thus one should expect each paroxysm to arrive two to three hours later or earlier than the one that preceded it.[20] Characteristically for Galen, the resulting argument seems designed to strike the reader as both straightforward and mathematically complex at the same time.[21]

Galen's decision to use hours as a structuring tool in this clinical narrative permits him to accomplish a number of things. First, it gives him a way

---

20 Galen seems untroubled by the nine-hour jump in the timing of the paroxysms on Day 9 and Day 11.

21 Barton has observed that, "despite [Galen's] idealization of the autodidactic doctor and his implication that anyone who follows his example can enjoy similar fame and success, he appears to relish the mastery of such knowledge more than he does attempting any real communication of it" (Barton 1994, 137).

TABLE 9.1 The pattern of paroxysms experienced by the patient within the first five days of his illness, and Galen's attribution of them to particular tertians. Shaded cells indicate seasonal hours of the night, unshaded cells seasonal hours of the day.

Hour	Day 1	Day 2	Day 3	Day 4	Day 5
4				(Tertian 2)	
5					
6					
7		Tertian 2			
8					
9					(Tertian 3)
10			Tertian 3		
11					
12					Tertian 1
1					
2			Tertian 1		
3					
4					
5	Tertian 1				

to describe the behavior of an intermittent fever that is more precise and information-rich than the terminology commonly employed during his day. Galen makes this claim explicitly in another of his fever treatises, *On the Differences Among Fevers*, in the midst of a discussion about how best to refer to tertians of varying paroxysmal lengths. Galen describes how other physicians often resort to vague adjectival modifiers to differentiate between tertians whose paroxysmal periods are exactly 24 hours and those that are longer by various degrees. These physicians call the former sort of tertian ἁπλῶς ... μεμηκυσμένον ("strict" or "simply lengthened"), and divide the latter category into the μεμηκυσμένον ἱκανῶς ("sufficiently lengthened"), ἐπιπλέον μεμηκυσμένον ("very lengthened"), and ἐπιπλεῖστον [μεμηκυσμένον] ("the most lengthened of all").[22] Galen observes that such labels are relative terms, at best, and do not necessarily help a doctor to predict a fever's specific behavior. Instead, he recommends a different method: ἀλλὰ καὶ δηλῶσαι βουλόμενος ἑτέρῳ τὸν τοῦ κάμνοντος πυρετὸν ὁποῖός τίς ἐστιν, ἀκριβέστερον δηλώσει τό τε τοῦ παροξυσμοῦ καὶ τὸ τοῦ διαλείμματος εἰπὼν μῆκος, ἢ ζητῶν ὄνομα σαφῶς καὶ ἀφωρισμένως

22 VII.372 Kühn.

TABLE 9.2 The paroxysms of what Galen identifies as Tertian 1. In this table, the days advance in units of two because the paroxysmal period of a tertian is approximately two days.

Hour	Day 1	Day 3	Day 5	Day 7	Day 9	Day 11	Day 13	Day 15	Day 17	Day 19
6										
7					x					
8										
9				x						
10										
11										
12			x							
1										
2		x								
3										
4						x	x			
5	x							x		
6									x	
7										
8										x

ἐνδείξασθαι, ταὐτὸν δυνάμενον (VII.373 Kühn) ("But if one should want to explain to someone else what sort of fever a patient has, it will be more precisely clear if one mentions the length of the paroxysm and of the interval than if one seeks a term that can indicate the same thing clearly and definitively"). In the example that Galen offers, he states the durations of both the paroxysm and remission period specifically in hours.[23]

Galen is concerned that disputes over fever terminology can lead to the propagation of harmful errors, both among students building up the foundations of their medical knowledge and among actual physicians attempting to treat patients. At this point in *On the Differences Among Fevers*, Galen essentially asks: if one physician cannot understand what another physician means

23 ὑποκείσθω γὰρ τόνδέ τινα πέντε καὶ δέκα μὲν ὥρας πυρέττειν, ἀπύρετον δὲ γίγνεσθαι τρισὶ καὶ τριάκοντα, καὶ τοῦθ' ἑξῆς αὐτῷ κατ' ἀναλογίαν ἀπαντᾷν (VII.372 Kühn) ("Let us assume that this particular man experiences fever for fifteen hours and becomes fever-free for thirty-three, and that this state of affairs persists for him in the successive days according to this progression").

by "tertian fever," then what good is that term? And how is medical progress ever to be achieved, if physicians are unable even to agree on the technical language common to their art? Galen avidly promoted the development of a universal "scientific language" which would allow practitioners to communicate with one another without any loss or garbling of information.[24] In the realm of fever typology, he seems to see the hour-unit as offering physicians a way to describe and identify fevers precisely, without resorting to more approximate or controversial terms.[25]

Secondly, by using hours in the case history from *On Crises* quoted above, Galen can produce a temporal map of these fevers that is so detailed as to reinforce the impression that Galen acquired his temporal data points through genuine autopsy. The reader pictures Galen by the sickbed, with one eye on his patient and the other on a clock. Galen enhances this image later in the passage with additional references to his own personal experience: e.g., ἡμῖν μὲν οὖν ἅπασαν τοῦ πυρετοῦ τὴν ἰδέαν ἐπισκεψαμένοις (IX.681 Kühn) ("to me, who examined the whole form of the fever") or ἐντεῦθεν δὲ παραφυλάττοντες ἅπαντας (IX.682 Kühn) ("from there, keeping an eye on everything"). Thus, Galen presents himself as a physician who grounds his theories in empirical research.[26] Meanwhile, by mapping out the progress of the triple tertian in a schematic way, Galen suggests to his readers that he has subjected his empirical data to systematic analysis. This exercise helps Galen support the claim that a coherent, logical framework underlies even the most erratic fever periods. Such a framework promises to enable doctors to identify and make predictions about any kind of febrile behavior. The balance between logic and empiricism that Galen displays here is the hallmark of his own scientific method, or ἀπόδειξις, a system which allowed Galen to walk the line between Rationalist and Empiricist approaches to medicine. Thus, by using hours to structure his

---

24 On Galen's desire for a universal scientific language, see also von Staden 2002, 109–40.

25 Galen does warn his readers that fever periodicities are not always sufficient, on their own, for producing accurate diagnoses and prognoses. A good doctor must consider the whole picture: the quality of the patient's internal heat; the presence or absence of other symptoms, such as shivering or inflammation; the characteristics of the pulse, urine, and stool. We see this in our *On Crises* passage, for example, when Galen observes that the patient's pulse, temperature, shivering, and sweating all corroborate the conclusion that Galen had tentatively drawn from the fever's paroxysmal periods—namely, that he is faced with a triple tertian. Yet, among these diagnostic features, Galen seems to give pride of place to fever periodicity, as measured specifically in hours.

26 Cf. (as one *comparandum* among many) *De diebus decr.* IX.772 Kühn, where Galen accuses ἔνιοι τῶν ἰατρῶν ("some doctors") of weighing in on medical discussions οὐδὲ παραγεγονέναι … οὐδεπώποτε ("despite never actually having attended") those who suffered from the illness at issue.

clinical narrative in this context, Galen could bolster his pet claim to be a meticulous practitioner of ἀπόδειξις.[27]

3 Galen's Approach in Context

Galen seems to have been neither the first nor the only physician to propose a theory of compound fevers to explain irregular paroxysmal cycles. Nor was he alone among his contemporaries in using hours to express fever periodicities. Galen stands out (amidst our limited extant sources) in two ways: first, in specifying the length of each paroxysm and abatement in terms of hours; and second, in distinguishing between "stationary" and "moving" paroxysms, a practice that allowed him to integrate precise and allegedly observational measurements into a simpler, more practical classification system.[28] While few fever treatises from among Galen's near-contemporaries have come down to us,[29] Galen alludes to the existence of such treatises in his work *On Periods*.

In this polemical text, Galen criticizes physicians who have adopted, in his opinion, overly-complicated and thoroughly unrealistic schemes for classifying fevers. Though he does not mention the affiliations of these doctors, it is possible that Galen means to target members of the Pneumatic sect.[30] This sect earned its name from its members' belief that the πνεῦμα was the fundamental principle of life. The health of the πνεῦμα could supposedly be determined by analyzing a person's pulse. We know from Galen that Pneumatic physicians developed elaborate systems for classifying pulses, often incorporating charts

27 Cf., for example, *De diff. febr.* VII.370 Kühn: ἡμεῖς ἐπειράθημεν εἰς ὅσον οἷοί τ' ἦμεν ἀκριβῶς ἐξεργάσασθαι τὸν τύπον ἅπαντα, χρόνῳ παμπόλλῳ δι' ἐμπειρίας ἅμα καὶ λόγου τὰς χρησίμους ἐξευρόντες ἐν αὐτῷ διαφοράς ("I have tried, as best I can, to express every fever type accurately, finding out the useful differences among them through both experience and logic over a long period of time").

28 Celsus says: *aliae tempore eodem postridie revertuntur, aliae vel serius vel celerius* (III.3.4) ("Some [paroxysms] recur at the same time the next day, while others recur either later or sooner"). He seems, however, to be merely making a general observation, not identifying discrete categories analogous to Galen's "stationary" and "moving" types.

29 The Pneumaticist Aretaeus of Cappadocia (c. 150–190 CE) wrote a treatise on fevers, fragments of which are preserved in Ps-Alexander of Aphrodisias's work *On Fevers* (16.1, 24.5, 30.1 in Tassinari 1994, 2019–34). The fever treatise by Ps-Alexander of Aphrodisias (150–200 CE) shares neither Galen's interest in fever temporality, nor his humoral view of anatomy and pathology. While Aretaeus occasionally mentions fevers in his extant works, *On Causes and Signs of Acute and Chronic Diseases* and *On Therapy of Acute and Chronic Diseases*, he does not discuss their periodicities (see, e.g., *De caus. sign. acut. morb.* 1.4). See Keyser and Irby-Massie 2008, 55–56 [Ps-Alexander] and 129–130 [Aretaeus].

30 On this sect, see, e.g., Gourevitch 1998, 115–17.

alongside comprehensive descriptions.[31] Galen's imagined interlocutors in *On Periods* seem to adopt a similar strategy with regard to periodic fever classification. It is also noteworthy that the two physicians Galen names within this text, Agathinus and Philistion, may be connected to the Pneumatic school.[32] Within the text, however, the only term that Galen uses to describe his interlocutors is οἱ νεώτεροι ("the younger ones," cf. VII.476 Kühn). Thus, his critique may, in fact, be aimed at common trends among the new generation of physicians, broadly speaking.

In *On Periods*, Galen charges these physicians, whoever they are, with a number of medical crimes: (1) incorrectly identifying fever types, thereby creating classificatory confusion; (2) producing tardy and unhelpful diagnoses; and, in general, (3) failing to adhere to the proper scientific method. In criticizing such behavior, Galen also creates—and seizes—the opportunity to present himself in contrasting terms. Galen, he himself would have us believe, (1) utilizes a precise classificatory system that is simultaneously nuanced and easy to use; (2) regularly produces rapid and accurate diagnoses, due to his great familiarity with fevers; and (3) obtains his prognoses via true ἀπόδειξις. I examine each of these claims more closely to see how Galen uses hourly timekeeping in their support.

In his other fever treatises, Galen often accuses doctors of creating classificatory confusion by being insufficiently precise. Toward the beginning of *On the Differences Among Fevers*, for example, Galen complains that most doctors οὐ διακρίναντες ἀπὸ τῶν χρησίμων τὰς ἀχρήστους (VII.274 Kühn) ("fail to distinguish the useless [fever categories] from the useful"), sometimes omitting important types or adding superfluous ones. In *On Periods*, Galen focuses his attack on doctors who do not limit themselves to the three simple fever types recognized by Galen (the quotidian, the tertian, and the quartan), and who adhere too strictly to the notion that all such types must be ὁλοκλήροις ἡμέραις τε καὶ νυξὶ περιγράφηται (VII.476 Kühn) ("defined in complete days and nights").[33] These

31 See Smith 1979, 100. A tantalizing reference in the writings of the Imperial-period physician Marcellinus suggests that a Greek doctor (Herophilus, in this case) might sometimes use a small clepsydra as one might use an egg-timer in order to measure the rate of a patient's pulse. Unfortunately, Marcellinus' testimony is the only secure evidence from the ancient world for such a practice to have come down to us. Therefore, we cannot infer from it how common this manner of pulse-timing may have been. However, our extant medical literature suggests that ancient Greek physicians overwhelmingly preferred qualitative, rather than quantitative, methods for describing the pulse's speed and strength.

32 On Agathinus and Philistion, see Keyser and Irby-Massie 2008, 42–43 and 649–650, respectively.

33 Cf. VII.489 Kühn: τύπον οὗτοι καλοῦσιν οὐχ ἅπασαν περίοδον, ἀλλ' ἥτις ἂν ἐξ ὁλοκλήρων ἡμερῶν τε καὶ νυκτῶν συγκέηται, καὶ καθ' ἕκαστόν γε πλῆθος ἡμερῶν ἴδιον εἶναι βούλονται τύπον, οὐδὲ

misguided physicians must resort to convoluted and, in Galen's view, utterly impractical calculations in order to account for fevers whose periods deviate from the simple types by just a few hours. Galen condemns these men as ones who τῶν μαθητῶν κατατρίψαντες τὸν χρόνον, ὅταν ἐπὶ τοὺς κάμνοντας ἀφίκωνται, χείρους εἰσὶν ἰδιώτου παντός (VII.476 Kühn) ("after wasting their own and their students' time, whenever they come to patients' bedsides, are worse than any layman"). For a layman, Galen insists, would not be confused by a quotidian fever whose paroxysms arrived two hours earlier than expected; despite this slight temporal variation, that layman would still φαντασίαν τινὰ ἔχει περιόδου τῆς καλουμένης ἀμφημερινῆς (VII.476–477 Kühn) ("have the impression of the period called 'daily'"). The offending doctors, however, assume that each manifestation of what Galen would call a "simple moving quotidian" must actually constitute a highly complex compound:

> οἱ δὲ ταῖς εἰρημέναις ὑποθέσεσι δουλεύοντες, ἐὰν ἐντύχωσιν ἀῤῥώστῳ δυοῖν ὥραιν ἑκάστης ἡμέρας τὸν παροξυσμὸν ἔχοντι προληπτικόν, εἴκοσι μὲν εἶναι καὶ δυοῖν ὥραιν φασὶ τὴν περίοδον, τὸν δ' ὅλον τύπον οὐχ ἁπλοῦν, ἀλλά τινα σύνθετον ὑπάρχειν νομίζουσιν· εἶτα ζητήσαντες ἱκανῶς καὶ λογισάμενοι κατὰ σφᾶς αὐτούς, οἱ μὲν ἐπὶ δακτύλων, οἱ δὲ καὶ ἐπὶ διαγράμματος ἔν τινι βιβλίῳ γεγραμμένου, δώδεκά φασιν εἶναι δωδεκαταίους τὸν τοιοῦτον τύπον. γελώντων δὲ ἐπὶ τούτοις τῶν ἀκουσάντων, ἀγανακτοῦσιν.
>
> VII.477 KÜHN

> Those who are slaves to the aforementioned hypotheses ... would say that [the fever's] period is 22 hours, and they would think that the whole type is not simple, but some sort of compound. Then, having investigated sufficiently and calculated on their own—some on their fingers, some with a diagram drawn in a book—they would say that such a fever type is twelve "dodecan" fevers (i.e., fevers whose paroxysms arrive on the twelfth day, and whose periods thus last eleven days). When those who hear them laugh at them, they are irritated.

The methods used to arrive at such a peculiar diagnosis will be considered shortly. For now, what matters is that, in Galen's view, these doctors misidentify

κ' ἢ λ' μόνον ἡμερῶν, ἀλλὰ καὶ τούτων ἐξωτέρω τὰς τῶν τύπων περιόδους ἐκτείνοντες ("These men do not call every period a 'type,' but whatever period is composed of complete days and nights, and they want there to be a type for each particular quantity of days. And [they want these types to include] not 20 or 30 days alone, but also the periods that extend beyond those of the types").

a moving quotidian—with a period of *approximately* 24 hours—as a different illness with a strict 22-hour period. To explain this strange period, the doctors must resort to a rather absurd combination of twelve so-called "dodecan" fevers.

Despite the abuse that Galen heaps upon his rivals, he seems to be fascinated by the mathematical methods they employ. Rather than simply dismiss these doctors' strategies out of hand, as too silly to warrant consideration, Galen instead devotes the greater part of *On Periods* to explaining and testing them—before ultimately subjecting them to crippling refutations. In order to elucidate the mathematical operations that underpin these doctors' diagnostic systems, Galen begins by laying out a table of his own, which he advises any interested reader either to commit to memory or to copy out onto a notecard for easy consultation during the discussion that follows.[34] In the left-hand column of the table, Galen lists the different fever types recognized by these doctors, which are not limited to quotidians, tertians, and quartans, but go all the way up, day by day, to "quinquagintans," fevers whose paroxysms occur every 50th day. In the right-hand column, Galen gives the period of each of these fever types, listed not in units of whole days but in the equivalent number of equinoctial hours. Thus, the number 48 appears next to the type labeled "tertian," the number 120 next to "sextan," and the number 936 next to "quadragintan." Galen invites us to employ these figures in this mathematical formula (the variables were assigned by this author, not Galen):

$$P/p = N$$

In this equation, P = the period, in hours, of a fever type listed on the table; p = the period, in hours, of an observed fever; and N = the number of fevers (of the type corresponding to period P) that have combined to create the compound. Galen explicates this formula using the following example:

> ἔστω τινὰ παροξύνεσθαι μὲν ὥραις στ', ἐξανίεσθαι δὲ δυοῖν, ὥστε τὴν ὅλην περίοδον ὑπάρχειν ὡρῶν ὀκτώ· κατὰ ταύτην τὴν ὑπόθεσιν ἐπισκέψῃ, τί μέρος ἐστὶ τὰ η' τῶν κδ', εἶθ' εὑρὼν ὅτι τρίτον, ἀμφημερινοὺς ἐρεῖς ἐπιπεπλέχθαι τρεῖς· ἀμφημερινοὺς μὲν, ὅτι τοῦ τῶν κδ' ἀριθμοῦ μόριόν ἐστιν ὁ τῶν ὀκτώ· τρεῖς δὲ, ὅτι τρίτον.
>
> VII.493–494 KÜHN

34 VII.489–493 Kühn.

> Let it be the case that someone suffers a paroxysm for six hours and experiences abatement for two hours, with the result that the whole period of hours is eight. In accordance with this hypothesis, consider what fraction of 24 is 8. Then, having discovered that it is one third, you will say that three quotidian fevers have combined. Quotidians, because eight is a fraction of the number 24; three, because that fraction is one third.

If we plug these figures into the equation, we get: 24 hours / 8 hours = 3 fevers (each of length 24 hours). An observed paroxysmal period of 16 hours, Galen continues, would indicate a combination of three tertians, because 48 hours / 16 hours = 3 fevers (each of length 48 hours).

Since, however, the only given in any particular diagnostic situation is p, how, Galen asks, should one decide which values to assign to P and N? In order to ensure that both of these are whole numbers, Galen ultimately advises using the Rule of the Greatest Common Denominator expressed by Euclid in chapter 7 of his *Elements*.[35] Galen offers, as one of several examples, the case of a fever with an observed 15-hour period. He recommends first finding the lowest denominator common to both 15 and 24 (the number of hours in a day); this number is 3, which goes into 15 five times and into 24 eight times. Then: ἐὰν οὖν τὸν ε′ πολλαπλασιάσῃς ἐπὶ τὸν τῶν κδ′, ἐάν τε τὸν η′ ἐπὶ τῶν ιε′, τὸν αὐτὸν ἀριθμὸν εὑρήσεις γινόμενον, ὥσπερ ἑκατὸν καὶ εἴκοσιν ὡρῶν, ἡμερῶν δὲ δηλονότι ε′ (VII.511–512 Kühn) ("If you cross-multiply that five by 24 and that eight by 15, you will discover that the same number is produced, namely 120 hours, which is clearly 5 days"). Thus, you are plainly dealing with eight "sextan" fevers whose paroxysms arrive every sixth day inclusive. Galen devotes several pages to similar calculations, before he abruptly concludes the treatise by declaring κάλλιον γὰρ ἦν αὐτῶν ὅλως καταφρονεῖν μεμνημένους τοῦ Πυθικοῦ παραγγέλματος, ὡς φείδεσθαι προσήκει χρόνου (VII.512 Kühn) ("It would be better to despise these things all together, recalling the pronouncement of Pythian Apollo that it is best not to waste time").

Galen seems to relish the opportunity to impress his readers with this virtuosic display of mathematics.[36] Yet, Galen does not hesitate to point out the

---

35 VII.511 Kühn.

36 Though it should be noted that Galen does not consistently use the formula that he himself lays out. For instance, shortly after introducing the formula, Galen says: ἐπειδὴ γὰρ ὅταν ἡ κατὰ τοὺς παροξυσμοὺς περίοδος ὡρῶν ᾖ δώδεκα, δύνανται μὲν εἶναι καὶ δύο ἀμφημερινοί, δύνανται δὲ καὶ τριταῖοι τέτταρες, καὶ τεταρταῖοι δ᾽ οὐδὲν ἧττον ἕξ, καὶ ὀκτὼ πεμπταῖοι, καὶ δέκα ἑκταῖοι, καὶ δώδεκα ἑβδομαῖοι, καὶ δεκατέσσαρες ὀγδοαῖοι, καὶ δεκαὲξ ἐνναταῖοι, καὶ δεκαοκτὼ δεκαταῖοι (VII.497–498 Kühn) ("Since, whenever the paroxysmal period is twelve, it can be two quotidians but it can also be four tertians, and—still further—six quartans, eight

significant problems with this fever classification system, in his signature vitriolic style.[37] To begin with, he notes that this diagnostic method does not yield one decisive diagnosis for each observed period. There are instead many numbers of which a given *p* can be a factor. He asks, τί γὰρ μᾶλλον ἄχρι τῶν ὀκτωκαίδεκα δεκαταίων, ἢ τῶν εἴκοσι δωδεκαταίων, ἢ τῶν εἰκοσιδύο δωδεκαταίων προαίρεσθαι προσῆκεν, ἤ τινος τῶν ἐφεξῆς περιόδων; (VII.498 Kühn) ("Why is it more suitable to proceed as far as 18 decans or 20 dodecans or 22 dodecans, or one of the periods that come after?"). In other words, with so many possible fever types and combinations, how does one know where to stop? Such a question, Galen points out, matters on the practical level because, at least according to his own theory of pathology, each fever type is produced by a different humor or combination of humors, and therefore demands a different regimen of allopathic treatment (VII.485–488 Kühn). In Galen's opinion, classification systems that are unable to help a doctor determine how best to treat his patient are nothing more than γυμνάσματα τοῖς νέοις γράψαι ταῦτα παραπλήσια τοῖς περὶ δεῖπνον αἰνίγμασιν (VII.487 Kühn) ("exercises for young men, similar to the riddles that are posed around the dinner table"). In short, they are utterly useless. Elsewhere in his corpus, Galen recommends that physicians pay attention not simply to the periodicity of the paroxysms, but rather to a whole suite of critical signs, including the quality of the patient's urine, the speed of his pulse, the consistency of his stool, etc.[38] In *On Periods*, Galen seems to suggest that a similar practice would help the physician to adjudicate in cases where, based on periodicity alone, multiple fever configurations could be viable.

Galen's second damning criticism of this system is that, if one allows *homogeneous* compounds of fever types that extend beyond the quartan all the

---

quintans, ten sextans, twelve septans, fourteen octans, sixteen nonans, and eighteen decans"). This sequence has its own mathematical logic: as *P* increases by one day, *N* increases by two. Galen does not offer an explanation of this system. On the diffusion of mathematics during the Greco-Roman period, see Cuomo 2001, 143–211.

37 VII.497 Kühn: ἀρκεῖ καὶ μέχρι τοσούτου προεληλυθέναι τῆς ματαιοτεχνίας αὐτῶν, ἵνα καὶ διὰ βραχέων κεφαλαίων ἐπιδεῖξαι δυνηθῶμεν ἕκαστον τῶν ἑπομένων ἁμαρτημάτων ("[I]t suffices to proceed in the useless art of these men up to such a point, in order that we might be able to demonstrate, through a short summary, each of the errors that follow").

38 Galen lists some of these critical signs in *On Sects for Beginners*: θαυμάζω γὰρ αὐτῶν, εἰ μήθ' ἱδρῶτάς ποτε μήτ' οὖρα μήτ' ἐμέτους μήτε διαχωρήματα πλείω τῶν κατὰ φύσιν εἶδον ἐν ταῖς νόσοις χρηστῶς κενούμενα, καὶ τὸ πάντων ἀτοπώτατον, εἰ μηδ' ἐκ ῥινῶν αἱμορραγίαν ποτὲ κρίνασαν ἐθεάσαντο (I.94 Kühn) ("I marvel at those [physicians] if they did not ever look at the sweats, the urine, the vomit, or the excrement beyond what is natural, though these evacuations are useful in diseases. And strangest of all, [I marvel] if they never saw that the bleeding of the nose is decisive"). Further discussion of these signs can be found in, e.g., *Method of Medicine*.

way up to the "quinquagintan," then one must also be prepared to consider all the *heterogeneous* compounds that could be formed from those types. After pointing out that numerous heterogeneous combinations are available even to physicians who limit themselves to mixing quotidians, tertians, and quartans, Galen exclaims:

> ἐὰν οὖν ἐπιπλέκῃς ἅπασαν τὴν τοῦ πεμπταίου ποικιλίαν τῇ κατὰ τὸν ἑκταῖον, ἐννοεῖς, οἶμαι, πόσον ἔσται πλῆθος· ἐὰν δὲ μίξῃς αὐταῖς καὶ τὸν ἑβδομαῖον, ἢ ὀγδοαῖον, ἤ τινα τῶν ἐφεξῆς, εἶτα πάλιν ἐκ τεττάρων τύπων οὕτως ἀλλήλοις ἐπιπεπλεγμένων ἑτέρας ποιήσῃς συζυγίας, εἶτ' ἐκ πέντε, κᾄπειτ' αὖθις ἐκ πλειόνων, οἶμαί σε λογίζεσθαι πόσος ἀριθμὸς ἔσται συζυγιῶν, ἢ πόσων δεήσει βιβλίων εἰς τὴν διδασκαλίαν αὐτῶν.
>
> VII.505 KÜHN

> If you then combine every complex of the quintan with that of the sextan, you will know, I think, what the quantity will be. And if you also mix with them the septan or octan or one of the ones that come after, and then you make other compounds, first out of the four types thus combined with one another, then out of five types, and then again out of more, I think that you will appreciate how great a number of combinations there will be—or how many books you will need in order to teach about them!

As someone who generally supports integrating mathematics into the art of medicine, Galen can appreciate these physicians' desire to create a fever classification system that is mathematically grounded.[39] The problem, in his opinion, is that they get it wrong; they fail to subject their results to the logical and empirical tests that would alert them to the impracticality of their method.[40]

For Galen, this method can result in the following errors. First, since it does not impose any limit on how long a paroxysmal cycle can be, physicians may find themselves waiting indefinitely to see whether another paroxysm is forthcoming, or whether they can, in fact, declare a patient cured. As Galen puts it, τοῖς δ' εἰς πολὺ πλῆθος ἡμερῶν ἐκτείνουσι τὰς περιόδους ἀδύνατον προγνῶναι τὸν ἐσόμενον παροξυσμόν (VII.481 Kühn) ("It is impossible for those who extend their periods into a great number of days to make a prognosis about the paroxysm that will come"). Second, this method allows for fever types that defy experiential wisdom. For instance, Galen recounts a debate between himself

39 On Galen's fascination with mathematics, see Lloyd 2006, 110–30.

40 Galen describes such physicians as τοῖς ἀλόγωι τριβῆι μεταχειριζομένοις τὴν τέχνην (VII.478 Kühn) ("those who practice the [medical] art with illogical method").

and a rival physician over how to characterize a fever whose paroxysms last five hours, and whose periods of abatement last two hours (yielding a total cycle of seven hours). Galen tells us: ἀπεκρινάμην οὖν εὐθέως αὐτῷ, μηδ' οἷόν τε γενέσθαι τοῦτο. τοὺς γὰρ ὀλιγοχρονίους οὕτω παροξυσμοὺς μακρὰς ποιεῖσθαι τὰς παρακμάς, καὶ τοὐπίπαν γε καὶ εἰς ἀπυρεξίαν ἀφικνεῖσθαι (VII.479 Kühn) ("I replied straight away that this is not even possible. For paroxysms that are of such short duration produce abatements that are long, and on the whole, they result in a fever-free state"). He then presents his interlocutor as someone who μηδὲν … σκόπει τό γε νῦν, πότερον ἀδύνατός ἐστιν ἡ ὑπόθεσις, ἢ δυνατή, γυμνασίας δ' ἕνεκα τὰ τοιαῦτα συγχώρει ζητεῖσθαι (VII.479 Kühn) ("does not … even consider whether a hypothesis is impossible or possible, but concedes that such things are sought for the sake of exercise"). Finally, this method is further flawed because, as we have just seen, it does not offer clear strategies for distinguishing between several different fever combinations that, mathematically, could all yield the same paroxysmal period.

Galen, of course, would have his readers believe that he himself is exempt from such criticisms. Not only has he succeeded in identifying the errors that these other doctors have built into their systems, but Galen has also proposed a different system which—despite its own complications and defects, characteristically unacknowledged by Galen—he presents as being cleaner and more effective. We recall that Galen recognizes only three simple fever types, each of which is assigned to its own humor and associated method of treatment. These can combine to explain irregular patterns, but Galen's system avoids excessive complexity (at least in terms of the types and numbers of fevers) by introducing the category of the "moving" fever, whose paroxysms can arrive a few hours early or late.

Furthermore, Galen suggests that, unlike the doctors he is critiquing, he himself knows how to test medical theories using the proper blend of logic and empiricism. Galen's favorite term for this approach, ἀπόδειξις, appears only once in this treatise, but it carries great significance; Galen describes the entire exercise that he has performed in writing *On Periods* as an instance of ἀπόδειξις.[41] Galen brings his argument to a close by saying, ἔχετε γοῦν ἤδη καὶ ταύτην ἀπόδειξιν τῆς μεθόδου, καίτοι γ' οὐ βουληθέντος ἐμοῦ διδάσκειν ὑμᾶς ἄχρηστα πράγματα (VII.512 Kühn) ("So now you also have this demonstration of their method, despite the fact that I did not want to teach you useless things"). Here, ἀπόδειξις does not refer to a formal logical proof, but rather to the rigorous testing and refutation of other doctors' hypotheses that constitutes the

41 The related term ἐνδείκνυμι, "to point out, to prove," makes several appearances, at VII.479, 498, 499, 505 K.

bulk of *On Periods*. Galen's apodeictic strategy in this treatise is: first, to identify the premises upon which these rival doctrines are constructed, and to pursue them to their logical conclusions; second, to assess the viability of these conclusions by comparing them against his personal experiences of what works in the sick room. In doing so, Galen displays his own logical aptitude and commitment to empiricism, while also demonstrating his adroitness at making tables and manipulating mathematical formulae. Ultimately, the results of these rival diagnostic methods speak for themselves: Galen is able (or so he claims) to identify fevers quickly and offer patients accurate prognoses; the other doctors, in contrast, must wait for long periods of time before they can even feel confident offering a diagnosis, much less a prognosis, and their conclusions are usually wrong.

Of course, as Wittern has observed, there is little reason to believe that Galen's system was, in fact, substantially more practical or empirically-grounded than the ones he critiques.[42] Galen does, however, offer us a fascinating perspective on the "proper" role of "exactness" or "precision" in ancient sciences. By Galen's time, this concept (often rendered in Greek as ἀκρίβεια) was articulated most frequently in two ways: through simple quantification and/or through complex διαίρεσις, or sub-division of categories.[43] As Heilen has already illustrated in his contribution to this volume, intellectual elites of the Roman Imperial period were often more interested in cultivating the *perception* of this kind of ἀκρίβεια than in reflecting the actual exactitude that their tools and processes could yield. Elsewhere, Galen too is guilty of this charge. In his literature on pulse theory, for example, as Barton has aptly demonstrated, Galen creates a superabundance of categories and sub-categories, a strategy that grants him substantial interpretive flexibility and the chance to show off his refined intellectual sensibilities.[44] In *On Periods*, however, Galen reverses this approach. Rather than privilege extreme διαίρεσις, Galen places a premium on (comparative) simplicity and ease of use. He clearly expected this strategy to give him the upper hand in the intellectual battle, or ἀγών, over fever periodicities and help him draw potential pupils away from his rivals. Thus, this Galenic text indicates that, in some medical circles, there was a backlash against excessive precision.

42 Wittern 1989, 21.

43 On the concept of ἀκρίβεια in the Classical period, see Kurz 1970.

44 Barton 1994, 133–68.

4 Conclusion

This chapter explored how Galen used hour-units to participate in an ongoing medical debate about classifying irregular periodic fevers, and how, in the process, he could portray himself as a faultless practitioner of ἀπόδειξις. For Galen, numbered hours seemed to offer physicians a lexicon for describing periodic fevers that was more quantitative and broadly comprehensible than the language typically employed. Because Galen was willing to chart the exact hours at which a fever's paroxysms arrived, he was also able to distinguish between stationary and moving fevers, a distinction that he considered useful for resolving seemingly irregular patterns. Furthermore, by incorporating systematic hourly timekeeping into his fever theory, Galen could demonstrate to his readers that he, in contrast to many of his contemporaries, possessed the qualities of an exemplary scientist. He was meticulous about recording and organizing observational data in keeping with specific temporal schemes; he was adept at interpreting this data with the aid of logical reasoning and mathematical formulae; and, thanks to his scientific method, he was capable of catching and correcting any potential errors. Galen's example illustrates that some Imperial-period physicians, at least, considered hourly timekeeping a useful tool, whether for practical or rhetorical purposes.

References

Barton, Tamsyn. 1994. *Power and Knowledge: Astrology, Physiognomics, and Medicine under the Roman Empire*. Ann Arbor, MI: University of Michigan Press.

Bates, Don G. 1981. "Thomas Willis and the Fevers Literature of the Seventeenth Century." *Medical History* 25 (Supplement 1): 45–70.

Burke, Paul F. 1996. "Malaria in the Greco-Roman World: A Historical and Epidemiological Survey." In *Aufstieg und Niedergang der römischen Welt*, edited by Wolfgang Haase and Hildegard Temporini, II.37.3: 2252–81. Berlin: de Gruyter.

Cooper, Glen M. 2011. "Galen and Astrology: A Mésalliance?" *Early Science and Medicine* 16: 120–46.

Cuomo, Serafina. 2001. *Ancient Mathematics*. New York: Routledge.

De Zulueta, Julian. 1973. "Malaria and Mediterranean History." *Parassitologia* 15: 1–15.

Gourevitch, Danielle. 1998. "The Paths of Knowledge: Medicine in the Roman World." In *Western Medical Thought from Antiquity to the Middle Ages*, edited by Mirko D. Grmek and Bernardino Fantini, translated by Antony Shugaar, 104–38. Cambridge: Harvard University Press.

Grmek, Mirko D. 1989. *Diseases in the Ancient Greek World*. Translated by Leonard Muellner and Mireille Muellner. Baltimore, MD: Johns Hopkins University Press.

Jones, Alexander. 1991. "The Adaptation of Babylonian Methods in Greek Numerical Astronomy." *Isis* 82: 440–53.

Jones, Alexander, ed. 1999. *Astronomical Papyri from Oxyrhynchus (P. Oxy. 4133–4300a). Edited with Translations and Commentaries*. 2 vols. Philadelphia: American Philosophical Society.

Jones, William H.S. 1907. *Malaria: A Neglected Factor in the History of Greece and Rome*. Cambridge: Macmillan & Bowes.

Jones, William H.S. 1909. *Malaria and Greek History*. Manchester: Manchester University Press.

Jouanna, Jacques, ed. 2013. *Hippocrate, Pronostic*. Paris: Les Belles Lettres.

Keyser, Paul T., and Georgia L. Irby-Massie, eds. 2008. *The Encyclopedia of Ancient Natural Scientists: The Greek Tradition and Its Many Heirs*. New York: Routledge.

Kühn, Karl G., ed. 1825. *Claudii Galeni Opera Omnia*. Vol. IX. Leipzig: Knobloch. [Repr. Hildesheim: Olms, 2001, and Cambridge: Cambridge University Press, 2012.]

Kurz, Dietrich. 1970. *AKPIBEIA. Das Ideal der Exaktheit bei den Griechen bis Aristoteles*. Göppingen: Kümmerle.

Langholf, Volker. 1990. *Medical Theories in Hippocrates. Early Texts and the "Epidemics."* Berlin: de Gruyter.

Littré, Maximilien Paul Émile. 1839. *Œuvres complètes d'Hippocrate. Traduction nouvelle … Accompagnée d'une introduction, de commentaires médicaux, de variantes et de notes philologiques … par E. Littré.* Paris: Les Belles Lettres.

Lloyd, Geoffrey E.R. 1979. *Magic, Reason and Experience*. Cambridge: Cambridge University Press.

Lloyd, Geoffrey E.R. 2006. "Mathematics as a Model of Method in Galen." In *Principles and Practices in Ancient Greek and Chinese Science*, edited by Geoffrey E.R. Lloyd, 110–30. Burlington, VT: Variorum.

Lonie, Iain M. 1981. "Fever Pathology in the Sixteenth Century: Tradition and Innovation." *Medical History* 25 (Supplement 1): 19–44.

Miller, Kassandra J. 2017. "A Doctor on the Clock: Hourly Timekeeping and Galen's Scientific Method." Ph.D. Thesis, Chicago: University of Chicago.

Miller, Kassandra J. 2018. "From Critical Days to Critical Hours: Galenic Refinements of Hippocratic Models." *TAPA* 148: 111–38.

Nutton, Vivan. 2012. *Ancient Medicine*. New York: Routledge.

Pampana, Emilio. 1963. *A Textbook of Malaria Eradication*. Oxford: Oxford University Press.

Sallares, Robert. 2002. *Malaria and Rome: A History of Malaria in Ancient Italy*. Oxford: Oxford University Press.

Scheidel, Walter. 2001. *Death on the Nile: Disease and Demography of Roman Egypt*. Leiden: Brill.

Smith, Wesley D. 1979. *The Hippocratic Tradition*. Ithaca, NY: Cornell University Press.

Tassinari, Piero. 1994. "Il trattato dello ps. Alessandro d'Afrodisia." In *Aufstieg und Niedergang der römischen Welt*, edited by Wolfgang Haase and Hildegard Temporini, 37.2: 2019–34. Berlin: de Gruyter.

Temkin, Owsei. 1973. *Galenism: Rise and Decline of a Medical Philosophy*. Ithaca, NY: Cornell University Press.

Toomer, Gerald J. 1985. "Galen on the Astronomers and Astrologers." *Archive for the History of Exact Sciences* 32: 193–206.

Van der Eijk, Philip J. 2014. "An Episode in the Historiography of Malaria in the Ancient World." In *Medicine and Healing in the Ancient Mediterranean World*, edited by Demetrios Michaelides, 112–17. Oxford: Oxbow Books.

Von Staden, Heinrich. 2002. "'A Woman Does Not Become Ambidextrous'. Galen and the Culture of the Scientific Commentary." In *The Classical Commentary: Histories, Practices, Theory*, edited by Roy K. Gibson and Christina S. Kraus, 109–40. Leiden: Brill.

Wittern, Renate. 1989. "Die Wechselfieber bei Galen." *History and Philosophy of the Life Sciences* 11: 3–22.

Yeo, In-Sok. 2005. "Hippocrates in the Context of Galen: Galen's Commentary on the Classification of Fevers in Epidemics VI." In *Medicine and Philosophy in Classical Antiquity. Doctors and Philosophers on Nature, Soul, Health and Disease*, edited by Philip J. Van der Eijk, 433–43. Cambridge: Cambridge University Press.

Index

Works can be found via authors' names. References to footnotes are in *italics.*

www.ingramcontent.com/pod-product-compliance
Lightning Source LLC
LaVergne TN
LVHW020634300926
849161LV00004B/695